Edited by
Ralf B. Wehrspohn,
Uwe Rau, and
Andreas Gombert

Photon Management in Solar Cells

Related Titles

Wehrspohn, R.B., Kitzerow, H., Busch, K. (eds.)

Nanophotonic Materials

Photonic Crystals, Plasmonics, and Metamaterials

2008
Print ISBN: 978-3-527-40858-0; also available in electronic formats

Abou-Ras, D., Kirchartz, T., Rau, U. (eds.)

Advanced Characterization Techniques for Thin Film Solar Cells

2011
Print ISBN: 978-3-527-41003-3; also available in electronic formats

Würfel, P.

Physics of Solar Cells

From Basic Principles to Advanced Concepts

Second Edition
2009
Print ISBN: 978-3-527-40857-3; also available in electronic formats

Pagliaro, M., Palmisano, G., Ciriminna, R.

Flexible Solar Cells

2008
Print ISBN: 978-3-527-32375-3; also available in electronic formats

Quaschning, V.V.

Renewable Energy and Climate Change

2010
Print ISBN: 978-0-470-74707-0; also available in electronic formats

Edited by
Ralf B. Wehrspohn, Uwe Rau, and Andreas Gombert

Photon Management in Solar Cells

Verlag GmbH & Co. KGaA

The Editors

Prof. Dr. Ralf B. Wehrspohn
Martin Luther University
Institute of Physics
Heinrich-Damerow-Str. 4
06120 Halle
Germany

and

Fraunhofer-Institute for Mechanics of
Materials IWM
Walter-Hulse-Strasse 1
06120 Halle
Germany

Prof. Dr. Uwe Rau
Research Center Jülich
IEF5-Photovoltaics
Leo-Brandt-Straße
52428 Jülich
Germany

Dr. Andreas Gombert
Soitec Solar GmbH
Bötzinger Str. 31
79111 Freiburg
Germany

All books published by **Wiley-VCH** are carefully produced. Nevertheless, authors, editors, and publisher do not warrant the information contained in these books, including this book, to be free of errors. Readers are advised to keep in mind that statements, data, illustrations, procedural details or other items may inadvertently be inaccurate.

Library of Congress Card No.: applied for

British Library Cataloguing-in-Publication Data
A catalogue record for this book is available from the British Library.

Bibliographic information published by the Deutsche Nationalbibliothek
The Deutsche Nationalbibliothek lists this publication in the Deutsche Nationalbibliografie; detailed bibliographic data are available on the Internet at <http://dnb.d-nb.de>.

© 2015 Wiley-VCH Verlag GmbH & Co. KGaA, Boschstr. 12, 69469 Weinheim, Germany

All rights reserved (including those of translation into other languages). No part of this book may be reproduced in any form – by photoprinting, microfilm, or any other means – nor transmitted or translated into a machine language without written permission from the publishers. Registered names, trademarks, etc. used in this book, even when not specifically marked as such, are not to be considered unprotected by law.

Print ISBN: 978-3-527-41175-7
ePDF ISBN: 978-3-527-66569-3
ePub ISBN: 978-3-527-66568-6
Mobi ISBN: 978-3-527-66567-9
oBook ISBN: 978-3-527-66566-2

Cover Design Grafik-Design Schulz, Fußgönheim, Germany
Typesetting Laserwords Private Limited, Chennai, India
Printing and Binding Markono Print Media Pte Ltd., Singapore

Printed on acid-free paper

Contents

Preface *XIII*
List of Contributors *XV*

1 **Current Concepts for Optical Path Enhancement in Solar Cells** *1*
Alexander N. Sprafke and Ralf B. Wehrspohn
1.1 Introduction *1*
1.2 Planar Antireflection Coatings *2*
1.3 Optical Path Enhancement in the Ray Optical Limit *4*
1.4 Scattering Structures for Optical Path Enhancement *5*
1.5 Resonant Structures for Optical Path Enhancement *7*
1.6 Ultra-Light Trapping *10*
1.7 Energy-Selective Structures as Intermediate Reflectors for Optical Path Enhancement in Tandem Solar Cells *13*
1.8 Comparison of the Concepts *16*
1.9 Conclusion *17*
References *17*

2 **The Principle of Detailed Balance and the Opto-Electronic Properties of Solar Cells** *21*
Uwe Rau and Thomas Kirchartz
2.1 Introduction *21*
2.2 Opto-Electronic Reciprocity *21*
2.2.1 The Principle of Detailed Balance *21*
2.2.2 The Shockley–Queisser Limit *22*
2.2.3 Derivation of the Reciprocity Theorem *24*
2.3 Connection to Other Reciprocity Theorems *29*
2.3.1 Emitter and Collector Currents in Transistors *29*
2.3.2 Tellegens's Network Theorem *30*
2.3.3 Differential Reciprocity Relations by Wong and Green *31*
2.3.4 Würfel's Generalization of Kirchhoff's Law *33*
2.3.5 Reciprocity Relation for LED Quantum Efficiency *33*
2.3.6 Shockley–Queisser Revisited *34*
2.3.7 Influence of Light Trapping *35*

2.4	Applications of the Opto-Electronic Reciprocity Theorem	*37*
2.4.1	Experimental Verifications	*37*
2.4.2	Spectrally Resolved Luminescence Analysis	*39*
2.4.3	Luminescence Imaging	*40*
2.5	Limitations to the Opto-Electronic Reciprocity Theorem	*43*
2.6	Conclusions	*44*
	References	*44*
3	**Rear Side Diffractive Gratings for Silicon Wafer Solar Cells**	*49*
	Marius Peters, Hubert Hauser, Benedikt Bläsi, Matthias Kroll,	
	Christian Helgert, Stephan Fahr, Samuel Wiesendanger, Carsten Rockstuhl,	
	Thomas Kirchartz, Uwe Rau, Alexander Mellor, Lorenz Steidl,	
	and Rudolf Zentel	
3.1	Introduction	*49*
3.1.1	Gratings for Solar Cells – Basic Idea and Challenges	*49*
3.1.2	A Short Literature Review	*50*
3.2	Principle of Light Trapping with Gratings	*52*
3.3	Fundamental Limits of Light Trapping with Gratings	*56*
3.4	Simulation of Gratings in Solar Cells	*58*
3.4.1	Optical Simulation Using RCWA/FMM	*58*
3.4.2	Optical Simulation Using the Matrix Method	*61*
3.4.3	Electro-Optically Coupled Simulation Using RCWA and Sentaurus Device	*65*
3.5	Realization	*67*
3.5.1	Electron-Beam Lithography	*68*
3.5.2	Self-Organizing Photonic Crystals	*72*
3.5.3	Fabrication of Rear Side Gratings via Interference Lithography and Nanoimprint Lithography	*75*
3.6	Topographical Characterization	*78*
3.6.1	Atomic Force Microscopy	*78*
3.6.2	Scanning Electron Microscopy	*80*
3.6.3	Focused Ion Beam Milling	*81*
3.7	Summary	*84*
	References	*84*
4	**Randomly Textured Surfaces**	*91*
	Carsten Rockstuhl, Stephan Fahr, Falk Lederer, Karsten Bittkau,	
	Thomas Beckers, Markus Ermes, and Reinhard Carius	
4.1	Introduction	*91*
4.2	Methodology	*93*
4.2.1	Structure of a Referential Solar Cell and Description of Available Substrates	*94*
4.2.2	Rigorous Methods	*96*
4.2.3	Scalar Methods	*97*
4.2.4	Properties of Interest	*98*

4.2.5	Near-Field Scanning Optical Microscopy	99
4.3	Properties of an Isolated Interface	100
4.3.1	Near-Field Properties	100
4.3.2	Far-Field Properties	102
4.4	Single-Junction Solar Cell	104
4.4.1	Absorption Enhancement	104
4.4.2	Design of Optimized Randomly Textured Interfaces	106
4.5	Intermediate Layer in Tandem Solar Cells	110
4.6	Conclusions	112
	Acknowledgments	113
	References	113
5	**Black Silicon Photovoltaics**	**117**

Kevin Füchsel, Matthias Kroll, Martin Otto, Martin Steglich, Astrid Bingel, Thomas Käsebier, Ralf B. Wehrspohn, Ernst-Bernhard Kley, Thomas Pertsch, and Andreas Tünnermann

5.1	Introduction	117
5.1.1	Fabrication Methods	117
5.1.2	Reactive Ion Etching	119
5.1.3	Laser Processing	122
5.1.4	Chemical and Electrochemical Etching	124
5.2	Optical Properties and Light Trapping Possibilities	126
5.2.1	Overview	126
5.2.2	ICP-RIE Black Silicon	128
5.2.3	Influence of Dielectric Coatings	130
5.2.4	Influence of the Substrate Thickness and Limiting Efficiency	132
5.3	Surface Passivation of Black Silicon	135
5.3.1	Requirements for Black Silicon Passivation	136
5.3.2	Possible Passivation Schemes	136
5.3.3	Passivation of Black Silicon Surfaces	139
5.3.3.1	Surface Damage and Sample Cleaning	139
5.3.3.2	Effective Passivation of ICP-RIE Black Silicon	140
5.4	Black Silicon Solar Cells	142
	References	144
6	**Concentrator Optics for Photovoltaic Systems**	**153**

Andreas Gombert, Juan C. Miñano, Pablo Benitez, and Thorsten Hornung

6.1	Fundamentals of Solar Concentration	153
6.1.1	Introduction	153
6.1.2	Concentration and Acceptance Angle	153
6.1.3	Optical Efficiency	157
6.1.4	Effect of Spatial and Spectral Non-uniformities on the Cell Illumination	158
6.2	Optical Designs	159
6.2.1	Classical Imaging Concentrators	160

6.2.2	Nonimaging Secondary Optics	161
6.2.3	Advanced Concentrator Designs	162
6.2.4	Freeform SMS Concentrators	163
6.2.5	Multifold Köhler Concentrators	164
6.2.6	Comparison	166
6.3	Silicone on Glass Fresnel Lenses	169
6.3.1	Physical Influence of Lens Temperature	170
6.3.2	Influence of Lens Temperature on Efficiency	172
6.4	Considerations on Concentrators in HCPV Systems	175
6.4.1	General Requirements on CPV Concentrator Optics	175
6.4.2	Design Considerations	176
6.4.3	Experiences with Concentrator Designs	178
6.5	Conclusions	179
	References	179

7 Light-Trapping in Solar Cells by Directionally Selective Filters 183
Carolin Ulbrich, Marius Peters, Stephan Fahr, Johannes Üpping, Thomas Kirchartz, Carsten Rockstuhl, Jan Christoph Goldschmidt, Andreas Gerber, Falk Lederer, Ralf Wehrspohn, Benedikt Bläsi, and Uwe Rau

7.1	Introduction	183
7.2	Theory	185
7.2.1	Radiative Efficiency Limit	185
7.2.2	Ultra-Light-Trapping	187
7.2.2.1	Universal Light-Trapping Limit for Completely Randomized Light	187
7.2.3	Annual Yield for Directionally Selective Solar Absorbers	190
7.3	Filter Systems	192
7.3.1	1D Layer Stack Rugate Filters	192
7.3.2	3D Photonic Crystal Opal Structures	193
7.4	Experimental Realization	197
7.4.1	Bragg Filter Covering a Hydrogenated Amorphous Silicon Solar Cell	197
7.4.2	Bragg Filter Covering a Germanium Solar Cell	201
7.5	Summary and Outlook	202
	References	203

8 Linear Optics of Plasmonic Concepts to Enhance Solar Cell Performance 209
Gero von Plessen, Deepu Kumar, Florian Hallermann, Dmitry N. Chigrin, and Alexander N. Sprafke

8.1	Introduction	209
8.2	Metal Nanoparticles	210
8.2.1	Optical Excitations in Metal Nanoparticles	210
8.2.2	Control of Optical Properties	212
8.2.2.1	Resonance Energies of Particle Plasmons	213

8.2.2.2	Linewidths of Particle-Plasmon Resonances	*215*
8.2.2.3	Peak Heights of Particle-Plasmon Resonances	*216*
8.2.2.4	Scattering Quantum Efficiencies	*216*
8.2.2.5	Light-Scattering Patterns	*217*
8.2.2.6	Near-Field Effects	*217*
8.2.2.7	Combinations of Effects	*218*
8.3	Surface-Plasmon Polaritons	*218*
8.4	Front-Side Plasmonic Nanostructures	*219*
8.5	Rear-Side Plasmonic Nanostructures	*221*
8.6	Further Concepts	*222*
8.7	Summary	*226*
	Acknowledgments	*226*
	References	*227*

9 Up-conversion Materials for Enhanced Efficiency of Solar Cells *231*
Jan Christoph Goldschmidt, Stefan Fischer, Heiko Steinkemper, Barbara Herter, Sebastian Wolf, Florian Hallermann, Gero von Plessen, Jacqueline Anne Johnson, Bernd Ahrens, Paul-Tiberiu Miclea, and Stefan Schweizer

9.1	Introduction	*231*
9.2	Up-Conversion in Er^{3+}-Doped ZBLAN Glasses	*232*
9.2.1	Samples	*232*
9.2.2	Optical Absorption	*233*
9.2.3	Up-Conversion	*234*
9.3	Up-Conversion in Er^{3+}-Doped β-$NaYF^4$	*237*
9.3.1	Device Measurements	*239*
9.4	Simulating Up-Conversion with a Rate Equation Model	*240*
9.5	Increasing Up-Conversion Efficiencies	*242*
9.5.1	The Up-Converter Material	*242*
9.5.1.1	Phonon Energy	*242*
9.5.1.2	Doping Concentration	*243*
9.5.2	The Environment around the Up-Converter	*245*
9.5.2.1	Plasmon Enhanced Up-Conversion	*245*
9.5.2.2	Modeling Dielectric Nanostructures	*246*
9.5.3	Spectral Concentration	*248*
9.6	Conclusion	*251*
	Acknowledgments	*252*
	References	*252*

10 Down-Conversion in Rare-Earth Doped Glasses and Glass Ceramics *255*
Stefan Schweizer, Christian Paßlick, Franziska Steudel, Bernd Ahrens, Paul-Tiberiu Miclea, Jacqueline Anne Johnson, Katharina Baumgartner, and Reinhard Carius

10.1	Introduction	*255*
10.2	Physical Background	*257*

10.2.1	Rare-Earth Ions	*257*
10.2.2	Glass Systems	*258*
10.2.2.1	Phonons	*259*
10.2.2.2	ZBLAN Glasses	*259*
10.2.2.3	Borate Glasses	*260*
10.3	Down-Conversion in ZBLAN Glasses and Glass Ceramics	*260*
10.3.1	Samples	*261*
10.3.2	Glass-Ceramic Cover Glasses for High Efficiency Solar Cells	*261*
10.3.2.1	Absorption	*262*
10.3.2.2	Short-Circuit Current	*264*
10.3.2.3	Internal Conversion Efficiency	*265*
10.3.2.4	Quantum Efficiency Increase	*266*
10.3.3	Influence of Multivalent Europium-Doping	*267*
10.3.3.1	X-Ray Absorption near Edge Structure	*268*
10.3.3.2	X-Ray Diffraction	*270*
10.3.3.3	Photoluminescence	*272*
10.3.4	Conclusion	*274*
10.4	Down-Conversion in Sm-Doped Borate Glasses for High-Efficiency CdTe Solar Cells	*275*
10.4.1	Samples	*275*
10.4.2	External Quantum Efficiency of CdTe Solar Cells	*276*
10.4.3	Optical Absorption and Fluorescence Emission	*276*
10.4.4	Efficiency Increase	*277*
10.4.5	Conclusion	*279*
10.5	Summary	*280*
	Acknowledgment	*281*
	References	*281*
11	**Fluorescent Concentrators for Photovoltaic Applications**	*283*
	Jan Christoph Goldschmidt, Liv Prönneke, Andreas Büchtemann, Johannes Gutmann, Lorenz Steidl, Marcel Dyrba, Marie-Christin Wiegand, Bernd Ahrens, Armin Wedel, Stefan Schweizer, Benedikt Bläsi, Rudolf Zentel, and Uwe Rau	
11.1	Introduction	*283*
11.2	The Theoretical Description of Fluorescent Concentrators	*285*
11.2.1	Detailed Balance Considerations	*285*
11.2.2	Photonic Structures to Increase Fluorescent Collector Efficiency	*286*
11.2.3	Possible System Configurations – Side-Mounted and Bottom-Mounted Solar Cells	*287*
11.2.3.1	Thermodynamic Efficiency Limits of Fluorescent Concentrators	*290*
11.2.3.2	Non-perfect Photonic Structure	*292*
11.2.3.3	Luminescent Materials in Photonic Structures	*293*
11.3	Materials for Fluorescent Concentrators	*296*

11.3.1	Systems Based on Organic Matrix Materials	*296*
11.3.1.1	The Matrix Material	*296*
11.3.1.2	The Luminescent Species	*299*
11.3.2	Completely Inorganic Systems Based on Rare Earths	*304*
11.4	Experimentally Realized Fluorescent Concentrator Systems	*307*
11.4.1	Systems with Side-Mounted Solar Cells	*307*
11.4.2	Systems with Bottom-Mounted Monocrystalline Silicon Solar Module	*308*
11.4.3	Increasing Efficiency with Photonic Structures	*310*
11.4.3.1	Systems with Side-Mounted III–V Solar Cell	*310*
11.4.3.2	Systems with Bottom-Mounted Amorphous Silicon Solar Cell	*312*
11.5	Conclusion	*314*
	Acknowledgments	*314*
	References	*315*
12	**Light Management in Solar Modules**	*323*
	Gerhard Seifert, Isolde Schwedler, Jens Schneider, and Ralf B. Wehrspohn	
12.1	Introduction	*323*
12.2	Fundamentals of Light Management in Solar Modules	*324*
12.2.1	Basic Physical Concepts of Light Management in Solar Modules	*324*
12.2.1.1	Optical Losses due to Reflection and Absorption	*324*
12.2.1.2	Optical Description of Textured Interfaces or Surfaces	*326*
12.2.1.3	Spectral Effects and Solar Concentration	*329*
12.2.1.4	Effects of Incidence Angle Variation and Diffuse Light	*331*
12.2.2	Assessment of the Optical Performance of Solar Modules	*331*
12.2.2.1	Experimental Techniques	*331*
12.2.2.2	Simulation Approaches and Studies	*333*
12.3	Technological Solutions for Minimized Optical Losses in Solar Modules	*334*
12.3.1	Minimization of Optical Losses in Front Glass Sheets	*334*
12.3.2	Anti-reflection (AR) Technologies for PV Module Front Surface	*336*
12.3.2.1	Nano-scale Technologies for Anti-reflective Treatment of PV Front Glass	*336*
12.3.2.2	Micro-scale Structures for Light-Trapping on PV Front Glasses	*338*
12.3.3	Material Selection and Optimization for Encapsulation Film	*338*
12.3.4	Minimization of Losses due to Metallization and Contact Tabs	*340*
12.3.5	Optical Optimization of Cell Front and Back Interface	*342*
12.3.6	Redirection of Light from Cell Interspaces	*343*
12.4	Outlook	*343*
	References	*344*

Index *347*

Preface

This edition on photon management in solar cells gives a comprehensive overview of the current state-of-the-art of the tricks in optics and photonics to increase the absorption of light in a solar cell. The ultimate aim is to have a really black solar cell. If we currently look around in the landscape, we still see reddish thin-film solar cell based on amorphous silicon or bluish solar cells based on multi-crystalline silicon. So there is still plenty of work to do to improve the current solar cell technology. And this becomes even more important for future silicon solar cells where the thickness of indirect band-gap absorber silicon shrinks to less than 100 μm. However, the spectrum that we see with our eye is not equivalent to the spectrum of the sun. There are, for example, about 20% of the photons in the infrared spectral range. If our eyes would see only the infrared spectral range of the sun below the band-gap of silicon, all the current silicon solar cells would look white! Thus, we are still far away from having really black silicon solar cells. The educated reader might ask if it is theoretically possible to have a completely black silicon solar cell and we would have to answer – no. We have to allow for a little bit of light by the radiative emission of the silicon solar cell. However, this can be managed in principle so that it only sees the small angle of the sun. Then, the solar cell radiative emission "sees" only itself and the sun. This is, in principle, the thermodynamic limit of the system solar cell and of the really black solar cell.

With the following contribution we wish to provide the reader with a lot of insights in the concept of photon management in solar cells, technical help for their current work, or just fun reading how a really black solar cell could be made. We like to thank A.N. Sprafke for helping us to edit the book.

Ralf B. Wehrspohn, Andreas Gombert, U. Rau

List of Contributors

Bernd Ahrens
Fraunhofer Center for Silicon
Photovoltaics CSP
Walter-Hülse-Str. 1
06120 Halle (Saale)
Germany

and

Martin Luther University of
Halle-Wittenberg
Centre for Innovation
Competence SiLi-nano®
Karl-Freiherr-von-Fritsch-Str. 3
06120 Halle (Saale)
Germany

Katharina Baumgartner
Forschungszentrum Jülich
GmbH
Institut für Energie- und
Klimaforschung (IEK-5)
52425 Jülich
Germany

Thomas Beckers
Imperial College London
Department of Physics and
Center for Plastic Electronics
South Kensington Campus
SW7 2AZ London
United Kingdom

Pablo Benitez
Universidad Politecnica de
Madrid
E.T.S. de Ingenieros de
Telecomunicacion
Cedint, Campus de
Montegancedo
28233, Pozuelo
Madrid
Spain

Astrid Bingel
Friedrich Schiller University Jena
Abbe Center of Photonics
Institute of Applied Physics
Max-Wien-Platz 1
07743 Jena
Germany

Karsten Bittkau
Forschungszentrum Jülich
GmbH
Institut für Energie- und
Klimaforschung (IEK-5)
52425 Jülich
Germany

Benedikt Bläsi
Fraunhofer Institute for Solar
Energy Systems
Solar Thermal and Optics
Heidenhofstraße 2
79110 Freiburg
Germany

Andreas Büchtemann
Fraunhofer-Institut für
Angewandte Polymerforschung
IAP
Postfach 600 651
14406 Potsdam
Germany

Reinhard Carius
Forschungszentrum Jülich
GmbH
Institut für Energie- und
Klimaforschung (IEK-5)
52425 Jülich
Germany

Dmitry N. Chigrin
RWTH Aachen University
Institute of Physics (IA)
Department of Physics
Templergraben 55
52056 Aachen
Germany

Marcel Dyrba
Fraunhofer Center for Silicon
Photovoltaics CSP
Walter-Hülse-Str. 1
06120 Halle (Saale)
Germany

and

Martin Luther University of
Halle-Wittenberg
Centre for Innovation
Competence SiLi-nano®
Karl-Freiherr-von-Fritsch-
Strasse 3
06120 Halle (Saale)
Germany

Markus Ermes
Forschungszentrum Jülich
GmbH
Institut für Energie- und
Klimaforschung (IEK-5)
52425 Jülich
Germany

Stephan Fahr
Friedrich-Schiller-Universität
Jena
Institute of Condensed Matter
Theory and Solid State Optics
Abbe Center of Photonics
Max-Wien-Platz 1
07743 Jena
Germany

Stefan Fischer
Fraunhofer Institute for Solar
Energy Systems
Solar Thermal and Optics
Heidenhofstraße 2
79110 Freiburg
Germany

Kevin Füchsel
Friedrich-Schiller-Universität
Jena
Institute of Condensed Matter
Theory and Solid State Optics
Abbe Center of Photonics
Max-Wien-Platz 1
07743 Jena
Germany

and

Fraunhofer Institute of Applied
Optics and Precision
Engineering IOF
Albert-Einstein-Strasse 7
07745 Jena
Germany

Andreas Gerber
Forschungszentrum Jülich GmbH
Institut für Energie- und Klimaforschung (IEK-5)
52425 Jülich
Germany

Jan Christoph Goldschmidt
Fraunhofer Institute for Solar Energy Systems
Solar Thermal and Optics
Heidenhofstraße 2
79110 Freiburg
Germany

Andreas Gombert
Soitec Solar GmbH
Bötzinger Str. 31
79111 Freiburg
Germany

Johannes Gutmann
Fraunhofer Institute for Solar Energy Systems
Solar Thermal and Optics
Heidenhofstraße 2
79110 Freiburg
Germany

Florian Hallermann
RWTH Aachen University
Institute of Physics (IA)
Department of Physics
Templergraben 55
52056 Aachen
Germany

Hubert Hauser
Fraunhofer Institute for Solar Energy Systems
Solar Thermal and Optics
Heidenhofstraße 2
79110 Freiburg
Germany

Christian Helgert
Friedrich-Schiller-Universität Jena
Abbe Center of Photonics
Institute of Applied Physics
Max-Wien-Platz 1
07743 Jena
Germany

Barbara Herter
Fraunhofer Institute for Solar Energy Systems
Solar Thermal and Optics
Heidenhofstraße 2
79110 Freiburg
Germany

Thorsten Hornung
Fraunhofer Institute for Solar Energy Systems
Solar Thermal and Optics
Heidenhofstraße 2
79110 Freiburg
Germany

Jacqueline Anne Johnson
University of Tennessee Space Institute
Department of Mechanical Aerospace and Biomedical Engineering
411 B.H. Goethert Parkway
Tullahoma, TN 37388
USA

Thomas Käsebier
Friedrich Schiller University Jena
Abbe Center of Photonics
Institute of Applied Physics
Max-Wien-Platz 1
07743 Jena
Germany

Thomas Kirchartz
Imperial College London
Department of Physics and
Center for Plastic Electronics
South Kensington Campus
SW7 2AZ London
United Kingdom

Ernst-Bernhard Kley
Friedrich Schiller University Jena
Abbe Center of Photonics
Institute of Applied Physics
Max-Wien-Platz 1
07743 Jena
Germany

Matthias Kroll
Friedrich Schiller University Jena
Abbe Center of Photonics
Institute of Applied Physics
Max-Wien-Platz 1
07743 Jena
Germany

Deepu Kumar
RWTH Aachen University
Institute of Physics (IA)
52056 Aachen
Germany

Falk Lederer
Friedrich-Schiller-Universität
Jena
Institute of Condensed Matter
Theory and Solid State Optics
Abbe Center of Photonics
Max-Wien-Platz 1
07743 Jena
Germany

Alexander Mellor
Universidad Politécnica de
Madrid
Instítuto de Energía Solar
Avenida Complutense 30
28040 Madrid
Spain

Paul-Tiberiu Miclea
Fraunhofer Center for Silicon
Photovoltaics CSP
Walter-Hülse-Str. 1
06120 Halle (Saale)
Germany

Juan C. Miñano
light prescriptions innovators
(LPI)
2400 Lincoln Ave,
Altadena, CA 91001
USA

Martin Otto
Martin-Luther-Universität
Halle-Wittenberg
Institute of Physics
Heinrich-Damerow-Str. 4
06120 Halle
Germany

Christian Paßlick
Martin Luther University of
Halle-Wittenberg
Centre for Innovation
Competence SiLi-nano®
Karl-Freiherr-von-Fritsch-Str. 3
06120 Halle (Saale)
Germany

Thomas Pertsch
Friedrich Schiller University Jena
Abbe Center of Photonics
Institute of Applied Physics
Max-Wien-Platz 1
07743 Jena
Germany

Marius Peters
Fraunhofer Institute for Solar
Energy Systems
Solar Thermal and Optics
Heidenhofstraße 2
79110 Freiburg
Germany

Liv Prönneke
Universität Stuttgart
Institut für Photovoltaik
Pfaffenwaldring 47
70569 Stuttgart
Germany

Uwe Rau
Institut für Energie- und
Klimaforschung 5 - Photovoltaik
Forschungszentrum Jülich
GmbH
Wilhelm-Johnen-Straße
52425 Jülich
Germany

Carsten Rockstuhl
Karlsruher Institut för
Technologie
Institut für Theoretische
Festkörperphysik
Wolfgang-Gaede-Str. 1
76128 Karlsruhe
Germany

Jens Schneider
Fraunhofer Center for Silicon
Photovoltaics CSP
Otto-Eißfeldt-Street 12
06120 Halle
Germany

Isolde Schwedler
Fraunhofer Center for Silicon
Photovoltaics CSP
Otto-Eißfeldt-Street 12
06120 Halle
Germany

Stefan Schweizer
Fraunhofer Center for Silicon
Photovoltaics CSP
Walter-Hülse-Str. 1
06120 Halle (Saale)
Germany

and

South Westphalia University of
Applied Sciences
Department of Electrical
Engineering
Lübecker Ring 2
59494 Soest
Germany

and

Fraunhofer Application Center
for Inorganic Phosphors
Branch Lab of Fraunhofer
Institute for Mechanics of
Materials IWM
Lübecker Ring 2
59494 Soest
Germany

List of Contributors

Gerhard Seifert
Fraunhofer Center for Silicon
Photovoltaics CSP
Otto-Eißfeldt-Street 12
06120 Halle
Germany

Alexander N. Sprafke
Martin Luther University
Halle-Wittenberg
Institute of Physics
Heinrich-Damerow-Str. 4
06120 Halle
Germany

Martin Steglich
Friedrich Schiller University Jena
Abbe Center of Photonics
Institute of Applied Physics
Max-Wien-Platz 1
07743 Jena
Germany

Lorenz Steidl
Johannes Gutenberg University
Mainz
Institute of Organic Chemistry
Duesbergweg 10-14
55099 Mainz
Germany

Heiko Steinkemper
Fraunhofer Institute for Solar
Energy Systems
Solar Thermal and Optics
Heidenhofstraße 2
79110 Freiburg
Germany

Franziska Steudel
Fraunhofer Center for Silicon
Photovoltaics CSP
Walter-Hülse-Strasse 1
06120 Halle (Saale)
Germany

Andreas Tünnermann
Friedrich Schiller University Jena
Abbe Center of Photonics
Institute of Applied Physics
Max-Wien-Platz 1
07743 Jena
Germany

and

Fraunhofer Institute of Applied
Optics and Precision Engineering
IOF, Albert-Einstein-Strasse 7
07745 Jena
Germany

Carolin Ulbrich
Forschungszentrum Jülich
GmbH
Institut für Energie- und
Klimaforschung (IEK-5)
52425 Jülich
Germany

Johannes Üpping
Martin-Luther-University
Halle-Wittenberg
Institute of Physics
Heinrich-Damerow-Str. 4
06120 Halle
Germany

Gero von Plessen
Physikalisches Institut
RWTH Aachen
52056 Aachen
Germany

Armin Wedel
Fraunhofer-Institut für
Angewandte Polymerforschung
IAP
Postfach 600 651
14406 Potsdam
Germany

Ralf B. Wehrspohn
Martin Luther Universität
Halle-Wittenberg
Institute of Physics
Heinrich-Damerow-Str. 4
06120 Halle
Germany

and

Fraunhofer Institute for
Mechanics of Materials (IWMH)
Walter-Hülse-Strasse 1
06120 Halle
Germany

Marie-Christin Wiegand
Fraunhofer-Center für
Silizium-Photovoltaik CSP
Walter-Hülse-Str. 1
06120 Halle (Saale)
Germany

Samuel Wiesendanger
Friedrich-Schiller-Universität
Jena
Institut für Festkörpertheorie
und -optik
Max-Wien-Platz 1
07743 Jena
Germany

Sebastian Wolf
Fraunhofer Institute for Solar
Energy Systems
Solar Thermal and Optics
Heidenhofstraße 2
79110 Freiburg
Germany

Rudolf Zentel
Johannes Gutenberg University
Mainz
Institute of Organic Chemistry
Duesbergweg 10-14
55099 Mainz
Germany

1
Current Concepts for Optical Path Enhancement in Solar Cells
Alexander N. Sprafke and Ralf B. Wehrspohn

1.1
Introduction

The conversion efficiency of a solar cell, that is, the ratio of electrical power extracted from the cell to the power of solar photons flowing into the cell, is directly connected to the number of photons absorbed in the absorber material of the cell. Therefore, it is of critical importance to insert as many photons as possible into the cell and keep them inside the cell until they are finally absorbed. While achieving the first aspect is referred to as antireflection, the second aspect is commonly called optical path enhancement, also known as light Trapping, which is the focus of this chapter.

Because of its fundamental significance to the solar-to-electrical conversion mechanism, light trapping should be considered for any solar absorber material. However, light trapping is of particular importance for solar cells based on crystalline silicon (c-Si). Owing to its abundance and to the long-existing mature technologies in the electronic industry, commercial c-Si based solar cells are widely available and dominate the PV market today [1]. But since c-Si is an indirect semiconductor, it is actually a relatively bad light absorber. Figure 1.1 shows the absorption depth δ of c-Si. λ of an absorbing material is defined as the distance at which the intensity of light decreases to $1/e$ after it enters the material. For wavelengths $\lambda < 500$ nm, most of the light energy is absorbed within a micron. For longer wavelengths, λ increases rapidly (note the logarithmic y-axis) and reaches values in the range of centimeters for wavelengths in the spectral range of the bandgap of c-Si at around $\lambda \approx 1150$ nm.

To compensate for this, one could either use thick c-Si absorbers, or apply light-trapping schemes to optically thicken the absorber material. Owing to high costs and challenging requirements on material quality, using millimeter or centimeter thick c-Si absorbers is not an option. In fact, quite contrary is the case. The current trend in the PV industry as well as in the research community is to develop concepts to increase material efficiency by decreasing the Si wafers thickness below 100 μm. To stay competitive, the conversion efficiency of thin c-Si cells needs to be

Photon Management in Solar Cells, First Edition. Edited by Ralf B. Wehrspohn, Uwe Rau, and Andreas Gombert.
© 2015 Wiley-VCH Verlag GmbH & Co. KGaA. Published 2015 by Wiley-VCH Verlag GmbH & Co. KGaA.

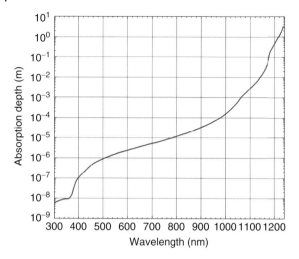

Figure 1.1 Absorption depth λ of crystalline silicon plotted against the wavelength λ of light. The optical properties to calculate λ were taken from Ref. [2].

higher or at least comparable to the efficiency of 180 μm thick, commercial solar cells available today.

In Figure 1.2 calculated absorption spectra of bare planar c-Si slabs of different thicknesses d are plotted. From the absorption at small wavelengths one can easily see that independent of the thickness light incoupling into silicon is rather weak despite the very high absorption coefficient of silicon in this spectral range (compare Figure 1.1). This is due to the strong refractive index contrast between silicon and air of $n_{Si}/n_{air} \approx 3,5$ in the visible spectral region and could be overcome by applying means of antireflection. Much more severe in the context of photovoltaic applications is the strong loss of absorption when decreasing the thickness of the c-Si slab. For a thickness of $d = 200$ μm, a typical thickness for commercially available c-Si solar cells, there is considerable absorption up to $\lambda \approx 950$ nm which then starts to decrease toward the spectral position of the band edge of c-Si. Thinning the c-Si slab leads to a drastic reduction of absorption and shifts the absorption edge to shorter wavelengths. For example, the integrated absorption for the $d = 0.5$ μm and the $d = 10$ μm slab is only 7.8% and 69% of that of the 200 μm slab, respectively (integrated from $\lambda = 400$ nm to 1200 nm).

These examples underly the importance that effective concepts for optical path length enhancement are inevitable for future solar cell designs. Here, we chose c-Si as an example material; however, in principle, these considerations apply to any solar cell material and have to be taken into account.

1.2
Planar Antireflection Coatings

From an optics point of view, the simplest solar cell consists of an homogeneous slab as introduced in the previous section, with an optical thickness of

1.2 Planar Antireflection Coatings

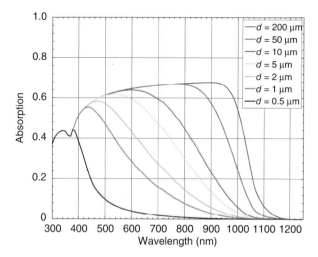

Figure 1.2 Calculated absorption spectra of planar c-Si slabs of varying thickness d suspended in air at normal light incidence. Note, that since the coherence length of sunlight is $< 1\,\mu m$, incoherent light was assumed in the calculations.

$d_{opt} = nd/\cos\theta_2$, where n and d are the refractive index and the geometrical thickness of the cell, respectively, and θ_2 is the angle of refraction (see Figure 1.3a for illustration). As pointed out above, a rather large amount of light energy does not reach the absorbing region but is reflected at the top surface. For example, the reflectivity of bare silicon is $R \approx 35\%$ for a wavelength of $\lambda = 600$ nm and normal incidence (compare Figure 1.2). Coupling a greater amount of light energy into the cell is commonly achieved by applying a planar antireflection coating (ARC) on top of the cell. This coating consists of a dielectric layer with a thickness of $\lambda/4$ such that the light wave reflected from the top surface and the wave reflected from the dielectric–semiconductor interface interfere destructively resulting in $R = 0$ in an ideal case for a particular design wavelength λ_{des}. In commercial photovoltaic applications mostly SiN deposited by chemical vapor deposition is applied as a coating material. A thickness of $d \approx 70$ nm minimizes reflection at $\lambda_{des} = 600$ nm, which is close to the peak of the solar spectrum and is responsible for the commonly known dark-bluish appearance of commercially available c-Si solar cells. A single-layer ARC minimizes light reflection of a particular design wavelength and angle of incidence only. This is in conflict with the broad solar spectrum and to the ever-changing altitude of the sun during the day. It is possible to minimize reflection losses for more than one wavelength by the use of double- or even more complex multi-layer antireflection coatings. However, another conceptual disadvantage of planar interference-based ARCs is their inherent symmetric mode of operation: They increase the flow of incoming light into the cell (at least for the design wavelength). However, photons that are not absorbed during their first pass through the cell and eventually reach the interface of the ARC again because of a scattering event or reflection at a mirror at the backside, are now efficiently coupled out of the solar cell. Thus, an ARC does not offer

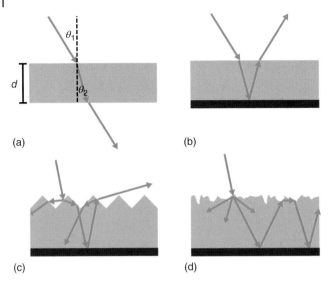

Figure 1.3 Light trapping concepts: (a) bare, planar slab of the absorber material, no light trapping applied, (b) backside reflector, (c) pyramidal texture, (d) random scattering texture. Sketched structures are not to scale.

any light trapping and additional or completely different concepts need to be considered to accomplish sufficient optical light path enhancement.

1.3
Optical Path Enhancement in the Ray Optical Limit

Light trapping in commercially available c-Si solar cells utilizes approaches that are based on ray optics. The most simple tool in this context would be a mirror at the backside of the absorber, which leads to a doubling of the optical light path (see Figure 1.3b); thus, the optical path enhancement is $\alpha_{opt} = 2$. A common approach to achieve real light trapping is to change the propagation direction of the light such that total internal reflection occurs (for a silicon–air interface, the critical angle is $\theta_c \approx 16°$). To achieve internal light trapping one has to move from planar surfaces to textured surfaces.

The frontside of commercial c-Si based solar cells are textured by alkaline or acidic chemical wet-etching [3, 4]. For example, anisotropic alkaline wet-etching of monocrystalline material (the <100> orientation is preferentially etched) results in randomly arranged pyramidal structures with feature sizes in the $\approx 10\,\mu m$ range. Using lithographic steps for inverted pyramids are also possible. In these structures light reflected from the surface gets a second chance to enter the absorber because of the tilted surfaces of the pyramids thereby leading to less reflection back into the upper halfspace (see Figure 1.3c). Additional ARCs may further decrease the reflection of the textured surface. Since the frontside is not planar anymore, light will be coupled into the silicon obliquely, not only leading to some optical path length enhancement, but also enabling total internal reflection for light rays coming back to the frontside. Thus, true optical path length enhancement is achieved. Today's world record single-junction silicon solar cell, the PERL cell [5] with a conversion efficiency of more than 24%, which uses pyramidal surface structures, holds the current world record as the single-junction silicon solar cell. In this case, inverted pyramids are arranged in a square lattice pattern (lattice constant ≈ 20 μm). To obtain high-efficiency c-Si solar cells with pyramidal front side textures one has to carefully optimize the geometric parameters of the pyramids to match the actual wafer thickness. Otherwise, light can be effectively coupled out of the cell because a large fraction of back-reflected light hits the pyramidal facets perpendicularly [3]. In solar cells based on materials that do not offer an easy way to obtain pyramidal structures , for example, thin GaAs membranes, texturing may be introduced in additional layers such as structured coatings [6].

1.4
Scattering Structures for Optical Path Enhancement

A very effective light trapping concept is the full randomization of the propagation directions of the incident sunlight after hitting the first absorber interface. A Lambertian scatterer at the front surface of the absorber material offers the highest degree of randomization [7]. In this case, the incoming light is scattered into propagation directions with a cosine distribution. The optically thicker medium, which is usually the absorber material, offers a larger phase space for light to be filled than does the adjacent medium (e.g., air). Thus, light will be preferably scattered into the absorber instead of out of it leading to a high confinement of light inside the desired volume. Additionally, an ideal Lambertian scatterer operates independently of the angle of incidence. This scenario was first elaborated by Eli Yablonovitch in 1982 [7]. He found the upper limit of light path enhancement for random scattering textures to be $\alpha_{opt} = 4n^2$, where n is the refractive index of the solar cell. For this limit, nowadays commonly referred to as the Yablonovitch or Lambertian limit, illumination from the frontside, a Lambertian front side texture and a perfect mirror at the backside are assumed (Figure 1.3d). Yablonovitch's theory applies in the limit of vanishing absorption, but has been generalized to apply for arbitrary absorption by Green [8]. For example, for a solar cell made of silicon $\alpha_{opt} \approx 50$ is theoretically possible.

A particularly interesting example of a material class that exhibits strong diffuse scattering close to the Lambertian case is *black silicon* (b-Si) [9–11]. Black Si surface textures consist of densely packed needle-like peaks and pits of irregular shape and high aspect ratios (see Figure 1.4(a)). Black silicon can be prepared by several methods such as metal-assisted wet-chemical etching, inductively coupled reactive ion etching (ICP-RIE), or short-pulse laser irradiation [11–15]. Another important advantage of b-Si over conventional pyramidal textures is the comparatively large feature sizes of the latter. Application of pyramidal textures will fail

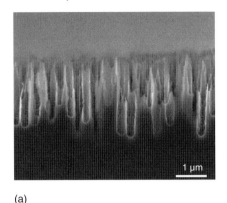 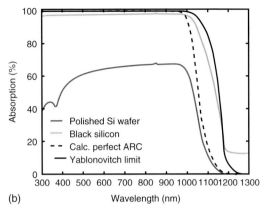

Figure 1.4 (a) SEM cross section of a typical b-Si sample fabricated by ICP-RIE. (b) Experimental absorption spectra of a b-Si structure (green line). For comparison the measured absorption spectra of a polished Si wafer (solid blue) as well as theoretical spectra of a perfect ARC (dashed black) and a perfect ARC with Lambertian scattering (Yablonovitch limit, solid black) are also plotted (adapted from Ref. [14]).

when the wafer thickness comes to values below 50 μm. For such thin wafers, b-Si is a suitable alternative.

A typical experimentally measured absorption spectrum of b-Si prepared by ICP-RIE is plotted in Figure 1.4(b) next to the spectrum of a bare c-Si wafer and calculated spectra of a wafer for which an ideal ARC but no light trapping is assumed, and the Yablonovitch limit, that is, with maximum light trapping. The b-Si sample reveals an almost complete suppression of the reflectivity over a broad spectral range making it far more effective than conventional ARCs. In the spectral region of the band gap of silicon, b-Si clearly outperforms a theoretical perfect ARC, which has to be attributed to its excellent light trapping properties. The absorption in the spectral region of the bandgap is very close to the Yablonovitch limit (see also Figure 1.10). The extraordinary antireflection and light trapping features of b-Si may be explained by an intuitive picture: On the one hand, the top of the texture consists of frayed tips of deep sub-wavelength dimensions that broaden slowly toward the wafer. Therefore, the surface can be treated as an effective medium with a smooth varying refractive index resulting in negligible back reflection, even for quasi-omnidirectional light incidence [16, 17]. On the other hand, the feature sizes at the bottom of the structure are in the range of a few hundred nanometers, that is, have dimensions of the wavelength of the incoming light. Thus, the bottom parts of the texture act as very effective light scatterers resulting in efficient randomization which is a condition for Lambertian light trapping. This makes black silicon an ideal candidate for photovoltaic applications, especially for ultrathin devices with a thickness below 50 μm.

Solar cell prototypes made of b-Si have already been demonstrated with efficiencies of up to 18.7% [11, 13, 18–21]. However, due to its strongly enlarged surface area and defect density, b-Si exhibits rather poor electronic surface quality, that is, high electronic recombination losses and degradation effects, which strongly affect solar cell device performance especially at short wavelengths [22, 23]. Nevertheless, it has been shown recently that an appropriate passivation of the b-Si surface by a conformally deposited Al_2O_3 layer can effectively compensate for these effects [14].

A thorough review on b-Si for photovoltaics will be given in Chapter 5 [24]. This section has also not covered scattering structures for thin-film photovoltaics such as amorphous and microcrystalline solar cells. These are treated in Chapter 4.

1.5
Resonant Structures for Optical Path Enhancement

Taking the simple planar slab model of the solar cell as introduced in Figure 1.3a, the maximum optical path enhancement is obviously attained when the impinging light is redirected to propagate parallel to the boundaries. In the case of normal light incidence this would mean a directional change of 90°.

Structures that are able to accomplish such a rather large redirection of the optical path are diffractive elements, such as gratings. A possible light path involving a diffraction grating in a solar cell may be described as sketched in Figure 1.5. Sunlight enters the absorber material of the cell. A reflection grating placed at the backside of the cell diffracts the light into at least one diffraction order m other than the 0th order as long as the grating constant is larger than the considered wavelength of light (for normal incidence). The amount of light energy going into a specific diffraction order is given by the diffraction efficiency, which crucially depends on the precise geometry of the actual grating [25]. Only light of a particular wavelength λ_g will be diffracted into a grazing angle as demanded, that is, into a direction parallel to the boundaries (also known as Rayleigh anomaly [26]). For $\lambda < \lambda_g$ the diffracted light travels back to the front surface but with a much larger angle of incidence compared to light that would have been reflected by a simple

Figure 1.5 Possible light path inside a solar cell with a backside grating. After entering the cell the light is diffracted by a grating at the backside of the cell. Due to the shallow angle of the diffraction order total internal reflection is possible. Upon the second diffraction the light is possibly diffracted into the direction of the former 0th diffraction order and leaves the cell (adapted from Ref. [27]).

backside mirror. Thus, total internal reflection at the front side is enabled. Hereafter, the light is redirected back onto the grating structure and is diffracted again. Because of the reciprocity theorem the light now may be either diffracted into the direction of the former 0th diffraction order and thus eventually leave the cell, or it may be diffracted into a diffraction order which is again trapped in the cell by total internal reflection. Again, the redistribution ratios of the light energy into the different diffraction orders are determined by the precise diffraction efficiencies. Also, because of the different angle of incidence onto the grating in the second pass through the cell, additional diffraction channels may be possible, which may lead to further light trapping by total internal reflection.

For a more detailed analysis instead of considering perfectly collimated light the divergence angle of the sunlight $\Theta_S = 4.7$ mrad, or the even more realistic angular divergence of the circumsolar radiation $\Theta_C = 44$ mrad, must not be ignored. Then, the conservation of étendue has to be taken into account. Assuming that light rays impinging on the grating structure perpendicularly will experience a change of propagation direction of 90° and that every other ray, which has a slightly oblique incidence due to the angular divergence, follows the conservation of étendue, the maximum optical path length enhancement is found to be

$$\alpha_{opt} \approx \frac{2n\Theta}{1 - \cos \Theta} \tag{1.1}$$

for $\Theta \ll 1$ as derived in Ref. [27]. In case of silicon, this gives an enhancement factor of $\alpha_{opt} = 2980$ and $\alpha_{opt} = 318$ considering Θ_S and Θ_C, respectively. Strictly speaking, these results hold for rotational symmetry of the problem and unity diffraction efficiency for all wavelengths and angles of incidence only. Thus, this approach constitutes a theoretical upper limit for any diffraction-based structure and emphasizes the high potential to effectively trap light. To obtain realistic characteristics of grating structures integrated into solar cells, rigorous numerical calculations that take the wave nature of light fully into account, such as the finite-difference time-domain method [28], have to be conducted.

On the downside it has to be noted that the spectral width of diffractive elements might be too narrow for certain applications. This especially accounts for 2D gratings. Therefore, more complex 2D gratings have been suggested that exhibit a broader spectral response, but these structures suffer from lower maximum path enhancements [29]. Since diffraction-based light trapping concepts rely on the coupling of light into modes that are bound to the absorber material by total internal reflection, the ultimate lower limit for the absorber film thickness is dictated by the monomodal regime. An approach suggested by Yu et al. [29] to circumvent this situation is to decouple the active layer from the light-guiding layer.

Several concepts are discussed in the literature to realize resonant structures for light trapping including dielectric structures [30–32] as well as metallic structures such as metallic gratings and metal nanoparticle arrays, which additionally exploit plasmonic effects [33, 34]. Early works were performed by Morf and coworkers [31] who applied binary gratings to the backside of a solar cell. In particular, blazed gratings are of interest since they exhibit enhanced diffraction efficiencies for certain diffraction orders in comparison to symmetric gratings. Morf and coworkers report path length enhancements of about $\alpha_{opt} \approx 5$ in these structures. The enhancement factor was determined by the change in reflectance which is a rather imprecise measure. An improved determination of the enhancement factor is to measure the spectrally resolved external quantum efficiency (EQE) enhancement

$$\alpha_{EQE}(\lambda) = \frac{EQE_w(\lambda)}{EQE_{w/o}(\lambda)}, \quad (1.2)$$

where the subscripts w and w/o indicate the EQE with and without the enhancing structure, respectively. It should be noted that in the limit of low absorption $\alpha_{EQE} = \alpha_{opt}$ holds. The first work using $\alpha_{EQE}(\lambda)$ as a measure was conducted by Zaidi et al. [35]. They found an enhancement factor of $\alpha_{EQE}(\lambda) = 2.7$ for a rectangular grating at the backside of a c-Si solar cell in the spectral region of low absorption at the band gap. Measurements on amorphous silicon–based solar cells revealed similar results [36]. Kimerling and coworkers investigated the combination of a grating and a dielectric Bragg mirror at the backside [37, 38]. They obtained $\alpha_{EQE}(\lambda) = 7$ and $\alpha_{EQE}(\lambda) = 135$ for a wavelength of $\lambda = 1100$ nm and $\lambda = 1200$ nm, respectively.

A systematic theoretical comparison of gratings, the combination of gratings with distributed Bragg reflectors (DBR), and 3D photonic crystals was conducted by the group of Joannopoulos [32]. 3D photonic crystal structures combine the properties of a grating (diffraction) and a mirror (reflection) and hence may reveal a similar or possibly higher potential. Joannopoulos et al. found that the efficiency of a 2 µm silicon solar cell can be increased by 25% by adding an optimized 1D grating to a DBR, 31.3% by a 2D grating instead of a 1D grating, and 26.3% by replacing the DBR with 2D photonic crystal. Using a 3D photonic crystal instead of any of these structures resulted in a slightly better enhancement of 26.5%.

Experiments using 3D photonic crystals on the backside of a silicon solar cell have been carried out very rarely. For example, in [39] it was found, that the short circuit current of a thin (1.5 µm) crystalline solar cell could be enhanced by 10% with a 3D photonic crystal on the backside in comparison to a reference cell without the crystal. The challenge is not only to create high-quality 3D photonic crystals such as inverted opals with a photonic band gap in the spectral range of the band gap of silicon, but also to transfer them to and integrate them into the solar cell.

Up to now the achieved optical path enhancements are still far from the limit given by the conservation of étendue (up to $\alpha_{opt} = 2980$ for silicon). Therefore, there is still a lot of room for more research on gratings and photonic crystals for light trapping in solar cells. A thorough review on the theory and technological realization of diffractive grating structures are given in Chapter 3.

An application field of photonic crystals that emerged during the recent years is that of integral photon management. Here, the absorber itself is structured as a 1D, 2D, or 3D photonic crystal which is very different from the approaches given above that leave the absorber material unaffected. This gives way to the exploitation of effects not considered before for light trapping, for example, slow light phenomena to increase the interaction duration with the absorber material, manipulating light propagation at particular frequency ranges and into certain propagation directions [40–42]. This may shed new light on the well-known properties of photonic crystals.

1.6
Ultra-Light Trapping

As pointed out in the previous section, the sunlight illuminates the solar cell with a very small divergence Θ_S; the circumsolar radiation divergence Θ_C is just a little bit larger. At the same time light is able to escape from the cell into the entire half-space above the cell, that is, the angular escape cone is $\Theta_e = 2\pi$. The difference of the small cone of incidence and the large escape cone constitutes an avoidable loss mechanism. Because of light path reversibility the acceptance cone and the escape cone are essentially equal. Therefore, shrinking the acceptance cone to match the angular divergence of the incoming irradiation from the sun would also lead to a restriction of the escape cone to a much smaller value. A concept to narrow down the acceptance/escape cone is the use of angular-selective reflectance filters and implementation of a photonic light trap [27, 43].

Figure 1.6 sketches the working principle of this approach. Sunlight passes through an angular selective filter that features high transmittance for an acceptance cone matched to the angular divergence of the sun light and low transmittance but high reflectance for all other angles of incidence. Then, the light enters the absorbing region of the solar cell that is equipped with a scattering light trapping mechanism at the frontside. While part of the isotropically scattered light is trapped by total internal reflection, the residual light escapes from the active region into the space above to hit the angular selective filter again. Only light that is scattered into the small acceptance range of the filter is able to leave the system. All other light is reflected back into the active region of the cell. Because of the small acceptance range of the filter a tracking system which follows the course of the sun is necessary. In combination with this tracking system this concept is called ultra-light trapping.

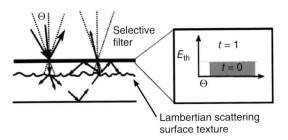

Figure 1.6 Schematic representation of the ultra-light trapping concept. Inset: Spectral range of an ideal angular selective filter (adapted from Ref. [43]).

In an ideal case, the angular-selective filter has unity transmittance for an angle of incidence smaller than the desired acceptance cone and unity reflectance for larger angles. Placed above a solar cell with a Lambertian-scatterering frontside and a perfect mirror at the backside, the maximal possible path length enhancement of this system is given by Goetzberger [44], Minano [45]

$$\alpha_{opt} = \frac{4n^2}{\sin^2\Theta}. \tag{1.3}$$

Inserting Θ_C results in a very large factor of optical path length enhancement of $\alpha_{opt} \approx 26 \times 10^3$ for silicon; inserting Θ_S even leads to $\alpha_{opt} \approx 2 \times 10^6$. The same values are derived for approaches involving concentration optics following the conservation of étendue. Additionally, for ultra-light trapping also the Shockley–Queisser limit for concentrating systems holds [46]. Ultra-light trapping and light concentration share a deep analogy to each other. This analogy is based on the fact that both concepts change the angular range of the acceptance cone. While the ultra-light trapping concept narrows down the acceptance cone to match the angular divergence of the sunlight, concentration systems work vice versa, that is, the angular divergence of the sunlight is enlarged to match the acceptance cone of the cell [47].

Until now, no experimental realization of ultra-light trapping has been conducted successfully to increase the efficiency of a solar cell. This is due to the fact that very few filters exist that exhibit the desired characteristics. Nevertheless, promising attempts have been conducted by Höhn et al. [48]. A comprehensive overview on the effects of different angular selective structures on silicon solar cells has been given by Ulbrich et al. [43, 49] and is also described in Chapter 7 of this book.

In addition to ideal filters, manufacturable rugate stacks and 3D photonic crystals were investigated in their work [43]. Rugate stacks show suppressed ripple formation in their optical spectra and no harmonic reflections as compared to Bragg stacks [50]. Iso-frequency surfaces for a frequency in the spectral region of the band gap of silicon are plotted in Figure 1.7(a) and (b) for a rugate filter and an inverted opal in Γ–X direction, respectively. Readers who are not familiar with this kind of plot may note that $k_x, k_y,$ and k_z give the direction of the light impinging on the structure whereas the surface basically stands for available photonic states, that is, light of directions that are part of the surface will be able to enter the structure, otherwise it will be reflected. The acceptance cone of the rugate filter is given by the second photonic band (Figure 1.7(a)). The photonic density of states vanishes for increasing angles because the stop gap shifts to higher frequencies; thus, light impinging from outside the acceptance cone is reflected. This holds true until for even higher angles the lower edge of the stop gap reaches the frequency that is being plotted, that is, light is transmitted by the first band and more importantly light can escape from the device into these directions. For the 3D photonic crystal the inverted opal structure was chosen due to its well-known properties. On first sight it seems feasible to use the inverted opal in the $\Gamma - L$ direction with the stop gap pointing to the sun and using the spectral shift of the

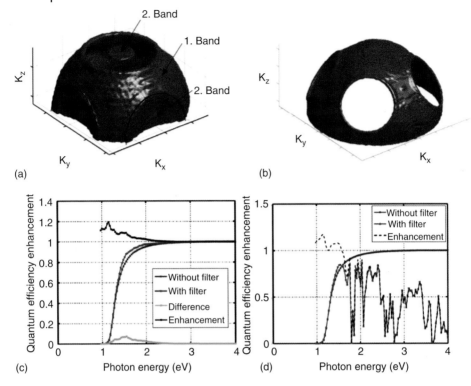

Figure 1.7 3D isofrequency surfaces of a rugate filter (a) and an inverted opal in Γ − X direction (b). (c): Calculated quantum efficiency for an inverted opal grown in Γ − X direction on top of a silicon solar cell. For photon energies above the photonic stop gap unity transmission is assumed. (d): Same as (c) but with realistic transmission properties (adapted from Ref. [43]).

stop gap similar to the mechanism of the rugate filter [51]. But it turns out that the use of the inverted opal in Γ − L direction results in unwanted strong reflections from higher-order stop gaps. Using the inverted opal in Γ − X direction enables to assign the stop gap itself for ultra-light trapping instead of the spectral shift of the stop gap. Figure 1.7(b) shows that the inverted opal in Γ − X direction does not match its acceptance cone to that of the angular divergence of the sun as tight as the rugate filter, but it apparently reflects a larger fraction of the hemisphere back.

Assuming perfect transmission of the entire solar spectrum through an inverted opal in Γ − X direction the calculated spectrally resolved quantum efficiency for a thin-film silicon solar cell is shown in Figure 1.7(c) [43]. Near the band gap of amorphous silicon the effect of the ultra-light trapping is visible. However, when the realistic transmission of the inverted opal is included, the efficiency of the device is clearly decreased for higher photon energies (see Figure 1.7(d)). Anyhow, the ultra-light trapping effect close to the band gap is still visible. This behavior originates from strong parasitic back-reflections caused by higher-order flat photonic bands. However, it is still possible that other 3D photonic crystal structures are more suitable for this application, for example, graded index ones. Up to now, 1D periodic structures such as rugate filters are found to have the highest potential [43].

1.7
Energy-Selective Structures as Intermediate Reflectors for Optical Path Enhancement in Tandem Solar Cells

The concepts described in the previous sections all aim toward maximizing the number of photons inside the absorbing region of the solar cell and thus primarily apply to single-junction solar cells. In these kind of cells every photon with an energy $E_\gamma \geq E_g$ absorbed by the absorber material with the band gap E_g initially generates an electron–hole pair of the energy E_γ, but the difference of the two energies, $E_\gamma - E_g$, is transferred as heat due to thermalization on a very short time scale ($\approx 10^{-12}$ s) [52]. Therefore, only E_g per absorbed photon is available. By the use of multiple layers of different E_g thermalization losses can be reduced. In this complex solar cell approach light trapping is needed, too, but now an additional task must be complied: Depending on their quantum energy the photons have to be distributed to the according layer. Furthermore, in case of stack-cell designs another important aspect has to be included. Stacked cells are multi-junction cells in which the active layers are electrically connected in series. Thus, the electrical current delivered by the structure is limited by the layer that generates the lowest current. Therefore, current-matching is necessary for optimum efficiency. To achieve current-matching the photon flux between the different layers has to be steered such that the generated currents are balanced.

Multi-junction cells achieve efficiencies above the single-junction limit [52, 53], while thin-film solar cells currently offer the best cost-effectiveness but at rather low efficiency. The micromorph tandem solar cell has the potential to combine the advantages of both cell concepts [54]. A micromorph tandem solar cell consists of a microcrystalline silicon (µc-Si) bottom cell of a few micrometer thickness and a hydrogenated amorphous silicon (a-Si:H) top cell of a thickness < 0.3 µm (see Figure 1.8). Ideally, the top cell ($E_g \approx 1.7$ eV) absorbs the high-energy photons while the low-energy photons are transmitted to the bottom cell ($E_g \approx 1.1$ eV). In the real world, the combined efficiency of these two layers suffers from the difference in absorbed photon flux between a-Si:H and µc-Si under AM1.5 irradiation. This leads to the above-mentioned electrical current mismatch between the two layers, which is one of the major limitations to this solar cell design.

To balance the currents and to operate the tandem solar cell at its maximum power point, intermediate reflective layers (IRLs) have been suggested [55]. These would be placed between the a-Si:H top cell and the µc-Si:H bottom cell. To tailor the photon distribution as desired, IRLs have to comply with the following criteria: High reflectance in the spectral region of low absorption of the top cell, high transmission in the spectral region of negligible absorption of the top cell, and high conductivity to avoid ohmic losses. In literature, three types of dielectric

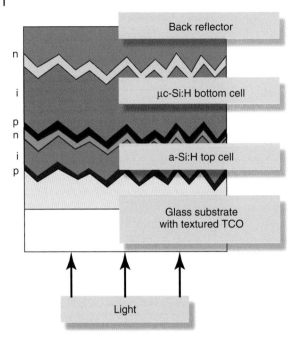

Figure 1.8 Schematic diagram of a state-of-the-art micromorph tandem solar cell without an intermediate reflector.

IRLs are predominantly discussed [56, 57]. The first type of IRL is a simple dielectric thin film that works similarly to planar interference-based ARCs as described above. They achieve a theoretical spectrally integrated absorption enhancement of $\eta_{j_{SC},theo} = 7\%$ which originates from doubling of the optical path length in the top cell [56]. The second type of IRLs are conductive Bragg reflectors, that is, is a multi-layer system that consists of two alternating conductive thin films of different refractive index. It is also based on interference but may achieve better reflectance properties than one layer only. A maximal theoretical enhancement of $\eta_{j_{SC},theo} = 20\%$ is expected [56]. 3D photonic crystal IRLs (3D-IRL) are a new class of IRL. They combine reflective and diffractive properties in a single functional layer, that is, in addition to an energy-selective redistribution an angular-selective redistribution of light is possible, which facilitates additional light-trapping mechanisms such as total internal reflection for optical path enhancement. It was shown theoretically by Bielawny et al. that 3D-IRLs can enhance the spectrally enhanced absorption by $\eta_{j_{SC},theo} = 28\%$ [56].

In experimental realizations of IRLs in the micromorph tandem cell ZnO or SiO$_x$ thin-films have been successfully integrated [58–60]. However, since thin-film solar cells are commonly applied to textured substrates any advanced IRL concept, Bragg or 3D photonic crystal based, needs to be compatible with textured substrates [61]. The maximum theoretical absorption enhancements $\eta_{j_{SC},theo}$ given above are all restricted to flat substrates. Nevertheless, it was shown by Fahr

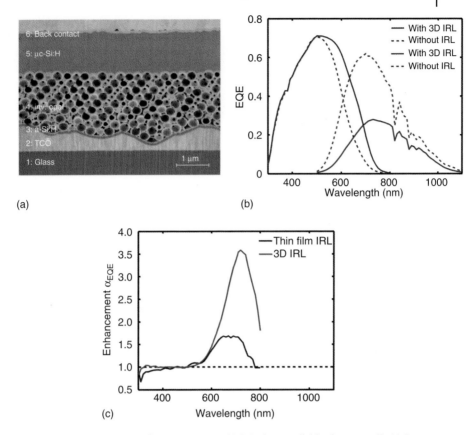

Figure 1.9 (a) Cross-sectional SEM image of the micromorph tandem cell with a 3D-photonic crystal IRL. The frontside of the cell is at the bottom. (b) Measured EQE of the micromorph tandem cell with and without a 3D-photonic crystal intermediate reflector (IRL) (red: top cell, blue:bottom cell). (c) Spectrally resolved enhancement of the EQE (α_{EQE}) of the a-Si:H top cell for a thin-film IRL (purple and for the 3D photonic crystal IRL (orange) (taken from Ref. [63]).

et al. that the given trend is still valid for textured substrates [62]. A 3D-IRL on a textured substrate fully integrated into a micromorph tandem cell has been successfully realized by Üpping *et al.* [63]. Figure. 1.9 (a) shows a cross-sectional SEM image of a micromorph tandem solar cell including the 3D-IRL. Experimentally obtained EQE measurements of the a-Si:H top cell (red) and the μc-Si:H bottom cell (blue) in a tandem solar cell with (solid) and without (dashed) a 3D photonic crystal IRL are presented in Figure 1.9 (b). In the wavelength range of 500–800 nm the inverted opal clearly enhances the EQE of the top cell while reducing that of the bottom cell due to the redistribution of the photon flux. The short circuit current density of this top cell is $j_{SC} = 9.4$ mA cm^{-2} without IRL and $j_{SC} = 11.7$ mA cm^{-2} with the inverted opal IRL. This corresponds to an enhancement of $\eta_{jSC} = 24.5$ % which is in very good agreement with previously published numerical calculations [56]. Further simulations reveal that the maximum of the absorption of the inverted opal is around 650 nm (not shown), which is in excellent agreement with the red shift in the absorption edge of the a-Si:H top cell (see Figure 1.9) [63].

A benchmark for the impact of the IRL is the enhancement ratio $\alpha_{EQE}(\lambda)$ that was introduced earlier. Note that α_{EQE} also displays the optical path enhancement. In Figure 1.9c experimentally determined values of $\alpha_{EQE}(\lambda)$ for a thin-film-based IRL (purple) and of our 3D photonic crystal IRL (orange) are plotted. $\alpha_{EQE}(\lambda)$ of the 3D photonic crystal IRL lies always above that of the thin-film IRL and has its maximum at about 720 nm exhibiting $\alpha_{EQE} = 3.6$ which is a factor of 2.25 larger than that of the thin-film-based IRL.

1.8
Comparison of the Concepts

The theoretical maximum path enhancements for solar (angular divergence $\Theta_S \approx 4.7$ mrad) and circumsolar (angular divergence $\Theta_C \approx 44$ mrad) irradiation, the spectral width, and the angular dependence of the light trapping concepts introduced in this chapter are summarized in Table 1.1. Generally, the light path enhancement α_{opt} of advanced light trapping concepts is increased compared to Yablonovitch-limited concepts but at the expense of spectral and/or angular acceptance. Therefore, application of these concepts to photovoltaics is highly dependent on the specific requirements and demands of the solar cell or the solar module and the area of application.

In Figure 1.10 the theoretical upper limit of the EQE enhancement α_{EQE} ($= \alpha_{opt}$) is plotted as a function of the refractive index n of the absorber material for circumsolar irradiation. It is clear from this figure that the advanced light trapping concepts presented here possess a high potential for a large α_{EQE}. Concepts based on light randomization already virtually achieve the Yablonovitch-limit experimentally, for example, by b-Si structures (see Figure 1.4(b)). The highest enhancement reported in applying a resonant light trapping scheme has been achieved by the Kimmerling group with $\alpha_{EQE} = 135$ (indicated by the red circle

Table 1.1 Overview of the different light trapping concepts for solar cells given in this book.

	Non-resonant	Resonant	Angular selective	Energy-selective
Max. path enhancement (Θ_S)	$4n^2$	$853n$	$181078n^2$	$> 2n$
Max. path enhancement (Θ_C)	$4n^2$	$91n$	$2067n^2$	$> 2n$
Spectral width	Very broad	Narrow	Broad	Broad
Angular dependence	None	Large	Very Large	Very small
Physical principle	Lambert–Beer law	Conservation of étendue	Conservation of étendue	Interference

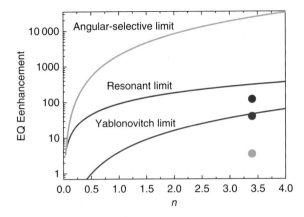

Figure 1.10 Theoretical upper limit of the integrated EQE enhancement for the Yablonovitch (blue), the resonant (red), and the angular-selective (green) light trapping concepts as a function of the refractive index n of the absorber material for circumsolar irradiation (Θ_c). The corresponding circles indicate the highest experimentally achieved enhancement found in literature so far [11, 38, 49].

in Figure 1.10) [38]. But still there is plenty of room to further investigate and optimize resonant light path enhancement concepts. For the experimental implementation of angular-selective structures, feasible concepts of integration into a solar cell are still lacking. Nevertheless, angular-selective light trapping structures hold the largest theoretical upper limit for enhancement and the first promising results have been achieved [49].

1.9 Conclusion

During the past 15 years, concepts to enhance the optical path length in solar cells have evolved significantly. This concerns theoretical understanding as well as experimental realization. In particular, with black silicon the Yablonovitch limit has been experimentally obtained for solar cells. More recent concepts such as resonant optical path enhancement have shown good progress in particular by the integration of photonic crystal approaches. The most demanding concept numerically as well as experimentally is probably the ultra-light trapping concepts by angular selective filters. Robust but also experimentally feasible concepts are missing. Moreover, some new ideas such as the use of slow light have just emerged, which might be of particular interest for 3D nanostructured solar cells.

References

1. ITRPV (2013) International Technology Roadmap for Photovoltaics 2013.

2. Green, M.A. (2008) Self-consistent optical parameters of intrinsic silicon at 300

K including temperature coefficients. *Sol. Energy Mater. Sol. Cells*, **92** (11), 1305–1310.
3. Campbell, P. and Green, M.A. (1987) Light trapping properties of pyramidally textured surfaces. *J. Appl. Phys.*, **62** (1), 243.
4. Campbell, P. (1993) Enhancement of light absorption from randomizing and geometric textures. *J. Opt. Soc. Am. B: Opt. Phys.*, **10** (12), 2410.
5. Zhao, J., Wang, A., Altermatt, P., and Green, M.A. (1995) Twenty-four percent efficient silicon solar cells with double layer antireflection coatings and reduced resistance loss. *Appl. Phys. Lett.*, **66** (26), 3636.
6. Chang, T.-H., Wu, P.-H., Chen, S.-H., Chan, C.-H., Lee, C.-C., Chen, C.-C., and Su, Y.-K. (2009) Efficiency enhancement in GaAs solar cells using self-assembled microspheres. *Opt. Express*, **17** (8), 6519.
7. Yablonovitch, E. (1982) Statistical ray optics. *J. Opt. Soc. Am. A*, **72** (7), 899.
8. Green, M.A. (2002) Lambertian light trapping in textured solar cells and light-emitting diodes: analytical solutions. *Prog. Photovoltaics Res. Appl.*, **10** (4), 235–241.
9. Yoo, J., Yu, G., and Yi, J. (2009) Black surface structures for crystalline silicon solar cells. *Mater. Sci. Eng., B*, **159**160, 333.
10. Koynov, S., Brandt, M.S., and Stutzmann, M. (2006) Black nonreflecting silicon surfaces for solar cells. *Appl. Phys. Lett.*, **88** (20), 203107.
11. Otto, M. et al. (2014) Black Silicon Photovoltaics. *Adv. Opt. Mater.* doi:10.1002/adom.201400395.
12. Li, X., Xiao, Y., Yan, C., Zhou, K., Schweizer, S.L., Sprafke, A., Lee, J.-H., and Wehrspohn, R.B. (2013) Influence of the mobility of Pt nanoparticles on the anisotropic etching properties of silicon. *ECS Solid State Lett.*, **2** (2), P22–P24.
13. Koynov, S., Brandt, M.S., and Stutzmann, M. (2007) Black multicrystalline silicon solar cells. *Phys. Status Solidi RRL*, **1** (2), R53.
14. Otto, M., Kroll, M., Käsebier, T., Salzer, R., Tünnermann, A., and Wehrspohn, R.B. (2012) Extremely low surface recombination velocities in black silicon passivated by atomic layer deposition. *Appl. Phys. Lett.*, **100** (19), 191603.
15. Gimpel, T., Höger, I., Falk, F., Schade, W., and Kontermann, S. (2012) Electron backscatter diffraction on femtosecond laser sulfur hyperdoped silicon. *Appl. Phys. Lett.*, **101** (11), 111911.
16. Stephens, R.B. and Cody, G.D. (1977) Optical reflectance and transmission of a textured surface. *Thin Solid Films*, **45**, 19–29.
17. Kroll, M., Kasebier, T., Otto, M., Salzer, R., Wehrspohn, R., Kley, E.-B., Tünnermann, A., and Pertsch, T. (2010) Optical modeling of needle like silicon surfaces produced by an ICP-RIE process. *Proc. SPIE*, **7725**, 772505.
18. Yuan, H.-C., Yost, V.E., Page, M.R., Stradins, P., Meier, D.L., and Branz, H.M. (2009) Efficient black silicon solar cell with a density-graded nanoporous surface: optical properties, performance limitations, and design rules. *Appl. Phys. Lett.*, **95** (12), 123501.
19. Fuechsel, K., Schulz, U., Kaiser, N., Kasebier, T., Kley, E.-B., and Tünnermann, A. (2010) Nanostructured SIS solar cells. SPIE Photonics for Solar Energy Systems.
20. Oh, J., Yuan, H.C., and Branz, H.M. (2012) An 18.2%-efficient black-silicon solar cell achieved through control of carrier recombination in nanostructures. *Nat. Nanotechnol.*, **7**, 1–6.
21. Repo, P., Benick, J., Vähänissi, V., Schön, J., Von Gastrowa, G., Steinhauser, B., Schubert, M.C., and Hermle, M. (2013) n-type black silicon solar cells. *Energy Procedia*, **38**, 866–871.
22. Ziegler, J., Otto, M., Sprafke, A.N., and Wehrspohn, R.B. (2013) Activation of Al_2O_3 passivation layers on silicon by microwave annealing. *Appl. Phys. A*, **113** (2), 285–290.
23. Song, J.-W., Jung, J.-Y., Um, H.-D., Li, X., Park, M.-J., Nam, Y.-H., Shin, S.-M., Park, T.J., Wehrspohn, R.B., and Lee, J.-H. (2014) Degradation mechanism of Al_2O_3 passivation in nanostructured Si solar cells. *Adv. Mater. Interfaces*, **1** (5), 1400010. doi: 10.1002/admi.201400010.

24. Wehrspohn, R., Sprakfe, A., Gombert, A., and Rau, U. (eds) (2015) *Photon-management in Solar Cells*, Wiley-VCH Verlag GmbH.
25. Petit, R. (ed.) (1980) *Electromagnetic Theory of Gratings*, Springer-Verlag.
26. Rayleigh, L. (1907) Note on the remarkable case of diffraction spectra described by Prof. Wood. *Philos. Mag.*, **14** (79), 60.
27. Peters, M., Bielawny, A., Bläsi, B., Carius, R., Glunz, S.W., Goldschmidt, J.C., Hauser, H., Hermle, M., Kirchartz, T., Löper, P., Üpping, J., Wehrspohn, R.B., and Willeke, G. (2010) Photonic concepts for solar cells, in *Physics of Nanostructured Solar Cells* (eds V. Badescu and M. Paulescu), NOVA Science Publishers, Inc.
28. Yee, K. (1966) Numerical solution of initial boundary value problems involving maxwell's equations in isotropic media. *IEEE Trans. Antennas Propag.*, **14** (3), 302.
29. Yu, Z., Raman, A., and Fan, S. (2010) Fundamental limit of nanophotonic light trapping in solar cells. *Proc. Natl. Acad. Sci. U.S.A.*, **107** (41), 17491.
30. Gale, M.T., Curtis, B., Kiess, H.G., and Morf, R.H. (1990) Design and fabrication of submicron grating structures for light trapping in silicon solar cells. *Proc. SPIE*, **1272**, 60–66.
31. Heine, C. and Morf, R.H. (1995) Submicrometer gratings for solar energy applications. *Appl. Opt.*, **34** (14), 2476.
32. Bermel, P., Luo, C., Zeng, L., Kimerling, L.C., and Joannopoulos, J.D. (2007) Improving thin-film crystalline silicon solar cell efficiencies with photonic crystals. *Opt. Express*, **15** (25), 16986.
33. Ferry, V.E., Munday, J.N., and Atwater, H. (2010) Design considerations for plasmonic photovoltaics. *Adv. Mater.*, **22** (43), 4794.
34. Atwater, H. and Polman, A. (2010) Plasmonics for improved photovoltaic devices. *Nat. Mater.*, **9** (3), 205.
35. Zaidi, S.H., Gee, J.M., and Ruby, D.S. (2000) Diffraction grating structures in solar cells. IEEE Photovoltaic Specialists Conference, p. 395.
36. Stiebig, H., Haase, C., Zahren, C., Rech, B., and Senoussaoui, N. (2006) Thin-film silicon solar cells with grating couplers – an experimental and numerical study. *J. Non-Cryst. Solids*, **352** (920), 1949.
37. Michel, J. and Kimerling, L.C. (2007) Design of highly efficient light-trapping structures for thin-film crystalline silicon solar cells. *IEEE Trans. Electron Devices*, **54** (8), 1926.
38. Zeng, L., Yi, Y., Hong, C., Liu, J., Feng, N., Duan, X., Kimerling, L.C., and Alamariu, Ba. (2006) Efficiency enhancement in Si solar cells by textured photonic crystal back reflector. *Appl. Phys. Lett.*, **89** (11), 111111.
39. Varghese, L.T., Xuan, Y., Niu, B., Fan, L., Bermel, P., and Qi, M. (2013) Enhanced photon management of thin-film silicon solar cells using inverse opal photonic crystals with 3D photonic bandgaps. *Adv. Opt. Mater.*, **1** (10), 692–698.
40. Chutinan, A. and John, S. (2008) Light trapping and absorption optimization in certain thin-film photonic crystal architectures. *Phys. Rev. A*, **78** (2), 1.
41. Deinega, A. and John, S. (2012) Solar power conversion efficiency in modulated silicon nanowire photonic crystals solar power conversion efficiency in modulated silicon nanowire photonic crystals. *Appl. Phys. Lett.*, **112**, 074327.
42. Gomard, G., Meng, X., Drouard, E., El Hajjam, K., Gerelli, E., Peretti, R., Fave, A., Orobtchouk, R., Lemiti, M., and Seassal, C. (2012) Light harvesting by planar photonic crystals in solar cells: the case of amorphous silicon. *J. Opt.*, **14** (2), 024011.
43. Ulbrich, C., Fahr, S., Üpping, J., Peters, M., Kirchartz, T., Rockstuhl, C., Wehrspohn, R., Gombert, A., Lederer, F., and Rau, U. (2008) Directional selectivity and ultra-light-trapping in solar cells. *Phys. Status Solidi A*, **205** (12), 2831.
44. Goetzberger, A. (1981) Optical confinement in thin Si-solar cells by diffuse back reflectors. IEEE Photovoltaic Specialists Conference, p. 867.
45. Minano, J.C. (1990) Optical confinement in photovoltaics, in *Physical Limitations to the Photovoltaic Solar Energy Conversion* (eds A. Luque and G.L. Araujo), Hilger, Bristol.
46. Shockley, W. and Queisser, H.J. (1961) Detailed balance limit of efficiency of

p-n junction solar cells. *J. Appl. Phys.*, **32** (3), 510.
47. Araujo, G. and Marti, A. (1994) Absolute limiting efficiencies for photovoltaic energy conversion. *Sol. Energy Mater. Sol. Cells*, **33**, 213.
48. Höhn, O., Kraus, T., Bauhuis, G., Schwarz, U.T., and Bläsi, B. (2014) Maximal power output by solar cells with angular confinement. *Opt. Express*, **22** (S3), A715.
49. Ulbrich, C., Peters, M., Bläsi, B., Kirchartz, T., Gerber, A., and Rau, U. (2010) Enhanced light trapping in thin-film solar cells by a directionally selective filter. *Opt. Express*, **18** (S2), A133.
50. Bovard, B.G. (1993) Rugate filter theory: an overview. *Appl. Opt.*, **32** (28), 5427.
51. Fahr, S., Ulbrich, C., Kirchartz, T., Rau, U., Rockstuhl, C., and Lederer, F. (2008) Rugate filter for light-trapping in solar cells. *Opt. Express*, **16** (13), 9332.
52. Würfel, P. (2005) *Physics of Solar Cells*, Wiley-VCH Verlag GmbH.
53. Conibeer, G. and Green, M.A. (2003) Third-generation photovoltaics. *Mater. Today*, **10** (11), 42–50.
54. Fischer, D., Dubail, S., Vaucher, N.P., Kroll, U., Tomes, P., Keppner, H., Wyrsch, N., and Shah, A. (1996) The "micromorph" solar cell: extending a-Si:H technology towards thin film crystalline silicon. IEEE Photovoltaic Specialists Conference, p. 1053.
55. Yamamoto, K., Yoshimi, M., Tawada, Y., Fukuda, S., Sawada, T., Meguro, T., Takata, H., Suezaki, T., Koi, Y., Hayashi, K., Suzuki, T., Ichikawa, M., and Nakajima, A. (2002) Large area thin film Si module. *Sol. Energy Mater. Sol. Cells*, **74**, 449.
56. Bielawny, A., Rockstuhl, C., Lederer, F., and Wehrspohn, R.B. (2009) Intermediate reflectors for enhanced top cell performance in photovoltaic thin-film tandem cells. *Opt. Express*, **17** (10), 8439.
57. O'Brien, P.G., Chutinan, A., Leong, K., Kherani, N.P., Ozin, G.A., and Zukotynski, S. (2010) Photonic crystal intermediate reflectors for micromorph solar cells: a comparative study. *Appl. Phys. Lett.*, **18** (5), 4478.
58. Buehlmann, P., Bailat, J., Domine, D., Billet, A., Meillaud, F., Feltrin, A., and Ballif, C. (2007) In situ silicon oxide based intermediate reflector for thin-film silicon micromorph solar cells. *Appl. Phys. Lett.*, **91** (14), 143505.
59. Das, C., Lambertz, A., Huepkes, J., Reetz, W., and Finger, F. (2008) A constructive combination of antireflection and intermediate-reflector layers for a-Si-μc-Si thin film solar cells. *Appl. Phys. Lett.*, **92** (5), 053509.
60. Soederstroem, T., Haug, F.-J., Niquille, X., Terrazzoni, V., and Ballif, C. (2009) Asymmetric intermediate reflector for tandem micromorph thin film silicon solar cells. *Appl. Phys. Lett.*, **94** (6), 063501.
61. Bielawny, A., Üpping, J., Miclea, P.T., Wehrspohn, R.B., Rockstuhl, C., Lederer, F., Peters, M., Steidl, L., Zentel, R., Lee, S.-M., Knez, M., Lambertz, A., and Carius, R. (2008) 3D photonic crystal intermediate reflector for micromorph thin-film tandem solar cell. *Phys. Status Solidi A*, **205** (12), 2796.
62. Fahr, S., Rockstuhl, C., and Lederer, F. (2010) The interplay of intermediate reflectors and randomly textured surfaces in tandem solar cells. *Appl. Phys. Lett.*, **97** (17), 173510.
63. Üpping, J., Bielawny, A., Wehrspohn, R.B., Beckers, T., Carius, R., Rau, U., Fahr, S., Rockstuhl, C., Lederer, F., Kroll, M., Pertsch, T., Steidl, L., and Zentel, R. (2011) Three-dimensional photonic crystal intermediate reflectors for enhanced light-trapping in tandem solar cells. *Adv. Mater.*, **23** (34), 3896.

2
The Principle of Detailed Balance and the Opto-Electronic Properties of Solar Cells

Uwe Rau and Thomas Kirchartz

2.1
Introduction

The celebrated detailed balance limit derived by Shockley and Queisser (SQ) in 1961 [1] describes the maximum power conversion efficiency that can be obtained with a solar cell using a single band gap absorber material, and has also taught generations of researchers how to use the thermodynamic principle of detailed balance to properly describe fundamental physics in solar cells. During the 50 years, hundreds of papers have been published that use the formal tools that were provided by SQ. Also within this book the SQ-limit serves as a reference in several chapters.

The present chapter uses the principle of detailed balance as briefly outlined (Section 2.2.1) to derive first the classical SQ-theory (Section 2.2.2). In Section 2.2.3 a generalization of the SQ-theory is developed that, on the one hand, contains the classical SQ-case as a limiting situation and, on the other, makes predictions about the opto-electronic properties of real world solar cells in terms of an opto-electronic reciprocity relation. Section 2.3 discusses relations of the opto-electronic reciprocity theorem to other reciprocity relations in electrical engineering and semiconductor physics. Section 2.4 gives some examples for applications of the reciprocity theorem for the analysis of solar cells and modules by spatially and/or spectrally resolved luminescence. The final Section 2.5 discusses the limitations of the reciprocity relations in thin-film solar cells.

2.2
Opto-Electronic Reciprocity

2.2.1
The Principle of Detailed Balance

The SQ-limit derives the maximum solar cell efficiency from macroscopic quantities of the solar cell, namely the absorptance and the temperature. The

Photon Management in Solar Cells, First Edition. Edited by Ralf B. Wehrspohn, Uwe Rau, and Andreas Gombert.
© 2015 Wiley-VCH Verlag GmbH & Co. KGaA. Published 2015 by Wiley-VCH Verlag GmbH & Co. KGaA.

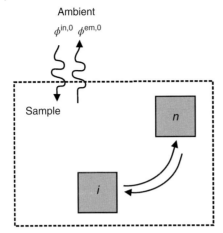

Figure 2.1 In thermal equilibrium the photon flux $\phi^{in,0}$ from the ambient and absorbed by a physical sample is compensated by the emitted flux $\phi^{em,0}$. Internal transport of charge carriers from one site (say i) to another site (n) is counterbalanced by its inverse process.

fundamental principle behind the SQ-limit, however, is a microscopic relation: the principle of detailed balance. If a sample is in thermal equilibrium, then all microscopic processes in the sample are exactly compensated by their respective inverse process. This principle is known since the first quarter of the twentieth century under various names, like "The Law of Entire Equilibrium" [2], "The Principle of Microscopic Reversibility" [3], or "The Principle of Detailed Balancing" [4]. Bridgman [5] formulated a definition that directly shows that the detailed balance is a consequence of the second law of thermodynamics. The definition reads: "No system in thermal equilibrium in an environment at constant temperature spontaneously and of itself arrives in such a condition that any of the processes taking place in the system by which energy may be extracted, run in a preferred direction, without a compensating reverse process."

Figure 2.1 illustrates the implications of the principle of detailed balance for the case of a solar cell. The sample is in thermal equilibrium with its ambient. Therefore, any incoming and absorbed (equilibrium) electromagnetic radiation flux $\phi^{in,0}$ is counterbalanced by the same amount $\phi^{em,0}$ of emitted radiation. In the same way each carrier transport process from site to site within the sample is compensated by the respective inverse process.

2.2.2
The Shockley–Queisser Limit

SQ used an ingeniously simplified concept of a solar cell to derive the maximum photovoltaic conversion efficiency as well as the current/voltage (J/V)-curve of such an idealized device. The basic ingredients are illustrated in Figure 2.2. In addition to the equilibrium radiation from the surrounding as shown in Figure 2.1, we have to consider the excess radiation ϕ^{sun} from the sun. Furthermore, all sites in

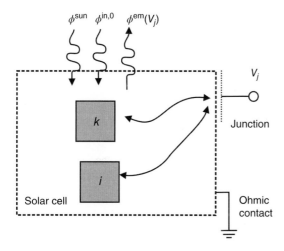

Figure 2.2 Illustration of the basic assumptions of the SQ-approach including the ambient photon flux $\phi^{in,0}$ and the excess flux ϕ^{sun} from the sun. All sites within the device are perfectly connected to the rectifying junction as well as to the Ohmic contact such that all photogenerated minority charge carriers are collected by the junction. In turn, the junction voltage V_j controls the luminescent emission $\phi^{em}(V_j)$.

the device are perfectly connected to the rectifying junction, such that each photogenerated minority charge carrier is collected by the junction and the flow of majority carriers to the Ohmic contact does not induce Ohmic losses. Considering a solar cell made from a semiconductor with a band gap energy E_g and assuming that all photons with energy $E \geq E_g$ are absorbed, we are readily able to calculate the short circuit current density J_{SC} of the solar cell via

$$J_{SC} = q \int_{E_g}^{\infty} \phi^{sun}(E) dE \tag{2.1}$$

where q denotes the elementary charge.

The perfect connection of the junction to the entire volume in the solar cell must be also reflected in the fact that the splitting $\mu = E_{F,n} - E_{F,p}$ of the quasi-Fermi levels $E_{F,n}$, $E_{F,p}$ of electrons and holes is controlled by the junction voltage V_j via $\mu = qV_j$. The emitted photon flux under the applied bias voltage V_j is therefore given by

$$\phi^{em}(V_j, E) = \frac{2\pi E^2}{h^3 c^2} \frac{1}{[\exp((E - qV_j)/kT) - 1]} \tag{2.2}$$

where h is the Planck constant, c the velocity of light in vacuum, and kT is the thermal energy. For voltages that are small compared with the emitted photon energies, that is, $(E - qV_j) \gg kT$, the Bose–Einstein term in Eq. (2.2) is well approximated by a Boltzmann distribution and we can simplify Eq. (2.2) to

$$\phi^{em}(V_j, E) = \phi_{bb} \exp\left(\frac{qV_j}{kT}\right) \tag{2.3}$$

where the black body emission spectrum is given by

$$\phi_{bb} = \frac{2\pi E^2}{h^3 c^2} \exp\left(-\frac{E}{kT}\right). \tag{2.4}$$

Subtracting the equilibrium emission from the cell's surface we arrive at the excess EL emission

$$\Delta\phi^{em}(V_j, E) = \phi_{bb}\left[\exp\left(\frac{qV_j}{kT}\right) - 1\right]. \tag{2.5}$$

This excess emission must be caused by the radiative recombination current J_{rad} of the charge carriers injected by the junction. Since in the radiative limit, there are no other possibilities to recombine, the entire dark current of the solar cell is given by

$$J_{rad} = q\int_{E_g}^{\infty} \Delta\phi(V_j, E)dE = q\int_{E_g}^{\infty} \phi_{bb}(E)dE \left[\exp\left(\frac{qV_j}{kT}\right) - 1\right]$$

$$= J_{0,rad}\left[\exp\left(\frac{qV_j}{kT}\right) - 1\right] \tag{2.6}$$

where $J_{0,rad}$ denotes the radiative saturation current (density). The current voltage characteristics in the SQ-limit is given by the superposition of the dark characteristics with the short circuit current defined in Eq. (2.1). Thus, the open circuit voltage V_{OC}^{rad} in the radiative limit follows from

$$V_{OC}^{rad} = \frac{kT}{q}\ln\left(\frac{J_{SC}}{J_{0,rad}} + 1\right). \tag{2.7}$$

Hence, the approach of SQ yields a complete diode theory not only for an ideal solar cell but also for an ideal LED. Note that a generalization of the fundamental SQ-theory instead of an abrupt increase of the absorptance A at the band gap energy E_g considers rather a continuous increase due to the real optical properties of the cell and of the absorber material [6].

It is important to note that the assumptions made above are valid for a solar cell that is connected to the ambient equally in all directions. As will be pointed out in Chapter 5 a restriction of the angle of acceptance (and consequently of emittance) leads to a substantial enhancement of the maximum efficiency achievable in the radiative limit by reducing $J_{0,rad}$ and, consequently, by enhancing $V_{OC,rad}$.

2.2.3
Derivation of the Reciprocity Theorem

The basic approach used to derive the opto-electronic reciprocity relation is sketched in Figure 2.3 defining a discretized model of a solar cell as used in Refs. [7, 8].

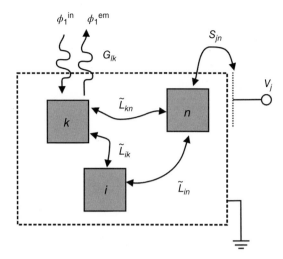

Figure 2.3 Scheme of the discrete solar cell model used for the derivation of the reciprocity relation from the principle of detailed balance.

Note that an analogous proceeding for the continuous case uses (self-adjoint) operators acting on function spaces and follows the same reasoning and yields analogous results [7]. Note further that the present approach, other than the original work [7], yields an expression that not only contains the EL but also the photoluminescent (PL) emission as outlined in Ref. [9]. Thus, we will show that application of the principle of detailed balance not only yields a reciprocity relation but also leads to a superposition principle of EL and PL.

At each site i in the absorber sketched in Figure 2.3, we define a concentration n of minority carriers. Minority carriers are generated optically by the incoming photon flux ϕ_l^{in} where the index l stands for the wavelength, location, and the angle of incidence of the incoming light. Incidence of photons via these optical "input channels" l leads to the photogeneration rate

$$g_i = \sum_l G_{il} \frac{\phi_l^{in}}{\phi_l^{in,0}} \tag{2.8}$$

at site i with the interaction coefficient G_{il} as a proportionality constant between g_i and the photon flux ϕ_l^{in} normalized to $\phi_l^{in,0} = \phi_l^{bb}$, that is, the equilibrium black body radiation of the surroundings. In turn, the total flux of photons ϕ_l^{em} leaving the output channel l is given by the sum

$$\phi_l^{em} = \sum_i G_{il} \frac{n_i}{n_i^0} \tag{2.9}$$

over the minority carrier density n_i at all sites i normalized to their equilibrium values n_i^0. Thus, in thermal equilibrium ($n_i = n_i^0$ and $\phi_l^{in} = \phi_l^{in,0}$) the number of carriers generated at any site i via the optical input channel l is the same as the number of photons that are emitted into channel l due to radiative recombination at site i.

In an analogous way we define the junction(s) to the device as electrical input channels such that through any junction j we have an injection (particle) current

$$J_i^{inj} = \sum_j S_{ij} \exp\left(\frac{qV_j}{kT}\right) \tag{2.10}$$

into the sites i. The currents extracted from all sites i into the junction j are given by the sum

$$J_j^{ext} = \sum_i S_{ij} \frac{n_i}{n_i^0} \tag{2.11}$$

using the same proportionality constant S_{ij} as in Eq. (2.4) such that in thermal equilibrium ($n_i = n_i^0$ and $V_j = 0$) any site i is in a detailed balance equilibrium with any junction j. Including the electrical and optical input/output channels into the transport equation of the solar cell leads to [10].

$$\sum_k \tilde{L}_{ik} \frac{n_k}{n_k^0} - \left(\sum_l G_{il} + \sum_j S_{ij}\right) \frac{n_i}{n_i^0} = -\sum_l G_{il} \frac{\phi_l^{in}}{\phi_l^{in,0}} - \sum_j S_{ij} \exp\left(\frac{qV_j}{kT}\right) - g_i^{th}. \tag{2.12}$$

In this steady state condition the non-diagonal elements of the matrix $\tilde{L}_{ik}$ are the net transfer rates from site k to site i originating from the normalized carrier density n_k/n_k^0 at site k [7, 10]. This implies direct transport of carriers as well as transfer via photon recycling, that is, the radiative recombination of a charge carrier at site k followed by the photogeneration of a carrier at site i. The symmetry $\tilde{L}_{ik} = \tilde{L}_{ki}$ warrants the principle of detailed balance [7, 10].

The diagonal elements $\tilde{L}_{ii}$ account for the losses at site i due to transitions to all other sites and to any kind of parasitic recombination described by a normalized non-radiative lifetime τ_i such that

$$\tilde{L}_{ii} = -\frac{1}{\tau_i} - \sum_{k \neq i} \tilde{L}_{ik} \tag{2.13}$$

with

$$\frac{n_i}{n_i^0} \frac{1}{\tau_i} = g_i^{th}. \tag{2.14}$$

Thus, parasitic recombination at all sites i is in equilibrium with the thermal generation rate g_i^{th}. Notably, the parasitic recombination term also accounts for radiative recombination followed by parasitic absorption of the generated photon, for example, in the substrate or the back contact of a solar cell.

2.2 Opto-Electronic Reciprocity

With the definition of the normalized excess carrier densities $\Delta v_k = n_k/n_k^0 - 1$ and photon fluxes $\Delta \varphi_l = \phi_l^{in}/\phi_l^{bb} - 1$ we obtain from Eq. (2.12)

$$\sum_k L_{ik} \Delta v_k = \sum_k \tilde{L}_{ik} \Delta v_k - \left(\sum_l G_{il} + \sum_j S_{ij} \right) \Delta v_i$$

$$= -\sum_l G_{il} \Delta \varphi_l - \sum_j S_{ij} \left[\exp\left(\frac{qV_j}{kT}\right) - 1 \right]. \quad (2.15)$$

with the matrix $L_{ik} = \tilde{L}_{ik}$ for $i \neq k$ and

$$L_{ii} = \tilde{L}_{ii} - \sum_l G_{il} - \sum_j S_{ij}. \quad (2.16)$$

From Eq. (2.9) we obtain the excess emission of photons

$$\Delta \phi_l^{em} = \sum_i G_{il} \Delta v_i \quad (2.17)$$

and from Eq. (2.11) the excess extracted current

$$\Delta J_j^{ext} = \sum_i S_{ij} \Delta v_i. \quad (2.18)$$

Equations (2), (2.17), and (2.18) simplify further upon describing the electrical and optical input/output formally with the coefficient H_{in} (being G_{il} or S_{ij}) and the excitation e_n (corresponding to $\Delta \phi_l$ or the exponential junction term). Then, instead of Eq. (2.9), we have

$$\sum_k L_{ik} \Delta v_k = \sum_n H_{in} e_n. \quad (2.19)$$

The optical emission or electrical extraction reads

$$\Delta I_n = \sum_i H_{in} \Delta v_i. \quad (2.20)$$

where $\Delta I_n = \Delta \phi_l^{em}$ for the optical channels and $\Delta I_n = \Delta J_j^{ext}$ for the electrical contacts.

Now we combine two solutions (a) and (b) of Eq. (2.19)

$$\sum_i \Delta v_i^{(a)} \sum_k L_{ik} \Delta v_k^{(b)} - \sum_i \Delta v_i^{(b)} \sum_k L_{ik} \Delta v_k^{(a)}$$

$$= \sum_i \Delta v_i^{(a)} \sum_n H_{in} e_n^{(b)} - \sum_i \Delta v_i^{(b)} \sum_n H_{in} e_n^{(a)}. \quad (2.21)$$

Because of the symmetry of L_{ik} the left hand side of Eq. (2) is zero. Combining Eq. (2.20) with the remainder of Eq. (2), we arrive at a general reciprocity relation connecting two arbitrarily combined input situations to the respective outputs

$$\sum_n \Delta I_n^{(a)} e_n^{(b)} = \sum_n \Delta I_n^{(b)} e_n^{(a)}. \quad (2.22)$$

For the description of EL/PL emission we have to consider a situation (sc) where the device is illuminated with light at an energy E' under short circuit conditions expressed as a normalized incoming photon flux at an energy E. We thus have at an input channel l in Eq. (2.22)

$$e_l^{(sc)} = \Delta\varphi_E^{(sc)} = \Delta\varphi^{(sc)}(E). \tag{2.23}$$

No other excitations are given in this situation (sc) which is not combined with a situation (bi) where a voltage bias V_j is applied to the junction, hence,

$$e_j^{(bi)} = \exp\left(\frac{qV_j}{kT}\right) - 1. \tag{2.24}$$

Additionally, the device is illuminated with a photon flux

$$e_{l'}^{(bi)} = \Delta\varphi_{E'}^{(bi)} = \Delta\varphi^{(bi)}(E'). \tag{2.25}$$

Hence, Eq. (2.22) transforms into

$$\Delta I_l^{(bi)} e_l^{(sc)} = \Delta I_{l'}^{(sc)} e_{l'}^{(bi)} + \Delta I_j^{(sc)} e_j^{(bi)} \tag{2.26}$$

and, finally, into

$$\Delta\phi_{(bi)}^{em}(E)\Delta\varphi_E^{(sc)} = \Delta\phi_{(sc)}^{em}(E')\Delta\varphi_{E'}^{(bi)} + \Delta J_j^{(sc)} \left[\exp\left(\frac{qV_j}{kT}\right) - 1\right]. \tag{2.27}$$

Reorganizing Eq. (2.27) leads to

$$\Delta\phi_{(bi)}^{em}(E) = \frac{\Delta\phi_{(sc)}^{em}(E')}{\Delta\phi^{in}(E)}\frac{\phi_{bb}(E)}{\phi_{bb}(E')}\Delta\phi_{(sc)}^{in}(E') + \frac{\Delta J_j^{(sc)}}{\Delta\phi^{in}(E)}\phi_{bb}(E)\left[\exp\left(\frac{qV_j}{kT}\right) - 1\right]. \tag{2.28}$$

With the definition of the (photovoltaic) external quantum efficiency

$$Q_e(E) := \frac{\Delta J_j^{(sc)}}{\Delta\phi^{in}(E)} \tag{2.29}$$

and the PL quantum efficiency (related to incoming photons of energy E' and outgoing photons at E)

$$Q_{PL}(E', E) := \frac{\Delta\phi_{(sc)}^{em}(E)}{\Delta\phi^{in}(E')} = \frac{\Delta\phi_{(sc)}^{em}(E')}{\Delta\phi^{in}(E)}\frac{\phi_{bb}(E)}{\phi_{bb}(E')} = Q_{PL}(E, E')\frac{\phi_{bb}(E)}{\phi_{bb}(E')} \tag{2.30}$$

we arrive at

$$\Delta\phi_{(bi)}^{em}(E) = Q_{PL}(E', E)\Delta\phi_{(sc)}^{in}(E') + Q_e(E)\phi_{bb}(E)\left[\exp\left(\frac{qV_j}{kT}\right) - 1\right] \tag{2.31}$$

stating that the combined PL/EL emission of a solar cell is a superposition of a pure EL part

$$\Delta\phi_{EL}^{em}(E) = Q_e(E)\phi_{bb}(E)\left[\exp\left(\frac{qV_j}{kT}\right) - 1\right] \tag{2.32}$$

which is driven by the junction voltage and a PL part

$$\Delta\phi_{\text{PL,sc}}(E) = Q_{\text{PL}}(E', E)\Delta\phi^{\text{in}}_{(\text{sc})}(E'). \tag{2.33}$$

driven by optical excitation of the solar cell under short circuit conditions. Thus, a rigorous application of the principle of detailed balance not only derives an RR between the EL emission and the external quantum efficiency of a solar cell (Eq. (2.32)) but also a superposition principle between EL and PL emission.

2.3 Connection to Other Reciprocity Theorems

The linear superposition of EL and PL emissions as described in Eq. (2.31) results from the linear character of the carrier transport equations allowing us to extrapolate all equilibrium rate constants toward a non-equilibrium situation. Thus, the reciprocity relation is a consequence of the principle of detailed balance and its linear extrapolation to a non-equilibrium situation and can be looked at as a special case of Onsager's reciprocity relation [11] or as a generalization of Casimir's application of Onsager's result to electrical conduction [12].

2.3.1 Emitter and Collector Currents in Transistors

From the general form given in Eq. (2.22) a large number of specific RRs can be derived by restriction to specific input/output channels. For instance, application of Eq. (2.26) to the situation of two junctions as in an npn transistor sketched in Figure 2.4 leads to the symmetry relation between emitter and collector currents. Using the notation of Shockley *et al.* [13], we compare the situation (*l*), where we

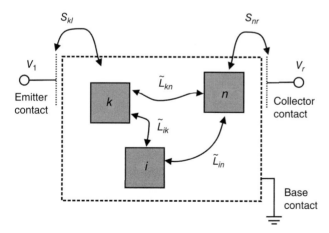

Figure 2.4 Modification of Figure 2.3 to illustrate the application of Eq. (2.26) to the situation considered by Shockley, Sparks, and Teal.

apply a voltage V_l to the left n-contact and collect an excess current $\Delta J_r^{(l)}$ at the right contact (cf. Figure 2.3 in Ref. [13]) to the complementary situation (r) with a voltage V_r applied to the right contact and a current $\Delta J_l^{(r)}$ collected at the left contact. Application of Eq. (2.26) to this situation and expressing the voltage and current terms explicitly leads to

$$\Delta J_l^{(r)} \exp\left(\frac{qV_l}{kT}\right) = \Delta J_r^{(l)} \exp\left(\frac{qV_r}{kT}\right) \tag{2.34}$$

which is equivalent to Eq. (6.18) in Ref. [13].

2.3.2
Tellegens's Network Theorem

Formally, Eq. (2.22) resembles Tellegen's RR [14]. This theorem (Eq. (2.1) in Ref. [14]) states that in an electrical network subject to Kirchhoff's current and voltage laws we have

$$\sum_n I_n V_n = \sum_b J_b U_b \tag{2.35}$$

where the sum on the left extends over all nodes n in the network and on the right over all branches b. The quantity I_n denotes the input/output currents at the nodes and V_n the electrical potential. Likewise, J_b are the currents U_b and the potential differences over the branches. It is important to notice that the currents I_n/J_b and the potentials V_n/U_b need not necessarily belong to the same physical situation but must only be currents obeying Kirchhoff's current law or potentials obeying Kirchhoff's voltage law within the same network topology. Therefore, we may compare to different bias situations (a) and (b) in a similar way as in Eq. (2.21)

$$\sum_n I_n^{(a)} V_n^{(b)} - I_n^{(b)} V_n^{(a)} = \sum_b J_b^{(a)} U_b^{(b)} - J_b^{(b)} U_b^{(a)}. \tag{2.36}$$

Equation (2.36) becomes especially interesting when comparing two different bias situations of a linear network as sketched in Figure 2.5. Here the current J_b over a branch is connected to the voltage drop U_b via $J_b = G_b U_b$ where G_b is the conductivity of branch b. Thus, Eq. (2.36) reads

$$\sum_n I_n^{(a)} V_n^{(b)} - I_n^{(b)} V_n^{(a)} = \sum_b J_b^{(a)} G_b J_b^{(b)} - J_b^{(b)} G_b J_b^{(a)} = 0, \tag{2.37}$$

that is, the contributions of the branches cancel out and the theorem interconnects only the input/output terminals in two different bias situations

$$\sum_n \delta I_n^{(a)} \delta V_n^{(b)} = \sum_n \delta I_n^{(b)} \delta V_n^{(a)}. \tag{2.38}$$

The formal similarity between Eqs. (2.26) and (2.37) results from the fact that in the derivation of Eq. (2.26) the normalized excess densities $\Delta\phi$ and $\Delta\nu$ for photons and charge carriers play the role of potentials whereas the fluxes $\Delta\phi$ and ΔJ correspond to the currents in Eq. (2.38). Using the linear dependence of the fluxes on the potentials, the mathematical derivation of Eq. (2.26) is identical to that of Eq. (2.38).

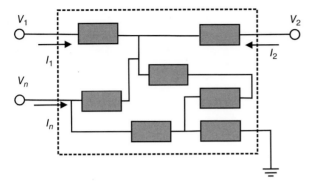

Figure 2.5 Sketch of a linear electrical network with voltages V_n applied to the electrical terminals and currents I_n entering into the terminals.

2.3.3
Differential Reciprocity Relations by Wong and Green

Recently, the opto-electronic reciprocity from Ref. [8] has been extended to situations where, due to non-linearities, the original theorem is no more valid in a global sense. Wong and Green [15] have demonstrated that in such cases the theorem may be retained in a differential form, that is, the theorem is valid for small perturbations around a bias situation. Such a non-linear situation is, for example, encountered when considering a spatially extended solar cell in terms of a distributed network model where local diodes are interconnected to the terminals by a network of resistors symbolizing, for example, the sheet resistance of the emitter or the bulk resistance of the base. The essential elements of such a network are sketched in Figure 2.6a. It is clear that such a network consisting of current sources and diodes (representing the local recombination and photo currents) and resistors is a non-linear system. However, given a bias point, small deviations can still be treated within a linear model.

In the following, we proof the theorem of Wong and Green using Tellegen's formalism. Note however, that mathematically the procedures given here and in Ref. [15] are almost identical except for the modification that in the present proof the diodes must not be grounded but can be anywhere in the network. This extends the theorem for cases of more complicated topology, for example, for the analysis of solar modules or even solar systems.

Similarly as in Eq. (2.23) we compare the situation (L) under illumination at a single location (cf. Figure 2.6b) with the situation where an additional voltage δV_0^b is applied to the terminal of the device (Figure 2.6c). As can be seen from Figure 2.6b, we have to consider three different nodes, namely the terminal as well as both sides of the local solar cell (locations i and i' in Figure 2.6b). Equation (2.38) then reads as

$$\delta I_0^L \delta V_0^b + \delta I_i^L \delta V_i^b + \delta I_{i'}^L \delta V_{i'}^b = \delta I_0^b \delta V_0^L + \delta I_i^b \delta V_i^L + \delta I_{i'}^b \delta V_{i'}^L. \tag{2.39}$$

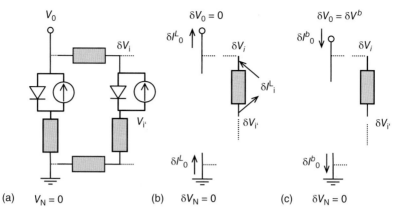

Figure 2.6 (a) Sketch of a non-linear network symbolizing a spatially extended solar cell. Only two of the many local diodes (and current sources) are shown. (b) Linearization of (a) around a given bias point. The diode and the current source of (a) is replaced by a resistor and the current source is replaced by a current input above and a current output below that resistor. (c) The linearized network model for the situation where a small change of the bias voltage δV_0^b is applied to the terminal of the device.

Because of $\delta I_{i'}^L = -\delta I_i^L$, $\delta V_0^L = 0$, $\delta I_{i'}^b = \delta I_i^b = 0$, and with the definition $\delta U_i^b := \delta V_i^b - \delta V_{i'}^b$ we obtain

$$\delta I_0^L \delta V_0^b = -\delta I_i^L \delta U_i^b. \tag{2.40}$$

The ratio $f_T = -\delta I_0^L / \delta I_i^L$ quantifies the share of the photo current (generated at site i) that is finally collected at the terminal. Therefore, f_T is denoted as the current transport [15] or current collection efficiency and fulfills

$$f_T(x,y) = -\frac{\delta j_0^L}{\delta j^L(x,y)} = \frac{\delta V_j^b(x,y)}{\delta V_0^b}, \tag{2.41}$$

where we have replaced the index i by the spatial coordinates (x,y), the currents δI by current densities δj, and the quantity δU_i^b by the junction voltage $\delta V_j(x,y)$. Thus, the current collection reciprocity theorem of Wong and Green relates the current collection efficiency $f_T(x,y)$ with the junction voltage $\delta V_j(x,y)$ that results from application of a small incremental bias voltage δV_0^b to the terminal of the device.

Equation (2.41) can be combined with Eq. (2.32) that reads in its differential and local form

$$\delta \phi_{EL}^{em}(E,x,y) = Q_e^{loc}(E,x,y) \phi_{bb}(E) \frac{\delta V_j^b(x,y)}{kT/q} \left[\exp\left(\frac{qV_j^b(x,y)}{kT}\right) - 1 \right] \tag{2.42}$$

Use of Eq. (2.41) yields

$$\delta \phi_{EL}^{em}(E,x,y) = Q_e^{loc}(E,x,y) f_T(x,y) \phi_{bb}(E) \frac{\delta V_0^b}{kT/q} \left[\exp\left(\frac{qV_j^b(x,y)}{kT}\right) - 1 \right] \tag{2.43}$$

2.3 Connection to Other Reciprocity Theorems

With the definition of the terminal quantum efficiency $Q_e^T(E, x, y) = Q_e^{loc}(E, x, y) f_T(x, y)$ and by reorganizing Eq. (2.43), Wong and Green obtain the differential form of the opto-electronic reciprocity as

$$\delta\phi_{EL}^{em}(E, x, y) \frac{kT/q}{\delta V_0^b} = Q_e^T(E, x, y)\phi_{bb}(E) \left[\exp\left(\frac{qV_j^b(x, y)}{kT}\right) - 1 \right] \quad (2.44)$$

Equations (2.41) and (2.44) are powerful, but yet to be exploited, tools for the analysis of EL experiments that use a differential approach.

2.3.4
Würfel's Generalization of Kirchhoff's Law

Strictly speaking, Kirchhoff's law [16] is not a reciprocity relation in the sense of input/output terminals. Neither is Würfel's generalization [17]. However, both are excellent examples for the application of the principle of detailed balance connecting the detailed balance pair light absorption/emission. In particular, Würfel's equation lies in the very heart of the interaction of light with semiconductors as it is used in this book. The radiation emitted from a semiconductor is given by

$$\phi^{em}(\mu, E) = a(E) \frac{2\pi E^2}{h^3 c^2} \frac{1}{[\exp((E-\mu)/kT) - 1]} \approx a(E)\phi_{bb} \exp\left(\frac{\mu}{kT}\right) \quad (2.45)$$

where $\mu = E_{F,n} - E_{F,p}$ is the split of electron and hole quasi-Fermi levels $E_{F,n}$ and $E_{F,p}$ of electrons and holes. The quantity a denotes the absorptance of the semiconductor. Since μ and a are local quantities, Eq. (2.45) is valid (and meaningful) only locally or for a thin semiconductor slab with constant μ and a. However, Eq. (2.45) is restricted neither to cases where the equilibrium rate constants apply nor to a linearization of the system in contrast to the reciprocity relations Eqs. (2.32) or (2.44).

2.3.5
Reciprocity Relation for LED Quantum Efficiency

Upto this point, we have considered only reciprocity relations for the solar cell in the short circuit situation compared to the radiative emission caused by application of a voltage bias. The open circuit situation of a real solar cell is given by the balance between absorbed radiation and losses by radiative emission (as in the SQ case) as well as by non-radiative recombination. In 1967, Ross [18] derived an equation for the attainable "potential difference μ caused by a radiation field in a photochemical system"

$$\mu = \mu_{max} - kT \ln(\kappa), \quad (2.46)$$

where μ_{max} is the maximal potential difference that is derived by equilibrating the incoming radiation with the rate of radiative decay. The quantity κ denotes the inverse of the luminescence yield. Thus, for $\kappa = 1$ we have $\mu = \mu_{max}$. Reference [18] uses thermodynamic terminology and arguments that are not straight forward to relate to photovoltaic device parameters. However, Smestad and Ries [19] derived a similar relation, now explicitly put into the context of the open circuit voltage V_{OC} of a solar cell. In the notation of Ref. [8] this relation reads

$$V_{OC} = V_{OC}^{rad} + \frac{kT}{q} \ln(Q_{LED}), \qquad (2.47)$$

where V_{OC}^{rad} is the open circuit voltage that would be attained if radiative losses, that is, losses by photon emission from the solar cell, were the only loss mechanism. The quantity Q_{LED} is the *external* LED quantum efficiency of the device. Similarly as for Eq. (2.46) we have $V_{OC} = V_{OC}^{rad}$ if $Q_{LED} = 1$. Though, the derivation of this relation is not complicated, a proper terminology prevents misunderstanding and facilitates experimental verification. Despite the fact that all versions of Eqs. (2.46) or (2.47) in Refs. [8, 18, 19] describe the same fact, we prefer the term external LED quantum efficiency [8] instead of luminescence yield [18] or luminescence efficiency [19] to emphasize the character of the equation as an external reciprocity relation connecting measurable input and output parameters. In view of the superposition principle of EL and photoluminescence [9] it is important to understand that Q_{LED} relates only to the voltage driven part of the luminescence that is seen from the outside. Green [20] has chosen the term *External Radiative Efficiency* (ERE) for the same quantity. Markvart [21] implemented Eq. (2.47) in his thermodynamic derivation of the open circuit voltage of a solar cell as an entropy generation term.

2.3.6
Shockley–Queisser Revisited

In the SQ-theory the ideal solar cell is at the same time an ideal diode. The reciprocity relations discussed above can be looked at as a generalization of the SQ-theory extending the very same detailed balance approach to the case of real world devices. This is seen when calculating from Eq. (2.32) the total radiative emission current

$$J_{0,rad} = q \int_{E_g}^{\infty} \overline{Q_e(E)} \phi_{bb}(E) dE \left[\exp\left(\frac{qV_j}{kT}\right) - 1 \right], \qquad (2.48)$$

where $\overline{Q_e(E)}$ is the angular weighted average of $Q_e(E)$ over the half sphere. Let us further consider the definition of the short circuit current density via the external quantum efficiency according to

$$J_{SC} = q \int_{E_g}^{\infty} Q_e(E) \phi^{sun}(E) dE. \qquad (2.49)$$

From these we obtain the constituting Eqs. (2.6) and (2.1) of the SQ-theory when choosing the limit

$$\overline{Q_e(E)} = Q_e(E) = A(E) \quad \begin{matrix} = 1 & \text{if } E \geq E_g \\ = 0 & \text{if } E < E_g, \end{matrix} \qquad (2.50)$$

that is, for idealized absorptance A and perfect carrier collection. Similarly, Würfel's Eq. (2.45) becomes identical to Eq. (2.2) of the SQ-theory if $qV_j = \mu$, that is, for flat Quasi-Fermi levels. The SQ-limit further requires the absence of non-radiative recombination to obtain $V_{OC} = V_{OC}^{rad}$. Because of Eq. (2.47) this requires $Q_{LED} = 1$. This implies that the ideal (SQ) solar cell is at the same time an ideal LED.

An important consequence of the reciprocity relations (2.32) and (2.47) is the fact that we can readily extract the open circuit voltage difference $V_{OC}^{rad} - V_{OC}$ from experimental data. Combining Eq. (2.47) with Eq. (2.7) yields

$$Q_{LED} = \exp\left(\frac{qV_{OC}}{kT}\right) \exp\left(\frac{-qV_{OC}^{rad}}{kT}\right) \approx \frac{J_{0,rad}}{J_{SC}} \exp\left(\frac{qV_{OC}}{kT}\right). \quad (2.51)$$

The quantities V_{OC} and J_{SC} are standard solar cell parameters and $J_{0,rad}$ can be calculated from the external quantum efficiency under the approximation $\overline{Q_e(E)} = Q_e(E)$. A comparison of the open circuit voltages of Cu(In,Ga)Se$_2$ and of Si solar cells with their respective limits using Eq. (2.51) has been performed in Refs. [22–24]. In this work, $J_{0,rad}$ is calculated from the EL emission spectra scaled with the measured $Q_e(E)$ data. More recently, Green [20] has calculated Q_{LED} (denoted as *ERE*) for a wide variety of state-of-the-art devices from published literature data using the external quantum efficiency $Q_e(E)$ alone. The range of quantum efficiencies Q_{LED} stretches over many orders of magnitude from 0.225 (for a GaAs device) down to appr. 3×10^{-9} for organic solar cells [20]. More specific examples for the application of the reciprocity relations will be delineated in the following section.

2.3.7
Influence of Light Trapping

It is important in the context of the present book to emphasize especially the optical part of the opto-electronic reciprocity relations. Light trapping has traditionally been seen to have a way to overcome poor carrier absorption, for example, in indirect semiconductors like silicon [25]. The maximum optical pathlength obtainable within geometric optics is $4n^2$ times the thickness of the absorber where n denotes the refractive index of the absorber material [26, 27]. This is known as the *Yablonovitch-limit* for light trapping and will be discussed in more detail in Chapter 5 of the present book. In addition, there is an influence of light trapping on the attainable open circuit voltage of the solar cell either because of the possibility of using thinner absorber layers with a correspondingly smaller share of non-radiative recombination [28] and/or by reducing losses due to parasitic light absorption. The latter is especially important for GaAs solar cells where, internally, radiative recombination is more important than non-radiative recombination and therefore light-management becomes especially important also for achieving maximum efficiency as pointed out recently by Miller *et al.* [29].

We will use the ideas of Ref. [29] as shown in Figure 2.7 to discuss the benefit of light-trapping in terms of reciprocity relations. The starting point is a flat solar cell without any surface texture and with a back contact (or a substrate) that is

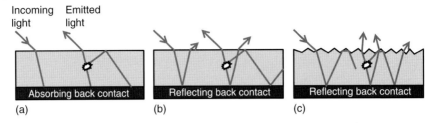

Figure 2.7 Incoupling of external light and outcoupling of EL emission in case of (a) a flat solar cell with a non-reflecting back contact, (b) a flat solar cell with a reflecting back contact, and (c) a solar cell with light-trapping and a reflecting back contact [29].

absorbing (Figure 2.7a). In this case, a portion of the incoming light is only weakly absorbed in the active photovoltaic material but will be parasitically absorbed in the back contact. This will lead to a poor external quantum efficiency in the spectral range of weak absorptance $a(E)$ although the carrier collection in the solar cell might be considered as being perfect. Let us assume $Q_e(E) = a(E)$. Similarly to EL emission the same problem arises as all light that is not emitted into the immediate escape cone is absorbed in the back contact, either directly or after one reflection at the front side due to total internal reflection. In the opto-electronic reciprocity relation (2.32) both facts, that is, the reduced Q_e as well as the reduced emission $\Delta\phi_{EL}^{em}$ are consistently contained. In addition, Eq. (2.47) tells us that the poor external LED quantum efficiency Q_{LED} leads to a reduced open circuit voltage. This is obvious, since parasitic absorption is a non-radiative loss, in the same way as non-radiative recombination.

This restriction changes when replacing the absorbing back contact by a perfectly reflecting one (Figure 2.7b). The fraction of absorbed light increases by a factor of two (in the weak absorption limit) as it does the portion of the emitted light, that is, Q_e increases in the same why as it does $\Delta\phi_{EL}^{em}$. A major change occurs in Q_{LED} because more light is emitted to the outside and, more importantly, the light emitted into the guided mode is no more a loss since, assuming a total eclipse of parasitic absorption, eventually it gets reabsorbed creating another electron–hole pair (photon-recycling). If we further assume that there is non-radiative recombination, we have $Q_{LED} = 1$ and, consequently, $V_{OC} = V_{OC}^{rad}$.

In the third situation (Figure 2.7c), we add light-trapping to the device. Now all optical modes in the absorber material are coupled to the outside world. Therefore, Q_e and $\Delta\phi_{EL}^{em}$ increase simultaneously as predicted by Eq. (2.23). Because of Eq. (2.48) the integral EL emission will lead to an increased increase $J_{0,rad}$. Because the black body spectrum contained in the integral of Eq. (2.48) in general puts more weight on the low absorption range (long wavelength) as it does the solar spectrum in Eq. (2.49) the increase in $J_{0,rad}$ will overcompensate the increase in J_{SC} leading to an overall decrease of the radiative open circuit voltage V_{OC}^{rad}. This is an effect complementary to the increase of V_{OC}^{rad} obtained from restricting the angle of acceptance of the incoming and outgoing light as discussed in Chapter 5. Despite the fact that the effect of light trapping on the radiative limit of the open circuit voltage is negative, in terms of final efficiency, light trapping is still beneficial due to the increase of the short circuit current density J_{SC}. This is especially true for absorber materials with a band gap larger than the optimum one (like GaAs) because increasing J_{SC} is especially valuable even when losing a little in open circuit voltage.

2.4
Applications of the Opto-Electronic Reciprocity Theorem

2.4.1
Experimental Verifications

The opto-electronic reciprocity connects absorption and charge carrier collection with the EL spectrum of a solar cell and Eq. (2.32) connects the open circuit voltage with the absolute amount of EL emission. Thus, there are three aspects of the reciprocity relations discussed above that can be tested in experiment. The earliest work used the comparison of absorption and emission of a semiconductor to analyze the optical properties of the material. Daub and Wurfel [30] showed that photoluminescence of a silicon wafer enables the measurement of ultralow absorption coefficients by making use of the generalized Planck's law. From Eq. (2.45), they obtained the absorptance from the PL spectrum. Because the reflection coefficients at the air–silicon interface are well known, it is straightforward to calculate the absorption coefficient from the absorptance. Given the fact that the Bose–Einstein Term in Eq. (2.45) improves the signal to noise ratio at low energies relative to those at high energies, the absorption coefficient from PL complements the absorption coefficient measured with alternative methods like transmission and reflection measurements. At a temperature $T = 295$ K, the absorption coefficient covered roughly 8 orders of magnitude while at $T = 90$ K, the dynamic range was even 18 orders of magnitude because of the larger effect of the Boltzmann term in Eq. (2.45). The excellent agreement between the absorption coefficient measured by transmission and the one measured with PL in the range where both overlap is evidence for the applicability of the generalized Planck's law and therefore for the optical part of the opto-electronic reciprocity.

In a next step, Trupke *et al.* [31] measured the EL of crystalline-Si solar cells to determine the absorptance of band to band transitions, which is difficult to distinguish of the absorptance of free carrier absorption and the absorptance of the metal back contact for flat and textured solar cells.

An early comparison of the measured quantum efficiency of a solar cell and the EL spectrum was again done by Trupke *et al.* on dye-sensitized solar cells [32]. This work was published 8 years before the theoretical derivation of the opto-electronic reciprocity. Thus, the external quantum efficiency was used here as an approximate measure of the absorptance and was compared with the absorptance derived from the EL using Eq. (2.45). Again, the agreement is excellent in the energetic range where the data overlap.

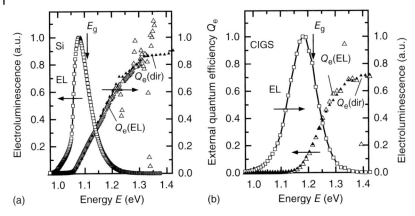

Figure 2.8 Comparison of measured photovoltaic external quantum efficiencies $Q_e(\text{dir})$ with the values $Q_e(\text{EL})$, calculated from the electroluminescence spectra using Eq. (2.36), for (a) a Si and (b) a Cu(In,Ga)Se$_2$ solar cell. (Data redrawn from Ref. [23].)

After the derivation of Eq. (2.32), more work was directed toward verification of the reciprocity relation [33]; upto the present, this includes crystalline Si [23, 24], Cu(In,Ga)Se$_2$ [22, 23], III/V based solar cells [34], as well as organic bulk heterojunction solar cells [35]. In all cases a good agreement between the experimental data and the predictions of Eq. (2.36) has been found. Figure 2.8 shows two examples, a silicon (a) and a Cu(In,Ga)Se$_2$ (b) solar cell, of a comparison between external quantum efficiency $Q_e(\text{EL})$ as it is calculated from the EL and $Q_e(\text{dir})$ that is measured directly. As can be seen, the prediction of Eq. (2.32) is accurate in the spectral range of sufficient overlap between the respective two data sets.

However, a problem of this kind of experimental analysis is that it is difficult to prove the charge collection aspect of the opto-electronic reciprocity. Does the EL actually contain information on charge carrier collection as suggested by Eq. (2.32) or is the agreement of the absorption and emission spectra mostly due to the fact that the differences between absorptance and quantum efficiency are always small and weakly energy dependent as compared to the energy dependence of the absorption coefficient? To study the impact of charge carrier collection it is instructive to compare the data obtained from EL imaging with spatially resolved quantum efficiency data. EL or PL imaging is a method to record the luminescence spatially resolved but without spectral information by photographing a solar cell or module which is biased to emit EL emission or illuminated to emit PL. If all ambient light (including the illumination source in case of PL) is blocked, the image will only contain the spatially resolved EL emission multiplied with the sensitivity of the camera. Usually, Si-CCD cameras are used for this purpose, but for low band gap semiconductors, InGaAs photodiode arrays are also used. When measuring the spatially resolved EL of an inorganic, crystalline semiconductor, the lateral variations of the EL will mostly originate from variations in the diffusion length of the minority carriers. A high-diffusion length will increase the carrier collection efficiency and therefore the quantum efficiency. From Eq. (2.32), we therefore expect that high- carrier diffusion lengths also lead to high EL emission. To study the local collection efficiency or quantum efficiency, it is possible to just measure the complete quantum efficiency on different macroscopic spots on the sample [24]. The alternative is to conduct a light beam induced current (LBIC) measurement, that is, to measure only at one or a few wavelengths while scanning the light source over the whole sample and obtain spatial resolution with limited spectral resolution. In both cases, the results are similar and indicate that in crystalline Si solar cells, the agreement between EL and local quantum efficiency is reasonable [24] and the information obtained from EL imaging and from LBIC scans is comparable [36]. The main advantage of EL imaging is that the measurement time is orders of magnitude shorter while the spatial resolution is usually better than in the case of an LBIC image.

The third aspect of the opto-electronic reciprocity is the connection between open-circuit voltage V_{OC} and LED quantum efficiency Q_{LED}. An interesting example where high-quality Si solar cells have been used as efficient LEDs was published by Green et al. [17, 37]. The authors took advantage of the high quality of light trapping and the low-defect densities and high open-circuit voltages in their solar cells to measure external LED quantum efficiencies of $Q_{LED} = 0.5\%$ at room temperature and $Q_{LED} = 1\%$ at around $T = 200$ K. These values are low compared to values for direct III/V semiconductors but represented a drastic improvement relative to the values obtained for the indirect semiconductor silicon before. A direct comparison between LED quantum efficiencies and open-circuit voltages was later done for crystalline Si [24] and for organic polymer:fullerene solar cells [35]. In both cases, the prediction of V_{OC} based on the LED quantum efficiency corresponded well to the actual open circuit voltage.

2.4.2
Spectrally Resolved Luminescence Analysis

The EL spectrum contains information on both charge carrier collection as well as the absorptance, which depends on the absorption coefficient, the thickness, and the reflective properties of the interfaces. Although, the effect of charge carrier collection, that is, the diffusion length, changes the spectral shape of the EL [38] and not only its amplitude, the EL offers no benefit compared to the standard measurement of the diffusion length in pn-junction solar cells using a normal internal quantum efficiency measurement [39, 40]. The spectrally resolved EL is a sensitive measure of the effect of light trapping on the increased pathlength in solar cells [24], but again this information can also be obtained from normal quantum efficiency measurements [41, 42].

EL spectroscopy provides a benefit as compared to normal quantum efficiency measurements, especially when quantum efficiency measurements are not easily possible in a particular situation or in a particular energy range and when the extra information contained in the absolute EL intensity is important. One example is the independent characterization of subcells in a multijunction solar cell. Because multijunction solar cells usually contain semiconductors with different band gaps, the emission from the different cells has a distinct energy and can be distinguished from those of the other layers. The measurement of a triple junction solar cell, for instance, will provide three distinct peaks whose absolute intensity reveals information on the quasi-Fermi level splitting. Using this principle it is possible to measure the EL as a function of injection current and reconstruct individual current/voltage curves of all subcells without having to contact them individually [34, 43].

In the context of organic solar cells, another interesting application of EL spectroscopy is the characterization of the interfacial charge transfer state at the interface between the electron donating and the electron accepting molecule or nanoparticle [44–46]. A direct characterization of the charge transfer state with quantum efficiency or related measurements of the absorption is possible but difficult because the absorption of the interfacial state is usually around 6 orders of magnitude weaker than the main absorption peak of the polymer [47–52]. In most cases the energetic offsets at the interface between the two molecules is sufficiently large to suppress the injection of electrons into the electron donating molecule and holes into the electron accepting molecule. In this case, the initially separated electrons and holes can only recombine at the donor–acceptor interface or at the electrodes, which makes the interfacial band gap at the donor–acceptor interface the major quantity controlling recombination and open circuit voltages in organic bulk heterojunction solar cells [53]. A study of the energies of the EL emission in a variety of polymer:fullerene solar cells verified that these are well correlated with the open circuit voltage [54].

A recent study [55] investigated the correlation between a disappearance of charge transfer emission and the loss in short circuit current density. The disappearance of charge transfer emission always implies an increase in emission from either the pure polymer or the pure fullerene. This emission is however an indication of insufficiently high band offsets at the donor–acceptor interface and that makes injection into the pure phases possible. These low band offsets also lead to increased geminate recombination of excitons that are not split at the heterointerface and therefore to a reduced generation and separation of free charge carriers and less photocurrent.

2.4.3
Luminescence Imaging

In case of spatially resolved techniques, the option to choose between techniques based on the detection of photocurrent and techniques based on emission of light shifts in favor of light emitting (i.e., luminescence based) techniques. This is due to the fact that the spatial resolution per measurement time that can be obtained by a camera is much higher than what is possible with techniques based on scanning a light source over a device. Spatial resolution at short measurement times is therefore a clear advantage of luminescence imaging and of high relevance for photovoltaic applications due to the large area of typically photovoltaic devices [56].

2.4 Applications of the Opto-Electronic Reciprocity Theorem

Luminescence images reveal three main types of features, namely (i) local mechanical problems as a crack or a disconnected contact finger, (ii) spatial variations of optical and electronic properties as for instance local variations of the minority carrier lifetime due to grain boundaries and dislocations in multicrystalline materials, and (iii) resistive effects that lead to a spatial gradient of voltage and therefore the luminescence.

Here, we want to focus on types (ii) and (iii) and the methods to calculate physical quantities from the EL images. In multicrystalline Si solar cells, the focus of the research has been on the determination of the local effective diffusion length, which includes the effects of bulk and surface recombination. The information on the effective diffusion length is included in the quantum efficiency of crystalline Si solar cells and therefore also in the EL spectrum. A simple measurement of the absolute EL intensity with a camera is the simplest approximation to a diffusion length determination [57]. A more advanced method uses different images taken with different filters to obtain some spectral information sufficient to more reliably determine the diffusion length [36, 58] and even to separate the influence of bulk recombination from the influence of surface recombination [59]. This method makes use of the fact that the diffusion length controls the spatial profile of the carrier concentrations. Since carriers closer to the back contact are more likely to be reabsorbed then carriers close to the front contact, the spatial carrier profile translates into a spectral change that can be detected with filters.

An alternative method to obtain the spatially resolved effective diffusion length from PL imaging was proposed by Hinken et al. [60]. The authors make use of the additional degree of freedom provided by PL imaging. While in EL imaging the voltage fixes the carrier concentration, in PL imaging one can change the carrier concentration using light and voltage bias at the same time. A comparison of a PL image taken at open and at short circuit is then used to obtain the effective diffusion length.

Resistive effects can be categorized into two groups, (i) series resistances that lead to a gradient in voltage and luminescence and (ii) shunts that lead to usually circular spots of negligible emission and radial luminescence gradients. The determination of series resistances from EL images is a frequently discussed topic with a different set of methods used in multicrystalline Si solar cells as opposed to thin-film solar cells. In the first case, usually the aim is to determine a spatially resolved series resistance from the local voltage and the local current density [61]. The local voltage can be obtained from the absolute EL emission and by comparison with the externally applied voltage the voltage drop from the contacts to one particular point on the device can be calculated. What is more difficult is to obtain the local current density to relate the voltage drop to a local series resistance. A possible way to solve that problem is to take an image at a low bias, where the currents are low and the resistive voltage drops are negligible relative to the applied voltage. In this case, the local current density should be proportional to the local emission and the solar cell behaves like a parallel connection of lots of perfect diodes. From the current density at low bias, the one at high bias is extrapolated and the series resistance is calculated [62]. Alternative methods are based on the comparison of two photoluminescence images taken at different light intensities and voltages [63].

In case of thin-film solar cells and modules, a spatially resolved determination of the series resistance is usually not necessary because the series resistance is controlled by the transparent conductive oxide layer, which has a spatially homogeneous conductivity when compared with multicrystalline silicon. In this case, a model can be fitted to the spatial gradient of the voltage obtained from the EL emission and local variations of the minority carrier lifetime or diffusion length can be neglected. With this method a simple determination of the ZnO : Al resistance in $Cu(In,Ga)Se_2$ based solar modules has been presented [64] and similar work has also been done on organic solar modules [65].

While the effect of the series resistance increases for higher applied voltages, the effect of shunts is mainly visible at low voltages. At the shunt, the local quasi-Fermi level splitting is low or zero and therefore there are hardly any excess carriers and the amount of radiative recombination is so low that it usually cannot be detected. In the unshunted regions around the shunt, the charge carriers flow toward the shunt and the radial voltage drop around the shunt depends on the resistivity of the transparent conductive oxide (in case of thin-film cells) or the emitter (in case of crystalline Si solar cells). The appearance on the EL image is that of a circular black spot with a radial gradient toward higher intensities. The relative influence of the shunt is highest when the ratio between the local conductivity of the shunt and the conductivity of the unshunted diodes around the shunt is highest. This is the case at low forward bias, because here the diode has a very low conductivity and the shunt has a high conductivity. At low forward bias, therefore, the whole cell might be black, while at higher voltages only the shunted region is black while the rest of the cell is virtually unaffected. To further study the properties of shunts with luminescence imaging, numerical modeling is usually applied using electrical network simulations usually based on SPICE [66–68].

This short overview of luminescence imaging is in no way complete. There are several other applications of EL and PL imaging that would go beyond the scope of this chapter but that should be briefly mentioned.

In addition to EL under forward bias, it is also possible to measure the EL under reverse bias and study the luminescence originating from hot spots that break down at reverse bias [69, 70]. Because the resulting EL originates from tiny spots, microscopic studies of EL are often used in context of reverse bias phenomena [71]. In these cases, the image covers only a tiny area and imaging has to be combined with scanning of the microscope to cover larger areas of interest. To study charge carrier lifetime with photoluminescence, the PL setup can be combined with confocal microscopes to get a high-resolution image of the cell [72–78]. For local high-intensity luminescence, scanning near field optical microscopes are also used for detection to increase the spatial resolution even more [79–82].

2.5 Limitations to the Opto-Electronic Reciprocity Theorem

The opto-electronic reciprocity strictly holds for linear systems where the recombination rate depends linearly on the concentration of minority carriers. Systems like that are usually the doped base of a pn-junction. In the case of crystalline Si solar cells, the reciprocity is particularly useful because most of the device thickness is made up of the base of the device and emitter and space charge region are thin relative to the base. Thus, it is reasonable to approximate the device as a base region with a junction that only serves as a boundary condition for the derivation of the reciprocity. In Cu(In,Ga)Se$_2$ solar cells, the base is still thick relative to the space charge region but for nearly all other thin-film solar cells, especially those made of disordered materials like amorphous Si or polymer based solar cells, this is no longer the case. In thin devices with thickness below 1 µm, it is usually not possible to define a large field free region where recombination is proportional to the minority carrier type. Instead, the devices will usually be designed such that the active layer is mostly intrinsic to optimize its electronic properties and then to have thin highly doped contact layers at both sides of the absorber. In the case of thin-film Si solar cells, these contact layers are made of around 20 nm thick doped layers while in the case of polymer:fullerene solar cells the layer at the cathode is usually just a low workfunction metal or combination of metals (as e.g., Ca/Al) and the anode layer is a p-type hole conductor (as, e.g., PEDOT:PSS, poly(3,4-ethylenedioxythiophene) poly(styrenesulfonate)). These devices aim to have an electric field in the whole absorber layer and not only at the periphery of it as in the case of crystalline Si or also dye cells to aid charge carrier collection in these low-mobility systems [10, 83].

The direct consequence of the lack of a field free region with well-defined minority carriers is the non-linearity of the recombination process. In this case, the Donolato theorem becomes invalid, that is, the relation between charge carrier injection in the dark and collection under illumination is violated and thus, the opto-electronic reciprocity does not hold anymore. However, while an exact quantitative relation does no longer exist, the qualitative relation between injection and extraction is still valid. Thus, also in disordered low-mobility semiconductors with non-linear recombination, there is a good agreement between a measure of charge collection (similar to LBIC) and a luminescence experiment that depends on charge carrier injection. A comparison of EL imaging and LBIC measurements was recently published for the case of polymer:fullerene solar cells and showed a good agreement [84].

Another complication arises in disordered semiconductors. While in crystalline semiconductors, most luminescence originates from radiative recombination of the free electron and hole populations in the conduction and valence bands, in disordered semiconductors a substantial part of the electron and hole population is localized and has a low or zero mobility relative to the free electrons and holes in the bands. For every material and for every temperature the luminescent emission may originate from different transitions between free and/or trapped carriers. In case trapped carriers are involved in recombination, one of the main assumptions used in the derivation of the opto-electronic reciprocity breaks down: The radiative recombination current does not scale with $\exp(\Delta E_f/kT)$ anymore [85]. Instead, the voltage dependence of the radiative recombination current will depend on the density of trapped states and the exact way depends on how probable the different transitions are with respect to each other [86]. Thus, the radiative recombination current could be written as $J_{rad} \sim \exp(\Delta E_f/(n_{id,rad}kT))$ where $n_{id,rad}$ is the radiative ideality factor [87]. In this case, all quantitative models based on the determination of the relative voltage from the luminescence emission would become more complicated because the radiative ideality factor $n_{id,rad}$ would have to be determined first by a separate measurement [88].

2.6
Conclusions

This chapter has summarized the recent progress in developing an understanding of solar cells in terms of the principle of detailed balance. The different reciprocity relations allow us to view and analyze solar cells and solar modules in terms of quantitative relations between their optical and electrical inputs and outputs. Thus, the fundamental principle of detailed balance has turned from a methodology to find the limitation of what we can achieve in the future into a method to analyze what we have already achieved by our present devices.

References

1. Shockley, W. and Queisser, H.J. (1961) Detailed balance limit of efficiency of pn-junction solar cells. *J. Appl. Phys.*, **32**, 510.
2. Lewis, G.N. (1925) A new principle of equilibrium. *Proc. Natl. Acad. Sci. U.S.A.*, **11**, 179.
3. Tolman, R.C. (1925) The principle of microscopic reversibility. *Proc. Natl. Acad. Sci. U.S.A.*, **11**, 436.
4. Lawrence, E.O. (1926) Transition probabilities: their relation to thermionic emission and the photo-electric effect. *Phys. Rev.*, **27**, 555.
5. Bridgman, P.W. (1928) Note on the principle of detailed balancing. *Phys. Rev.*, **31**, 101.
6. Tiedje, T., Yablonovitch, E., Cody, G.D., and Brooks, B.G. (1984) Limiting efficiency of silicon solar-cells. *IEEE Trans. Electron Devices*, **31**, 711.
7. Rau, U. and Brendel, R. (1998) The detailed balance principle and the reciprocity theorem between photocarrier collection and dark carrier distribution in solar cells. *J. Appl. Phys.*, **84**, 6412.
8. Rau, U. (2007) Reciprocity relation between photovoltaic quantum efficiency and electroluminescent emission of solar cells. *Phys. Rev. B*, **76**, 085303.
9. Rau, U. (2012) Superposition and reciprocity in the electroluminescence and photoluminescence of solar cells. *IEEE J. Photovoltaics*, **2**, 169.
10. Kirchartz, T. and Rau, U. (2008) Detailed balance and reciprocity in solar cells. *Phys. Status Solidi A*, **205**, 2737.
11. Onsager, L. (1931) Reciprocal relations in irreversible processes. I. *Phys. Rev.*, **37**, 405.
12. Casimir, H.B.G. (1945) On Onsager's principle of microscopic reversibility. *Rev. Mod. Phys.*, **17**, 343.

13. Shockley, W., Sparks, M., and Teal, G.K. (1951) Pn-junction transistors. *Phys. Rev.*, **83**, 151.
14. Tellegen, B.D.H. (1952) A general network theorem, with applications. *Philips Res. Rep.*, **7**, 259.
15. Wong, J. and Green, M.A. (2012) From junction to terminal: extended reciprocity relations in solar cell operation. *Phys. Rev. B*, **85**, 235205.
16. Kirchhoff, G. (1860) Ueber das Verhältniss zwischen dem Emissionsvermögen und dem Absorptionsvermögen der Körper für Wärme und Licht. *Ann. Phys.*, **185**, 275.
17. Wurfel, P. (1982) The chemical-potential of radiation. *J. Phys. C: Solid State Phys.*, **15**, 3967.
18. Ross, R.T. (1967) Some thermodynamics of photochemical systems. *J. Chem. Phys.*, **46**, 4590.
19. Smestad, G. and Ries, H. (1992) Luminescence and current voltage characteristics of solar-cells and optoelectronic devices. *Sol. Energy Mater. Sol. Cells*, **25**, 51.
20. Green, M.A. (2012) Radiative efficiency of state-of-the-art photovoltaic cells. *Prog. Photovoltaics*, **20**, 472.
21. Markvart, T. (2008) Solar cell as a heat engine: energy-entropy analysis of photovoltaic conversion. *Phys. Status Solidi A*, **205**, 2752.
22. Kirchartz, T. and Rau, U. (2007) Electroluminescence analysis of high efficiency Cu(In,Ga)Se-2 solar cells. *J. Appl. Phys.*, **102**, 104510.
23. Kirchartz, T., Rau, U., Kurth, M., Mattheis, J., and Werner, J.H. (2007) Comparative study of electroluminescence from Cu(In,Ga)Se-2 and Si solar cells. *Thin Solid Films*, **515**, 6238.
24. Kirchartz, T., Helbig, A., Reetz, W., Reuter, M., Werner, J.H., and Rau, U. (2009) Reciprocity between electroluminescence and quantum efficiency used for the characterization of silicon solar cells. *Prog. Photovoltaics*, **17**, 394.
25. Redfield, D. (1974) Multiple-pass thin-film silicon solar cell. *Appl. Phys. Lett.*, **25**, 647.
26. Yablonovitch, E. (1982) Statistical ray optics. *J. Opt. Soc. Am.*, **72**, 899.
27. Yablonovitch, E. and Cody, G.D. (1982) Intensity enhancement in textured optical sheets for solar-cells. *IEEE Trans. Electron Devices*, **29**, 300.
28. Brendel, R. and Queisser, H.J. (1993) On the thickness dependence of open-circuit voltages of P-N-junction solar-cells. *Sol. Energy Mater. Sol. Cells*, **29**, 397.
29. Miller, O.D., Yablonovitch, E., and Kurtz, S.R. (2012) Strong internal and external luminescence as solar cells approach the shockley-queisser limit. *IEEE J. Photovoltaics*, **2**, 303.
30. Daub, E. and Wurfel, P. (1995) Ultralow values of the absorption-coefficient of Si obtained from luminescence. *Phys. Rev. Lett.*, **74**, 1020.
31. Trupke, T., Daub, E., and Wurfel, P. (1998) Absorptivity of silicon solar cells obtained from luminescence. *Sol. Energy Mater. Sol. Cells*, **53**, 103.
32. Trupke, T., Wurfel, P., Uhlendorf, I., and Lauermann, I. (1999) Electroluminescence of the dye-sensitized solar cell. *J. Phys. Chem. B*, **103**, 1905.
33. Kirchartz, T., Helbig, A., Pieters, B.E., and Rau, U. (2011) Electroluminescence Analysis of Solar Cells and Solar Modules, in D. Abou-Ras, T. Kirchartz, and U. Rau (eds.) *Advanced Characterization Techniques for Thin Film Solar Cells*, Wiley-VCH Verlag GmbH & Co. KGaA, p. 61.
34. Kirchartz, T., Rau, U., Hermle, M., Bett, A.W., Helbig, A., and Werner, J.H. (2008) Internal voltages in GaInP/GaInAs/Ge multijunction solar cells determined by electroluminescence measurements. *Appl. Phys. Lett.*, **92**, 123502.
35. Vandewal, K., Tvingstedt, K., Gadisa, A., Inganas, O., and Manca, J.V. (2009) On the origin of the open-circuit voltage of polymer-fullerene solar cells. *Nat. Mater.*, **8**, 904.
36. Wurfel, P., Trupke, T., Puzzer, T., Schaffer, E., Warta, W., and Glunz, S.W. (2007) Diffusion lengths of silicon solar cells from luminescence images. *J. Appl. Phys.*, **101**, 123110.
37. Green, M.A., Zhao, J.H., Wang, A.H., Reece, P.J., and Gal, M. (2001) Efficient silicon light-emitting diodes. *Nature*, **412**, 805.

38. Kirchartz, T., Helbig, A., and Rau, U. (2008) Note on the interpretation of electroluminescence images using their spectral information. *Sol. Energy Mater. Sol. Cells*, **92**, 1621.
39. Arora, N.D., Chamberlain, S.G., and Roulston, D.J. (1980) Diffusion length determination in P-N-junction diodes and solar-cells. *Appl. Phys. Lett.*, **37**, 325.
40. Brendel, R. and Rau, U. (1999) Effective diffusion lengths for minority carriers in solar cells as determined from internal quantum efficiency analysis. *J. Appl. Phys.*, **85**, 3634.
41. Ulbrich, C., Peters, M., Blasi, B., Kirchartz, T., Gerber, A., and Rau, U. (2010) Enhanced light trapping in thin-film solar cells by a directionally selective filter. *Opt. Express*, **18**, A133–A138.
42. Peters, M., Goldschmidt, J.C., Kirchartz, T., and Blasi, B. (2009) The photonic light trap-Improved light trapping in solar cells by angularly selective filters. *Sol. Energy Mater. Sol. Cells*, **93**, 1721.
43. Roensch, S., Hoheisel, R., Dimroth, F., and Bett, A.W. (2011) Subcell I-V characteristic analysis of GaInP/GaInAs/Ge solar cells using electroluminescence measurements. *Appl. Phys. Lett.*, **98**, 251113.
44. Vandewal, K., Tvingstedt, K., Manca, J.V., and Inganas, O. (2010) Charge-transfer states and upper limit of the open-circuit voltage in polymer: fullerene organic solar cells. *IEEE J. Sel. Top. Quantum Electron.*, **16**, 1676.
45. Vandewal, K., Tvingstedt, K., Gadisa, A., Inganas, O., and Manca, J.V. (2010) Relating the open-circuit voltage to interface molecular properties of donor:acceptor bulk heterojunction solar cells. *Phys. Rev. B*, **81**, 125204.
46. Scharber, M.C., Lungenschmied, C., Egelhaaf, H.-J., Matt, G., Bednorz, M., Fromherz, T., Jia, G., Jarzab, D., and Loi, M.A. (2011) Charge transfer excitons in low band gap polymer based solar cells and the role of processing additives. *Energy Environ. Sci.*, **4**, 5077.
47. Vandewal, K., Gadisa, A., Oosterbaan, W.D., Bertho, S., Banishoeib, F., Van Severen, I., Lutsen, L., Cleij, T.J., Vanderzande, D., and Manca, J.V. (2008) The relation between open-circuit voltage and the onset of photocurrent generation by charge-transfer absorption in polymer: fullerene bulk heterojunction solar cells. *Adv. Funct. Mater.*, **18**, 2064.
48. Lee, J., Vandewal, K., Yost, S.R., Bahlke, M.E., Goris, L., Baldo, M.A., Manca, J.V., and Van Voorhis, T. (2010) Charge transfer state versus hot exciton dissociation in polymer-fullerene blended solar cells. *J. Am. Chem. Soc.*, **132**, 11878.
49. Benson-Smith, J.J., Goris, L., Vandewal, K., Haenen, K., Manca, J.V., Vanderzande, D., Bradley, D.D.C., and Nelson, J. (2007) Formation of a ground-state charge-transfer complex in polyfluorene/[6,6]-phenyl-C-61 butyric acid methyl ester (PCBM) blend films and its role in the function of polymer/PCBM solar cells. *Adv. Funct. Mater.*, **17**, 451.
50. Goris, L., Poruba, A., Hod'akova, L., Vanecek, M., Haenen, K., Nesladek, M., Wagner, P., Vanderzande, D., De Schepper, L., and Manca, J.V. (2006) Observation of the subgap optical absorption in polymer-fullerene blend solar cells. *Appl. Phys. Lett.*, **88**, 052113.
51. Goris, L., Haenen, K., Nesladek, M., Wagner, P., Vanderzande, D., De Schepper, L., D'Haen, J., Lutsen, L., and Manca, J.V. (2005) Absorption phenomena in organic thin films for solar cell applications investigated by photothermal deflection spectroscopy. *J. Mater. Sci.*, **40**, 1413.
52. Street, R.A., Song, K.W., Northrup, J.E., and Cowan, S. (2011) Photoconductivity measurements of the electronic structure of organic solar cells. *Phys. Rev. B*, **83**, 165207.
53. Scharber, M.C., Wuhlbacher, D., Koppe, M., Denk, P., Waldauf, C., Heeger, A.J., and Brabec, C.L. (2006) Design rules for donors in bulk-heterojunction solar cells – Towards 10% energy-conversion efficiency. *Adv. Mater.*, **18**, 789.
54. Tvingstedt, K., Vandewal, K., Gadisa, A., Zhang, F.L., Manca, J., and Inganas, O. (2009) Electroluminescence from charge transfer states in polymer solar cells. *J. Am. Chem. Soc.*, **131**, 11819.
55. Faist, M.A., Kirchartz, T., Gong, W., Ashraf, R.S., McCulloch, I., de Mello,

J.C., Ekins-Daukes, N.J., Bradley, D.D.C., and Nelson, J. (2012) Competition between the charge transfer state and the singlet states of donor or acceptor limiting the efficiency in polymer: fullerene solar cells. *J. Am. Chem. Soc.*, **134**, 685.

56. Trupke, T., Nyhus, J., and Haunschild, J. (2011) Luminescence imaging for inline characterisation in silicon photovoltaics. *Phys. Status Solidi RRL*, **5**, 131.

57. Fuyuki, T., Kondo, H., Yamazaki, T., Takahashi, Y., and Uraoka, Y. (2005) Photographic surveying of minority carrier diffusion length in polycrystalline silicon solar cells by electroluminescence. *Appl. Phys. Lett.*, **86**, 262108.

58. Giesecke, J.A., Kasemann, M., and Warta, W. (2009) Determination of local minority carrier diffusion lengths in crystalline silicon from luminescence images. *J. Appl. Phys.*, **106**, 014907.

59. Giesecke, J.A., Kasemann, M., Schubert, M.C., Wufel, P., and Warta, W. (2010) Separation of local bulk and surface recombination in crystalline silicon from luminescence reabsorption. *Prog. Photovoltaics*, **18**, 10.

60. Hinken, D., Bothe, K., Ramspeck, K., Herlufsen, S., and Brendel, R. (2009) Determination of the effective diffusion length of silicon solar cells from photoluminescence. *J. Appl. Phys.*, **105**, 104516.

61. Hinken, D., Ramspeck, K., Bothe, K., Fischer, B., and Brendel, R. (2007) Series resistance imaging of solar cells by voltage dependent electroluminescence. *Appl. Phys. Lett.*, **91**, 182104.

62. Haunschild, J., Glatthaar, M., Kasemann, M., Rein, S., and Weber, E.R. (2009) Fast series resistance imaging for silicon solar cells using electroluminescence. *Phys. Status Solidi RRL*, **3**, 227.

63. Kampwerth, H., Trupke, T., Weber, J.W., and Augarten, Y. (2008) Advanced luminescence based effective series resistance imaging of silicon solar cells. *Appl. Phys. Lett.*, **93**, 202102.

64. Helbig, A., Kirchartz, T., Schaeffler, R., Werner, J.H., and Rau, U. (2010) Quantitative electroluminescence analysis of resistive losses in Cu(In, Ga)Se(2) thin-film modules. *Sol. Energy Mater. Sol. Cells*, **94**, 979.

65. Seeland, M., Rosch, R., and Hoppe, H. (2012) Quantitative analysis of electroluminescence images from polymer solar cells. *J. Appl. Phys.*, **111**, 024505.

66. Kasemann, M., Grote, D., Walter, B., Kwapil, W., Trupke, T., Augarten, Y., Bardos, R.A., Pink, E., Abbott, M.D., and Warta, W. (2008) Luminescence imaging for the detection of shunts on silicon solar cells. *Prog. Photovoltaics*, **16**, 297.

67. Pieters, B.E. (2012) Spatial modeling of thin-film solar modules using the network simulation method and SPICE. *IEEE J. Photovoltaics*, **1**, 93.

68. Breitenstein, O., Bauer, J., Trupke, T., and Bardos, R.A. (2008) On the detection of shunts in silicon solar cells by photo- and electroluminescence imaging. *Prog. Photovoltaics*, **16**, 325.

69. Bothe, K., Ramspeck, K., Hinken, D., Schinke, C., Schmidt, J., Herlufsen, S., Brendel, R., Bauer, J., Wagner, J.M., Zakharov, N., and Breitenstein, O. (2009) Luminescence emission from forward- and reverse-biased multicrystalline silicon solar cells. *J. Appl. Phys.*, **106**, 104510.

70. Breitenstein, O., Bauer, J., Bothe, K., Kwapil, W., Lausch, D., Rau, U., Schmidt, J., Schneemann, M., Schubert, M.C., Wagner, J.M., and Warta, W. (2011) Understanding junction breakdown in multicrystalline solar cells. *J. Appl. Phys.*, **109**, 071101.

71. Schneemann, M., Helbig, A., Kirchartz, T., Carius, R., and Rau, U. (2010) Reverse biased electroluminescence spectroscopy of crystalline silicon solar cells with high spatial resolution. *Phys. Status Solidi A*, **207**, 2597.

72. Knabe, S., Gutay, L., and Bauer, G.H. (2009) Interpretation of quasi-Fermi level splitting in Cu(Ga,In)Se(2)-absorbers by confocally recorded spectral luminescence and numerical modeling. *Thin Solid Films*, **517**, 2344.

73. Gutay, L. and Bauer, G.H. (2007) Spectrally resolved photoluminescence studies on Cu(In,Ga)Se(2) solar cells with lateral submicron resolution. *Thin Solid Films*, **515**, 6212.

74. Bauer, G.H. and Gutay, L. (2007) Analyses of local open circuit voltages in polycrystalline Cu(In,Ga)Se-2 thin film solar cell absorbers on the micrometer scale by confocal luminescence. *Chimia*, **61**, 801.
75. Unold, T. and Gutay, L. (2011) Photoluminescence Analysis of Thin-Film Solar Cells, in D. Abou-Ras, T. Kirchartz, and U. Rau (eds.) *Advanced Characterization Techniques for Thin Film Solar Cells*, Wiley-VCH Verlag GmbH & Co. KGaA, p. 151.
76. Romero, M.J., Du, H., Teeter, G., Yan, Y.F., and Al-Jassim, M.M. (2011) Comparative study of the luminescence and intrinsic point defects in the kesterite Cu(2)ZnSnS(4) and chalcopyrite Cu(In,Ga)Se(2) thin films used in photovoltaic applications. *Phys. Rev. B*, **84**, 165324.
77. Gundel, P., Schubert, M.C., Heinz, F.D., Woehl, R., Benick, J., Giesecke, J.A., Suwito, D., and Warta, W. (2011) Microspectroscopy on silicon wafers and solar cells. *Nanoscale Res. Lett.*, **6**, 197.
78. Gundel, P., Heinz, F.D., Schubert, M.C., Giesecke, J.A., and Warta, W. (2010) Quantitative carrier lifetime measurement with micron resolution. *J. Appl. Phys.*, **108**, 033705.
79. Chappell, J., Lidzey, D.G., Jukes, P.C., Higgins, A.M., Thompson, R.L., O'Connor, S., Grizzi, I., Fletcher, R., O'Brien, J., Geoghegan, M., and Jones, R.A.L. (2003) Correlating structure with fluorescence emission in phase-separated conjugated-polymer blends. *Nat. Mater.*, **2**, 616.
80. Romero, M.J., Alberi, K., Martin, I.T., Jones, K.M., Young, D.L., Yan, Y., Teplin, C., Al-Jassim, M.M., Stradins, P., and Branz, H.M. (2010) Nanoscale measurements of local junction breakdown in epitaxial film silicon solar cells. *Appl. Phys. Lett.*, **97**, 092107.
81. Gutay, L., Lienau, C., and Bauer, G.H. (2010) Subgrain size inhomogeneities in the luminescence spectra of thin film chalcopyrites. *Appl. Phys. Lett.*, **97**, 052110.
82. Gutay, L., Pomraenke, R., Lienau, C., and Bauer, G.H. (2009) Subwavelength inhomogeneities in Cu(In,Ga)Se(2) thin films revealed by near-field scanning optical microscopy. *Phys. Status Solidi A*, **206**, 1005.
83. Kirchartz, T., Mattheis, J., and Rau, U. (2008) Detailed balance theory of excitonic and bulk heterojunction solar cells. *Phys. Rev. B*, **78**, 235320.
84. Hoyer, U., Pinna, L., Swonke, T., Auer, R., Brabec, C.J., Stubhan, T., and Li, N. (2011) Comparison of electroluminescence intensity and photocurrent of polymer based solar cells. *Adv. Energy Mater.*, **1**, 1097.
85. Pieters, B.E., Kirchartz, T., Merdzhanova, T., and Carius, R. (2010) Modeling of photoluminescence spectra and quasi-Fermi level splitting in mu c-Si:H solar cells. *Sol. Energy Mater. Sol. Cells*, **94**, 1851.
86. Gong, W., Faist, M.A., Ekins-Daukes, N.J., Xu, Z., Bradley, D.D.C., Nelson, J., and Kirchartz, T. (2012) Influence of energetic disorder on electroluminescence emission in polymer:fullerene solar cells. *Phys. Rev. B*, **86**, 024201.
87. Muller, T.C.M., Pieters, B.E., Kirchartz, T., Carius, R., and Rau, U. (2014) Effect of localized states on the reciprocity between quantum efficiency and electroluminescence in Cu(In,Ga)Se-2 and Si thin-film solar cells. *Sol. Energy Mater. Sol. Cells*, **129**, 95.
88. Tran, T.M.H., Pieters, B.E., Schneemann, M., Muller, T.C.M., Gerber, A., Kirchart, T., and Rau, U. (2013) Quantitative evaluation method for electroluminescence images of a-Si: H thin-film solar modules. *Phys. Status Solidi RRL*, **7**, 627.

3
Rear Side Diffractive Gratings for Silicon Wafer Solar Cells

Marius Peters, Hubert Hauser, Benedikt Bläsi, Matthias Kroll, Christian Helgert, Stephan Fahr, Samuel Wiesendanger, Carston Rockstuhl, Thomas Kirchartz, Uwe Rau, Alexander Mellor, Lorenz Steidl, and Rudolf Zentel

3.1
Introduction

3.1.1
Gratings for Solar Cells – Basic Idea and Challenges

Using very thin absorber layers for solar cells has several advantages: material consumption is reduced which leads to lower product costs. The path of the charge carriers to the contacts is shortened, leading to more efficient devices or lifted requirements on electronic material properties. In recent years, silicon wafers have become thinner and thinner, reaching thicknesses of 100 μm and below. The high potential for such ultrathin silicon wafer solar cells was shown, for example, in [1]. Furthermore, concepts were developed to deposit very thin layers of crystalline or amorphous silicon onto substrates [2, 3].

The disadvantage of thin layers is their low absorptance of photons with energies close to that of the semiconductor's band-gap. This is especially true for indirect semiconductors like crystalline silicon but also holds for direct semiconductors like amorphous silicon, although to a lower extent. For this reason, light trapping techniques, that is, ways to increase the absorptance without increasing the absorber thickness, have steadily gained importance for solar cells.

In this chapter we discuss how diffractive gratings can be applied for light trapping in solar cells. A grating is a periodic modulation of either an interface between two materials or of the permittivity itself. The periodicity can be along just one dimension or along two. The basic idea of this concept is sketched in Figure 3.1. light incident on a grating (either at the front or at the rear side of a solar cell, discussed here is the case of a rear side grating) will be diffracted into a number of secondary beams of different orders. The zeroth order corresponds to the direction of specular reflection, that is, the reflection that would occur if no grating was present. Light diffracted into higher orders travels along an oblique path and is totally internally reflected. The change of direction and the

Photon Management in Solar Cells, First Edition. Edited by Ralf B. Wehrspohn, Uwe Rau, and Andreas Gombert.
© 2015 Wiley-VCH Verlag GmbH & Co. KGaA. Published 2015 by Wiley-VCH Verlag GmbH & Co. KGaA.

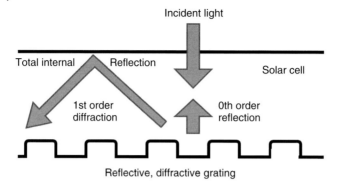

Figure 3.1 Sketch of the principle of light trapping in a solar cell with a diffractive grating. Incident light enters the solar cell and is diffracted at the rear side grating. Light diffracted into an angle greater than the critical angle is totally internally reflected and thus trapped in the solar cell. The increased path length due to this process results in an enhanced absorption within the solar cell and, eventually, in an increased current generation.

internal reflection provide the desired enhancement of the optical path length and absorption. The concept is discussed in more detail in Section 3.2. One reason for gratings having experienced a very high level of attention is that from a fundamental viewpoint, gratings have a very high potential for light trapping. Some theoretical considerations will be discussed in Section 3.3.

The grating concept, however, also faces several challenges. How much light is trapped by a grating depends on its geometrical shape, its period, and the properties of the solar cell it is used for. Optimizing gratings is a complex task and needs to be done on a case to case basis. This task requires sophisticated simulation and investigation methods, some of which are discussed in Section 3.4. Furthermore, the realization of gratings in solar cells requires high-precision texturing techniques in the micrometer and sub-micrometer range. How this texturing can be achieved will be addressed in Section 3.5.

Finally, in Section 3.6, we will give a short overview of the most important characterization techniques for solar cells with diffractive gratings. The chapter concludes with a summary and some recommendations for the design of solar cells with diffractive gratings.

3.1.2
A Short Literature Review

Because of the very large number of publications on gratings for PV (Pholovoltaics) applications, only a small excerpt of possible publications will be discussed here. The selection is confined to works of historical interest and to papers that played a role for some of the work discussed later on.

First ideas to use the diffractive properties of gratings for light trapping in solar cells were formulated by Sheng *et al.* in 1983 [4]. In this work, theoretical studies were performed for a 0.5 μm thick solar cell made of amorphous silicon (a-Si:H) with a 70 nm thick anti reflection (AR) coating and planar front and textured rear side. Compared to a solar cell without grating, a gain in absorbed photo current of 2 mA cm^{-2} was predicted for hemispherical incident light for a grating with a period of 250 nm and a depth of 80 nm. A comparable gain was found, in case of a restricted angle of incidence of $\pm 5°$. Furthermore, a gain in absorbed photocurrent of 3.5–4 mA cm^{-2} was predicted for a 2D cross-hatched grating.

In their widely recognized publication from 1995 [5], Heine and Morf performed a theoretical and experimental investigation of a crystalline silicon solar cell with a linear (1D) rear side grating. In this work, Heine and Morf assumed that the optimum grating parameters are those for which an order of diffraction appears under very shallow angles close to 90°. For a silicon solar cell with a thickness of 40 µm, this criterion resulted in a grating period of 310 nm (for the first order of diffraction) or 620 nm (for the second order of diffraction). Using the latter period, a path length enhancement of a factor of 3 was found experimentally (based on reflection measurements) for a binary grating and a factor of 5 for a blazed grating. The investigation, furthermore, included considerations about the use of front surface gratings as AR coatings. Here, a weighted average reflection of less than 0.6% was predicted. All results were restricted to the optical performance of the solar cell, and the impact of texturing on the electrical characteristics was not considered. Earlier works of this group were published in conferences already in 1990 [6] and earlier.

An extended study by the same authors together with Kiess et al. was published in Ref. [7] in 1997. An additional grating geometry of staircase shape was investigated. In this study, measured quantum efficiencies were also presented. An increase in external quantum efficiency (EQE) for a 40 µm thick solar cell with rear side grating of up to 25% was reported. The internal quantum efficiency (IQE), a measure for electrical losses inside the solar cell, remained unchanged. This was taken as an indication that the quality of the solar cell surface was not altered significantly by the introduction of a grating.

With the beginning of the new millennium, an ever increasing number of groups published results about diffraction gratings for different kinds of solar cells. Zaidi et al. [8] investigated gratings of different geometries for thin (1–20 µm thick) silicon solar cells and demonstrated path length enhancements in the range of two for 1D gratings. Even higher values were predicted for 2D gratings. Niggemann et al. [9] showed a significant increase of absorptance due to a diffractive grating in an organic solar cell. The impact of diffractive structures on thin-film silicon solar cells was investigated, for example, by Haase and Stiebig [10]. A further use of photonic and diffractive properties of photonic crystals was discussed by Bermel et al. [11]

Fundamental limits of light trapping were investigated by various groups. Theoretical are found, for example, in the publications by Kirchartz et al. [12] and Mellor et al. [13] revealed the high potential of diffraction with path length enhancements exceeding the Lambertian limit for light under normal incidence. Structures that could achieve such super-Lambertian light trapping were proposed in theoretical studies by Yu et al. [14] and Gjessing et al. [15].

The question whether periodic or random structures are superior was addressed by Battaglia et al. [16]. In this publication random and periodic structures for silicon thin-film solar cells were compared. The conclusion of this paper was that periodic structures can, under certain conditions, optically outperform random structure but that electrical characteristics also need to be considered. Optical and electrical demands define a trade-off and to not at this point no advantage of periodic structures could be found.

Concepts for the industrial realization of grating structures were investigated recently but are still in a very early stage. Parameter studies were published by Peters et al. [17]. The realization of gratings via lithography processes were investigated, for example, by Ferry for thin-film silicon solar cells [18], by Chou for organic solar cells [19] and by Bläsi et al. [20] for wafer-based silicon solar cells.

3.2
Principle of Light Trapping with Gratings

At first it has to be mentioned that for the sake of generality, this section is not restricted to gratings integrated into the rear side of a solar cell. The optical effects discussed further below do also occur for gratings integrated into the front side of the solar cell. Gratings have been used in the past in such configurations in solar cells as well and will be used for sure for that purpose in the future.

Generally spoken, the optical functionality of a grating is primarily governed by the lattice constant, also called the *grating period*. We will denote this parameter herein as Λ. Depending on the ratio between this period and the effective wavelength λ_{eff} (vacuum wavelength λ_0 divided by the refractive index n of the material in which the light propagation is considered) quite different optical effects can be observed when light interacts with a grating.

In order to simplify the discussion, we will, at first, consider only a corrugated interface that separates an upper half space made of air from a lower half space made of silicon. Moreover, the corrugation is assumed to be invariant in one lateral direction, which was chosen herein without loss of generality to be the x-direction. Along the other lateral direction, that is, the y-direction, it is thought to be composed of isosceles triangles with a height of 900 nm. Their base shall correspond to the lattice constant. This geometry is indicated in Figure 3.2a–c by the blue lines. As illumination, a plane wave, the electric field of which is polarized along the groves (y-direction) with a vacuum wavelength of 860 nm and normal incidence, is considered. The color in Figure 3.2a–c represents the normalized absolute value of the spatially resolved electric field, which can be calculated by applying one of the numerical methods indicated later in this chapter.

In Figure 3.2a, the grating period was set to 200 nm. In this case, the grating period is (much) smaller than the effective wavelength in both half spaces ($\lambda \ll \lambda_{eff}$). By reminding the reader that the absolute value of the electric field of a plane wave is a constant, we notice that in the space below the corrugation, that is, in the transmission region, the field amplitude as shown in Figure 3.2a shows exactly such a constant absolute; indicating that only a single plane wave

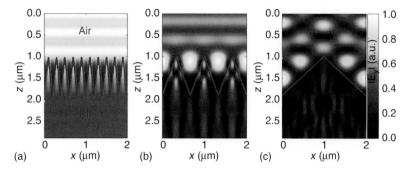

Figure 3.2 Normalized electric field amplitude for a plane wave with a vacuum wavelength of 860 nm illuminating a corrugated surface separating an upper air half space and a lower silicon half space in different parameter regimes. (a) The small period of 200 nm results in almost perfect reduction of reflection losses. (b) Strong scattering into silicon occurs for a period of 667 nm. (c) The large period of 2000 nm leads to scattering in both half spaces.

is transmitted. Above the corrugation, the electric field is a superposition of the incident and just one reflected plane wave. Since they are counter-propagating, the resulting field shows clearly a modulation of the absolute field strength in the vertical direction. Hence, the far-field reflected and transmitted by the grating can be safely described by a single plane wave each. No diffraction effects occur into the far-field, that is, the incident plane wave is not deflected by the periodic corrugation.

Such response resembles the response of a stratified medium. The geometry is said to be sub-wavelength and the incident light is not able to resolve the fine spatial details; it will experience an effective medium instead. In this sub-wavelength domain it is, therefore, possible to assign effective optical parameters to the grating structure. Since the structure is not invariant in the z-direction, these effective optical parameters will depend on z as well. Depending on the dimensionality of the grating and the polarization of the incident light, different rules exist [21, 22] for assigning these effective parameters. Details can be found in the indicated literature. Nonetheless, basic insight have been already obtained from heuristic considerations. A surface corrugation that continuously varies the filling fraction of the constituent materials along the surface normal can be considered in the limit of $\lambda \ll \lambda_{\text{eff}}$ as a medium characterized by a continuously varying effective refractive index profile. It continuously changes its index from that of the upper to that of the lower halfspace. This is obviously of major assistance to reduce Fresnel reflection losses and, hence, improves light coupling between the two half spaces [23, 24]. Since in this sub-wavelength domain the light is not able to resolve the spatial features, an exact periodicity is not required. Therefore, also randomly arranged nanorods [25–27] or nanosponges [28] are perfectly suited for the purpose of reducing Fresnel reflection losses and have been considered in the past for integration into solar cells. If the effective refractive index profile is discontinuous, as in the case of binary gratings, the reduction of reflection losses will not cover an as broad spectral range but still a good performance can be achieved for the relevant spectral range [5].

The grating period in Figure 3.2b was set to be 667 nm, that is, no higher diffraction orders than the zeroth exist. Nonetheless, the evanescent near field causes some modulation of the electric field in the air region close to the surface. However, in the transmitted region several diffraction orders propagate, which results in the highly modulated field strength. What the grating basically does is to change the propagation direction by deflecting the transmitted light into different orders of diffraction. For light under normal incidence, the angle of propagation increases with the order of diffraction. The angles under which the light propagates are solely given by the periodicity. The amount of light, however, that is distributed into the different orders depends on the exact shape of the surface profile. This occurs in a very deterministic sense and control over the shape of the geometry is the key. Calculation of these diffraction efficiencies can be done with the method detailed further below in this chapter. The modulation of the electric field is shown in Figure 3.2b in the transmission half space, and is a direct consequence of the interference among the different diffraction orders.

Since the local absorptance correlates with the local electric field strength, maximizing the field overlap with the absorbing semiconductor is the key task of optimizing the geometry of gratings employed in thin-film solar cells [29–38]. However, in absorber layers that are several wavelengths thick, the task is rather to excite solely the higher diffraction orders, since they correspond to the longest path lengths of the propagating light. Simple design strategies in this parameter regime consist of not in choosing a grating period to cause a diffraction order that propagates in the light absorbing material at an angle close to 90° for a wavelength close to the absorption edge, for example, in Ref. [5]. This is the spectral domain where the path length should be maximized in order to enhance the absorption. Finally, the grating has to be optimized to diffract the largest share of the incident light into that preferential diffraction order.

For Figure 3.2c the grating period is further increased to 2000 nm. In this regime, multiple diffraction orders exist for the silicon region as well as for the air half space. For even larger ratios of grating period to wavelength, the light propagation can be described with increasing accuracy by a ray optical picture. There, steep angles as in the case of V-shaped grooves [39–44] lead to multiple interactions of the incident light with the interface and, therefore, to enhanced transmittance. Moreover, light incident from the inside will be kept partially inside the absorber due to total internal reflection. The latter mechanism leads for purposefully structured surfaces to path lengths close to or even exceeding the Lambertian limit [41–43].

Besides the reduction of reflection losses and the enhancement of the path length due to scattering, the excitation of guided modes is also exploited [16]. The existence of guided modes requires a lower refractive index outside the guiding structure. Hence, the geometry considered in Figure 3.3 is modified by setting the material of the lower half space to air and by inserting an additional 500 nm thick silicon slab below the corrugation, as is shown in Figure 3.3b,c. It has to be

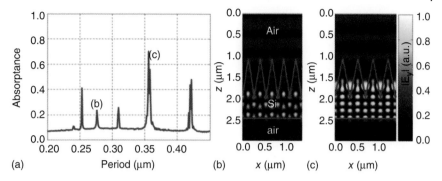

Figure 3.3 (a) Calculated absorptance as a function of the grating period for an illumination wavelength of 860 nm. Absolute electric field strength for a grating period in (b) of 276 nm and in (c) of 357 nm.

stressed that such a cell does not correspond to a real-world scenario but was only chosen for illustrative purposes. Figure 3.3a shows the calculated absorptance as a function of the grating period for a wavelength of 860 nm, at which light can propagate several tens of micrometers within crystalline silicon. Several distinct peaks can be observed above a smooth background absorptance of approximately 10%.

In order to understand these peaks in absorptance, it is instructive to consider first a non-absorbing slab with a high refractive index, which supports guided modes characterized by a real propagation constant. Such modes can be excited either by so-called butt-coupling, that is, illuminating from the side which leaves the waveguide modes unperturbed in all their properties, or by using a shallow grating coupler put on top of the waveguide. The latter can be considered as a small perturbation which only serves to compensate the mismatch between the lateral wave vector component of the incident plane wave and the propagation constant of the guided mode according to Eq. (3.13). Moreover, it can be safely assumed, that a small perturbation will have only a marginal effect on the real part of the propagation constant. Since the guided mode is coupled to the radiation modes outside the waveguide, this mode is called *leaky* and the propagation constant will be complex-valued in order to reflect the radiation losses. Nonetheless, using such a grating coupler will lead to sharp peaks in the reflection and transmission spectra [45], indicating the excitation of the guided modes by free space radiation.

Since the excitation of these modes leads to a field enhancement similar to the case of a Fabry–Perot-resonator with a high finesse, the presence of losses will coincide with peaks in the absorption spectrum. Introducing losses into the waveguide or increasing the corrugation will simply influence spectral position and line width of these resonances. The spatially resolved, absolute electric field strength for two such peaks, namely for a grating period of 276 and 357 nm, is shown in Figure 3.3b,c, respectively. The excitation of such guided modes has also been considered as the primary effect to enhance the absorption in solar cells;

most notably for rather thick solar cells which provide a high density of guided modes.

In a final comment, it has to be mentioned that for a real solar cell, the clear identification of these different physical effects is not always necessary while designing a specific grating. The huge spectral domain that needs to be considered in this design process and the complicated geometry of the solar cell which is often made of materials with largely disparate optical properties, makes this identification in each case challenging. Moreover, a full identification of all involved optical effects might not always be required, since in most cases one is rather inclined to optimize a certain number of free parameters of a chosen geometry such that, for example, the respective short circuit current density is maximized. Then, it is not necessary to be aware of the subtle mechanisms that finally lead to this enhancement. But the understanding and the recognition of the traces of these effects will provide guidelines for the design of advanced solar cells. For complicated geometries, where brute-force parameter scans are no longer possible since it would require computational resources in excess, the physical intuition born from the understanding of the regimes indicated above will allow to identify a sufficient optimal initial design for the grating to be integrated which can, nevertheless, always be fine-tuned using numerical efforts.

3.3
Fundamental Limits of Light Trapping with Gratings

To study the fundamental potential of diffraction gratings, we consider a perfect grating and calculate the path length enhancement generated by this grating for light incident from a small solid angle and for the case of light incident from the complete half sphere above the solar cell. We assume that a perfect grating redirects normally incident light such that it travels parallel to the surface of the cell.

The direction of all other incident rays is then controlled by the conservation of étendue ε. Incoming radiation shall have an étendue

$$d\varepsilon = n(\lambda)^2 d\Omega dA = d\Omega dA. \tag{3.1}$$

After refraction at the front surface, the angle changes according to Snell's law

$$\sin \theta_1 = n(\lambda) \cdot \sin \theta_2 \tag{3.2}$$

which defines the relation between the angle θ_1 of the incoming light beam and the angle θ_2 of the refracted light beam. The light beam impinging on the rear surface is then diffracted so that the angle of each ray is as high as possible without changing the étendue. Since the illuminated area and the refractive index stays the same, étendue conservation requires that the solid angle stays the same after the diffraction by the grating, which means that every light beam changing its direction by exactly 90°.

The path that the weakly absorbed light takes through the cell depends on the reflection at the front surface and on the critical angle $\theta_c = \arcsin(n^{-1})$ for total

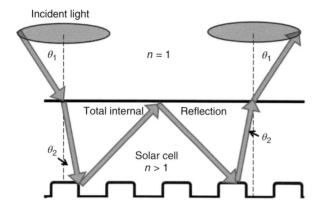

Figure 3.4 Path of the light inside a solar cell with a grating. The ideal grating changes the direction of light by 90° when it hits the rear contact. Due to the conservation of étendue it is not possible to just direct all light to an infinitely long path parallel to the surface.

internal reflection. For an angle $\pi/2 - \theta_2 > \theta_c$ the beam encounters total internal reflection, while for angles $\pi/2 - \theta_2 > \theta_c$, it is assumed that the light couples out of the cell. For a critical angle of total internal reflection $\theta_c > \pi/4$, that is, a refractive index $n < \sqrt{2}$, we have to distinguish between both cases – total internal reflection and outcoupling – depending on the angle of incidence. For smaller critical angles and larger refractive indices total internal reflection always occurs at the first reflection at the front side as depicted in Figure 3.4. Owing to the reciprocity of the light path, the light will couple out after the second hit of the rear surface. The light path is, therefore, symmetric to the surface normal at the point where the total internal reflection occurs and has the shape of a "W."[1]

For all critical angles, the first path from the front surface to the rear surface is $\cos^{-1}\theta$ in units of cell thickness and with θ being the angle to the surface normal. The path length from the first hit on the rear surface to the front surface is $\sin^{-1}\theta$ if we define the propagation direction θ relative to the surface. Now we have to distinguish between critical angles above or below 45°. We define an angle $\theta_b = \pi/2 - \theta_c$ which defines the border between total internal reflection and outcoupling. That means that for angles $0 < \theta < \theta_b$ the light has twice the path length as for angles $\theta_b < \theta < \theta_c$. The path length of unabsorbed light is, hence, given by

$$\frac{2\left[\int_0^{\theta_1} \frac{\sin\theta \cos\theta d\theta}{\cos\theta} + \int_0^{\theta_1} \frac{\sin\theta \cos\theta d\theta}{\sin\theta}\right] + \left[\int_{\theta_1}^{\theta_c} \frac{\sin\theta \cos\theta d\theta}{\cos\theta} + \int_{\theta_1}^{\theta_c} \frac{\sin\theta \cos\theta d\theta}{\sin\theta}\right]}{\int_0^{\theta_c} \sin\theta \cos\theta d\theta} = 4n^2$$

(3.3)

1) This argument is simplified and assumes that all light is diffracted into a single order and no further orders of diffraction exist. The argument can be generalized to more orders of diffraction with a 90° direction change toward the reflected order.

if the incident light has a Lambertian distribution. For the case of higher refractive indices $n > \sqrt{2}$ and smaller critical angles $\theta_c < \pi/4$, we get with a similar calculation

$$\frac{2\left[\int_0^{\theta_c} \frac{\sin\theta \cos\theta d\theta}{\cos\theta} + \int_0^{\theta_1} \frac{\sin\theta \cos\theta d\theta}{\sin\theta}\right]}{\int_0^{\theta_c} \sin\theta \cos\theta d\theta} = 4(n^2 + n - n\sqrt{n^2-1}) \leq 4n^2. \quad (3.4)$$

Interestingly, the path length enhancement becomes smaller than the Yablonovitch limit if total internal reflection always occurs. In this case not all angles are possible which is equivalent to the statement that not the whole phase space volume is occupied. However, for angles $\theta_c > \pi/4$ the whole phase space volume is occupied and the path length enhancement equals that of a Lambertian light-trapping scheme.

Let us now consider the case, when the irradiation of the sun impinges vertically on the system. For a refractive index of $n = 3.5$ and an angle of the sun of $\theta_s = 4.7$ mrad, the maximum path length for a directionally selective filter is $4n^2/\sin^2\theta_s \approx 2.2 \cdot 10^6$ [46]. The maximum path length for a non-Lambertian structure is then given by the "W" shape shown in Figure 3.4 and can be calculated by

$$\frac{2\left[\int_0^{\theta_c} \frac{\sin\theta d\theta}{\sqrt{1-\frac{\sin^2\theta}{n^2}}} + n\int_0^{\theta_s} d\theta\right]}{\int_0^{\theta_c} \sin\theta d\theta} = 2\left[1 + \frac{n\theta_s}{1-\cos\theta_s}\right] \approx 2973 \quad (3.5)$$

3.4
Simulation of Gratings in Solar Cells

3.4.1
Optical Simulation Using RCWA/FMM

In order to evaluate the performance of a specific grating design, it is eventually required to determine the solar cell efficiency. However, such a task would ask for a three-dimensional electro-optical treatment of the device in conjunction with complete knowledge of all material-specific parameters, such as doping concentration profiles, recombination rates, and so on. On the one hand, this treatment is cumbersome and numerically exhaustive. On the other hand, not all material characteristics are sufficiently well known. For example, the impact of surface recombination strongly depends on the grating morphology. Hence, the optimization of the grating morphology would require some *a priori* assumptions about surface effects which will affect for sure the predictive power.

A much simpler evaluation criterion, which is often fully sufficient, is the photo current density j_{ph}. For this, it is assumed that each photon absorbed in the semiconductor generates one electron-hole-pair and that all charge carriers are collected at the contacts. The integration of the absorptance of the semiconductor $a(\lambda)$ weighted with the solar spectral flux of incident photons $\Phi_{sun}(\lambda)$ over the spectral domain of interest then yields the absorbed photo current density

$$j_{ph} = e \cdot \int_0^\infty a(\lambda) \phi_{sun}(\lambda) d\lambda \tag{3.6}$$

with e being the elementary charge.

For the general case where, besides the semiconductor, other materials are absorbing as well, such as a metallic rear reflector or a transparent conductive oxide, the absorptance $a(\lambda)$ has to be determined via integration of the local absorptance over the relevant spatial domain. The local absorptance under monochromatic illumination is given by the time averaged divergence of the Poynting vector $\vec{\nabla} \cdot \langle \vec{S}(\vec{r}, \lambda) \rangle$ [47], with

$$\langle \vec{S}(\vec{r}, \lambda) \rangle = \frac{1}{2} \Re \left[\vec{E}(\vec{r}, \lambda) \times \vec{H}^*(\vec{r}, \lambda) \right] \tag{3.7}$$

Here, $\vec{E}$ and $\vec{H}$ depict the stationary complex electric and magnetic field vectors and the final question about how to determine these field components has to be solved.

The grating geometries of relevance will usually have sizes comparable to the wavelength, that is, approximated solutions are no longer sufficiently accurate and Maxwell's equations have to be solved on rigorous grounds. There exist several devoted numerical routines formulated either in time domain, such as the finite-difference time-domain (FDTD) method [48], or in frequency domain, such as the finite-element method (FEM), the rigorous coupled wave analysis (RCWA) or also called Fourier modal method (FMM) [49, 50], the Chandezon method, or the Korringa–Kohn–Rostocker (KKR) method [51] that can conveniently be used for grating structures. For the case of gratings, frequency domain solvers are recommended, since they avoid long computational run-times in case of exploiting resonances such as the guided modes discussed in Section 3.2 – Principles of light trapping with gratings. The authors describe in this section in detail the RCWA and, to some less extent, the KKR. The methods seem to be preferable since both provide a fast and accurate solution to the present class of problems and they seem to be the tools applied by most of the researchers that discuss the integration of gratings into solar cells.

Frequency domain solvers, in general, allow calculating the stationary field distributions, that is, all fields and sources have time dependence according to $\exp[-i\omega t]$ with ω being the angular frequency. Under the assumption that the media to be considered are characterized by linear, local, isotropic, and dispersive material properties that provide a link between the polarization and the electric field, the curl equations in frequency domain reduce to

$$\vec{\nabla} \times \vec{E}(\vec{r}, \omega) = +\frac{i\omega}{c}\sqrt{\frac{\mu_0}{\varepsilon_0}} \vec{H}(\vec{r}, \omega) \tag{3.8}$$

$$\vec{\nabla} \times \vec{H}(\vec{r},\omega) = -\frac{i\omega}{c} \varepsilon(\vec{r},\omega) \sqrt{\frac{\varepsilon_0}{\mu_0}} \vec{E}(\vec{r},\omega) \tag{3.9}$$

Within the considered grating geometry the permittivity is periodically modulated according to

$$\varepsilon(x, y, z) = \varepsilon(x + \lambda_x, y + \lambda_y, z) \tag{3.10}$$

with λ_x and λ_y being the lateral grating periods. For the RCWA, it is assumed that the grating structure can be divided into L layers, in which the permittivity is not z-dependent. For the KKR, the grating is also composed of L homogeneous layers, which may additionally contain non-overlapping, periodically arranged spheres with a radius dependent permittivity.

If such a periodic structure is illuminated by a plane wave characterized by the lateral wave vector components k_{x0} and k_{y0} the Floquet–Bloch-theorem [52] can be applied to each electromagnetic field component $f^{(l)}(x, y, z)$ in the lth layer

$$f^{(l)}(x + \lambda_x, y + \lambda_y, z) = f^{(l)}(x, y, z) \exp\left[i\left(k_{x0}\lambda_x + k_{y0}\lambda_y\right)\right] \tag{3.11}$$

Hence, $f^{(l)}(x, y, z)$ can be expressed as a pseudo Fourier expansion

$$f^{(l)}(x, y, z) = \sum_{m=-M}^{M} \sum_{n=-N}^{N} f^{(l)}(z) \exp\left[i\left(k_{xm}x + k_{yn}y\right)\right] \tag{3.12}$$

where k_{xm} and k_{yn} are given by

$$k_{xm} = k_{x0} + \frac{2\pi m}{\lambda_x}, \quad k_{yn} = k_{y0} + \frac{2\pi n}{\lambda_y} \tag{3.13}$$

In the frame of Fourier expansions, Eq. (3.12) is exact in the limit of $M, N \to \infty$. However, for a numerical implementation M and N will have to be limited. In the homogeneous half spaces above ($z < 0$) and below ($z > z_{L+1}$) the grating geometry the field components are written as a Rayleigh-expansion

$$f^{(0)}(x, y, z) = \sum_{m,n,\pm} f^{(0)}_{mn\pm} \exp\left[i\left(k_{xm}x + k_{yn}y \pm \gamma^{(0)}_{mn}z\right)\right], \quad z < 0 \tag{3.14}$$

$$f^{(L+1)}(x, y, z) = \sum_{m,n,\pm} f^{(L+1)}_{mn\pm} \exp\left[i\left(k_{xm}x + k_{yn}y \pm \gamma^{(L+1)}_{mn}z\right)\right], \quad z > z_{L+1} \tag{3.15}$$

where $\gamma^{(0,L+1)}_{mn}$ is given by

$$(\gamma^{(0,L+1)}_{mn})^2 = \frac{\omega^2}{c^2} \varepsilon^{(0,L+1)} - k^2_{xm} - k^2_{yn} \tag{3.16}$$

with the sign convention $\Re \gamma^{(0,L+1)}_{mn} > 0$ and $\Im \gamma^{(0,L+1)}_{mn} > 0$. This Rayleigh-expansion corresponds to the discrete diffraction orders including the evanescently decaying near field. The incident fields are indicated by $a+$ for ($z < 0$) and $a-$ for ($z > z_{L+1}$).

For the RCWA, the z-independence of the permittivity in the lth layer $\varepsilon^{(l)}(x, y)$ allows separating $f^{(l)}_{mn}(z)$ in Eq. (3.12) into

$$f^{(l)}_{mn}(z) = f^{(l)}_{mn} \exp\left[i\gamma^{(l)}z\right] \tag{3.17}$$

The two curl Eqs. (3.8) and (3.9) are then used to eliminate the z-components of the two fields. Fourier transforming the permittivity function $\varepsilon^{(l)}(x,y)$ in a specific layer by applying adopted factorization rules [49, 50] results in an eigenvalue problem with $U = 2(2M+1)(2N+1)$ eigenvectors $(\vec{E}_{xu}^{(l)}, \vec{E}_{yu}^{(l)})^T$ with the corresponding eigenvalue $\gamma_u^{(l)}$. The z-dependent electric field components $E_{jmn}^{(l)}(z)$ in Eq. (3.12) with $j \in \{x, y\}$ can then be written as

$$E_{jmn}(z)^{(l)} = \sum_{u=1}^{U} E_{jmnu}^{(l)} \left\{ a_u^{(l)} \exp\left[i\gamma_u^{(l)} \tilde{z}\right] + b_u^{(l)} \exp\left[-i\gamma_u^{(l)}(\tilde{z} - h_l)\right] \right\} \quad (3.18)$$

where $\tilde{z} = z - z_l$ depicts the relative z-coordinate in the lth layer with thickness h_l. An analogous expression can be found for the magnetic field components. The unknown coefficients $a_u^{(l)}$ and $b_u^{(l)}$ are connected with $a_u^{(l+1)}$ and $b_u^{(l+1)}$ and via a scattering matrix, which can be formulated by enforcing the continuity of the lateral field components at the respective layer interface. All scattering matrixes are then used to build a final scattering matrix that connects the incident field with all outgoing fields. Then, the electromagnetic field can be calculated everywhere in space. Note that the numerical effort scales linearly in the number of layers L and in the third power of U.

In contrast to the RCWA, where the modes within each layer have to be found numerically by solving an eigenvalue problem, the fields within each layer are already analytically known within the frame of the KKR. At all interfaces and especially in homogeneous layers the field is written as a Rayleigh-expansion similar to Eqs. (3.14)–(3.16). In the case of layers containing spheres the field is written using spherical Bessel and Hankel functions and spherical harmonics. This allows for deriving a simple scattering matrix that connects incident with scattered spherical waves similar to the scattering process at a single sphere [53]. The only computational burden is to translate the spherical waves outside the spheres into plane waves required for the formulation of the boundary conditions at the interfaces [51]. Again a scattering matrix is formulated for each layer, which finally allows connecting the illuminating plane wave with all outgoing fields.

3.4.2
Optical Simulation Using the Matrix Method

In the previous section, a method was presented that involves a fully coherent treatment of the entire system in which interference between all Fourier orders is considered. This implies that the thickness of the solar cell absorber region is smaller than the coherence length of the incident; not: an appropriate assumption for thin-film solar cells. The method can be used also for thick, wafer-based solar cells but at the cost of increased computing time as special care must be taken to remove undesired interference effects due to the thick substrate.

In this section a method that can be used to calculate the absorptance in a wafer-based crystalline silicon solar cell directly is described. We turn our attention to a system in which the thickness of the unstructured part of the absorber region is thicker than the coherence length of the incident solar radiation (≈ 1 µm). In this case, the Fourier orders in the grating region must be treated coherently as before, whereas the Fourier orders in the solar cell bulk must be treated incoherently, that is, they must not interfere with one another. To do this, the system is divided into two separate regions: the diffraction grating region (including the reflector and passivation layer where appropriate) and the solar cell bulk region.

The diffraction grating region is treated by either of the wave-optical methods described in Section 3.4.1 for a specially selected set of incident plane waves yielding a far-field scattering matrix for this region. The solar cell bulk is simulated using another matrix, which describes the absorptance of the scattered orders and, where appropriate, their escape through the front face of the solar cell. Combining these two matrices leads to a matrix equation that yields the brightness of all orders in the solar cell bulk, from which the absorption enhancement is calculated. This method is described in detail in Ref. [13] and the first results of the method can be found in [54]. We now give some of the mathematical details of the method followed by some exemplary results.

We begin by considering the solar cell to be illuminated by a plane wave. In the steady state, there will be a discrete set of N diffraction orders propagating in the homogeneous solar cell bulk, which are given by the Fraunhofer equation (3.12). These can be decomposed into s and p polarizations giving $2N$ polarized diffracted orders. The complex electric field amplitude of the ith polarized order immediately after diffraction from the grating (subscript "diff") is linearly related to that of all other polarized orders immediately before incidence on the grating (subscript "inc"):

$$E_i^{\text{diff}} = \sum_{j=1}^{2N} S_{i,j} E_j^{\text{inc}} \tag{3.19}$$

where $S_{i,j}$ are the complex valued scattering matrix elements of the diffraction grating; these are obtained by applying RCWA or FMM solely to the grating region. The number of Fourier components to be considered in the grating region when calculating the scattering matrix must be chosen by convergence testing. This will not necessarily be equal to N: the number of propagating orders in the solar cell bulk.

We now consider the incident light not as a plane wave, but as a pencil of rays which fill some solid angle $d\Omega$. The *brightness* (defined as power flux per unit area per unit solid angle – also known as *radiance*) in a ray of infinitesimal solid angle $d\Omega$ is related to the intensity I and, hence, electric field amplitude E by [55]

$$B = \frac{I}{d\Omega} = \frac{n\varepsilon_0^2 |E|^2}{2 d\Omega} \tag{3.20}$$

where ε_0 is the vacuum permittivity and n is the refractive index. Assigning an appropriate wave vector to each pencil, the brightness of diffracted rays can be calculated by substituting Eq. (3.19) into Eq. (3.20). On doing so, we take the modulus squared of the sum in Eq. (3.19) which contains a number of interference terms between the fields of the different propagating orders. Since the cell thickness is greater than the coherence length of the light inside the cell, the orders lose coherence between diffraction from and incidence on the grating. The statistical net contribution from all interference terms is, therefore, zero [55]. Hence, the brightness of incident and diffracted orders can be related by:

$$B_i^{\text{diff}} = \cos\theta_i \sum_{j=1}^{2N} \frac{1}{\cos\theta_j} |S_{i,j}|^2 B_j^{\text{inc}} \quad B_{\text{TE}} \text{ and } B_{\text{TM}} \qquad (3.21)$$

where θ_i and θ_j are the polar angles of the incident and diffracted rays. The cosines arise from the condition $d\Omega_i \cos\theta_i = d\Omega_j \cos\theta_j$, which is a result of the Fraunhofer equation and expresses that the diffraction grating diffracts incident pencils of rays into diffracted pencils of the same étendue.

Equation (3.21) can be expressed in matrix form:

$$\bm{B}^{\text{diff}} = \hat{R}\bm{B}^{\text{inc}} \qquad (3.22)$$

where real valued $\hat{R}_{2N\times 2N}$ is defined by Eq. (3.22) and describes how the grating maps the brightness in the orders when entering the grating to the brightness in the orders when leaving the grating. To complete the picture, we require a matrix $\hat{C}_{2N\times 2N}$ such that

$$\bm{B}^{\text{inc}} = \hat{C}\bm{B}^{\text{diff}} \qquad (3.23)$$

The matrix $\hat{C}$ describes how the brightness is lost by absorption in the solar cell bulk and transmission into the surroundings at the solar cell front surface, and how brightness is introduced into the system by the solar illumination. In order to take AR coatings into account, an angular dependent transmission at the front surface can also be included in $\hat{C}$. Since the solar cell bulk is a homogeneous slab, calculation of $\hat{C}$ requires no complex computing. However, it is mathematically cumbersome and, therefore, its exact form is not stated here, but can be found in Ref. [13]. Combining Eqs. (3.22) and (3.23), we have

$$\bm{B}^{\text{inc}} = \hat{C}\hat{R}\bm{B}^{\text{inc}} \qquad (3.24)$$

which can be solved to give the brightness in all propagating orders within the solar cell bulk. Direct illumination from the sun is incident on the solar cell as a cone of light. The cone can be divided into a number of constituent pencil rays. The above described formalism is then performed on each pencil to build up a full description of the angularly dependent brightness inside the solar cell bulk. The fraction of incident power absorbed in the bulk with thickness d can then be calculated by

$$A = \frac{\int \left[\left(\exp\left\{\frac{\alpha d}{\cos\theta}\right\} - 1\right) B^{\text{inc}} + \left(1 - \exp\left\{\frac{\alpha d}{\cos\theta}\right\}\right) B^{\text{diff}}\right] \cos\theta \sin\theta\, d\theta}{B_0 \int \cos\theta \sin\theta\, d\theta}$$

(3.25)

Treating the incident illumination as a cone instead of a plane wave avoids problems such as overly sharp peaks at wavelengths at which a diffracted order has a polar angle of 90°. It also allows us to study the operation under concentrated illumination, for which the direct illumination has a wide angular divergence. The necessary number of pencils in which to divide the incidence cone should be found by convergence testing. For direct solar illumination with no concentration, it is usually sufficient to consider the incidence cone as a single pencil. However, care must be taken when diffracted cones overlap, or where different parts of the incidence cone correspond to different numbers of diffracted orders; in these cases, the incidence cone must be divided accordingly [13].

We shall now use the method to simulate a particular device. The device in question is a 200 μm crystalline silicon wafer with a bi-periodic binary checkerboard diffraction grating etched into the rear face. The grating period is 990 nm and the depth is 300 nm. Behind the diffraction grating is a 450 nm thick SiO_2 passivation layer, followed by a planar perfect rear reflector. The front face of the wafer is untextured and has a 63 nm thick SiN AR coating. The device is illuminated by an isotropic cone of light with a semi-angle of 0.26°; this corresponds to direct on-axis illumination from the sun without concentration. The grating region is simulated by RCWA using the commercial software package GdCalc®. The fraction of incident power absorbed in the silicon wafer is plotted in Figure 3.5a as a function of wavelength (black curve). Also shown is the calculated absorptance in a planar reference wafer that has the same thickness and front side AR coating, but whose rear face has only a planar perfect reflector with no diffraction grating and no passivation layer. A significant absorption enhancement can be seen in the long wavelength range in which silicon absorbs weakly. It can be seen that by treating the solar cell bulk incoherently, no interference oscillations are observed in the absorptance spectrum.

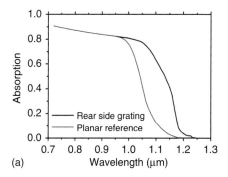
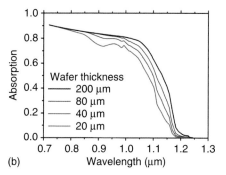

Figure 3.5 Absorptance spectra calculated using the far-field scattering matrix method. (a) Absorptance in a 200 μm thick crystalline silicon wafer equipped with a bi-periodic binary checkerboard diffraction grating compared to a planar reference. (b) Absorptance in wafers of different thicknesses all equipped with the same diffraction grating.

A further advantage of this simulation method arises when one wishes to apply the same diffraction grating to a variety of solar cell types. Because the time consuming wave-optical calculations are only applied to the diffraction grating region, parameters such as the solar cell thickness, absorption coefficient and AR coating can be changed without a costly re-simulation of the grating. To demonstrate this, the device structure described in the previous paragraph has been simulated with a range of wafer thicknesses. The calculated absorptance spectra are shown in Figure 3.5b. To calculate these curves, the diffraction grating was simulated once with RCWA over the range of wavelengths. For the 80 points shown, this required 2 h of computing time. Once this calculation was made, each curve shown in the figure was generated using the non-coherent part of the simulation; this required 30 s of computing time per wafer thickness.

3.4.3
Electro-Optically Coupled Simulation Using RCWA and Sentaurus Device

The RCWA method, described in the previous section, is one method to simulate the optical properties of a solar cell with a diffractive grating. The final result of this method is the spectrally resolved absorptance $a(\lambda)$ and with it the absorbed photocurrent density j_{ph} can be calculated (see Eq. (3.6)) As a standard, the AM1.5G spectrum as defined by IEC norm 60904-3 [56] is used. The absorbed photocurrent density j_{ph} can be said to be an upper limit for the short circuit current density j_{SC} generated by the solar cell. j_{ph}, however, does not consider any electrical characteristics of the solar. The spectrally resolved absorptance and j_{SC} are linked by

$$j_{SC} = \int_0^\infty d\lambda \; IQE(\lambda) a(\lambda) \phi(\lambda) = \int_0^\infty d\lambda \; EQE(\lambda) \phi(\lambda) \quad (3.26)$$

with $IQE(\lambda)$ and $EQE(\lambda)$ of the solar cell. Equation (3.26) can be used as a definition of IQE and EQE. The IQE combines all electrical losses within a PV device at short circuit conditions (see following text).

Knowing the spectrally and spatially resolved absorptance $a(\vec{r}, \lambda)$, the electrical characteristics of a solar cell can be calculated using a solar cell simulation program, by solving a set of continuity and transport equations self-consistently. Poisson's equation

$$\varepsilon \nabla^2 \varphi = -e(p - n + N_D - N_A) - \rho_{trap} \quad (3.27)$$

connects electrostatic potential φ and charge distribution. The electrostatic potential affects the charge distribution which again affects electrostatic potential. In Eq. (3.27) p represents the hole density, n the electron density, N_D the doping density of donors and N_A that of acceptors. ρ_{trap} is the trap density and ε the electric permittivity.

The drift-diffusion equations describe the carrier transport in the solar cell. At any point, the current is the sum of drift and diffusion current. They are related to electrostatic potential and carrier distribution. The equations are given by

$$\vec{j}_n = e\mu_n n\vec{E} + eD_n \vec{\nabla} n \quad (3.28)$$

$$\vec{J}_p = -e\mu_p p\vec{E} - eD_p \vec{\nabla} p \tag{3.29}$$

with μ_n and μ_p the electron and hole mobility, D_n and D_p the electron and hole diffusivities and $\vec{J}_n$ and $\vec{J}_p$ the electron and hole current.

Carrier conservation requires that, for an infinitesimal region, the net carrier flow is conserved. At equilibrium condition, the right hand side of the equations is zero as there is no net generation/recombination and no transient current. This requirement results in the following expressions

$$\vec{\nabla}\vec{J}_n = eG_{net} + e\frac{\partial n}{\partial t} \tag{3.30}$$

$$-\vec{\nabla}\vec{J}_p = eG_{net} + e\frac{\partial p}{\partial t} \tag{3.31}$$

with G_{net} being the net generation rate. The net generation rate consists of generation and recombination processes. The generation can be considered, in a first assumption, to be equal to the absorptance within the solar cell bulk material. Recombination is typically divided into four parts:

1) *Radiative recombination*: This is the direct reverse process to the generation of electron-hole pairs. An electron and a hole recombine radiatively via emission of a photon. This process is relevant in direct semiconductors like GaAs but plays only a minor role in silicon.
2) *Shockley Read Hall (SRH) recombination in the bulk*: SRH recombination summarizes all trap (defect)-assisted recombination processes. For the simulations of SRH recombination, a single trap-level in the center of the band-gap with a certain cross-section is usually assumed in crystalline Si solar cells while more complicated defect distributions are considered in thin-film Si devices.
3) *Surface recombination*: The description of surface recombination is similar to SRH recombination in the bulk; however, the trap density at the surface is typically higher than the one in the bulk and surface recombination can account for a significant part of the total recombination.
4) *Auger recombination*: This process is described as a three-particle interaction and is most relevant in strongly doped regions in the solar cell, that is, the emitter and the donor region.

For a more detailed description see, for example, Refs. [57] or [58]. Since a grating alters the surface of a solar cell, it can be assumed that it will have an effect on surface recombination and, hence, will affect the IQE.

Figure 3.6a shows the absorptance in a planar solar cell with a thickness of 40 µm with a planar front surface, an AR coating and a rear side binary grating with a period of 990 nm and a depth of 160 nm. An implementation of the RCWA [59], by Lalanne and Jurek [60] has been used to calculate this result. Details on the results can be found in Ref. [17] and details on the method can be found in Ref. [61]. Compared to a solar cell with planar rear side, the grating increases the calculated absorbed photo-current density by $\Delta j_{ph} = 1.94\,\text{mA}\,\text{cm}^{-2}$. Additionally, the

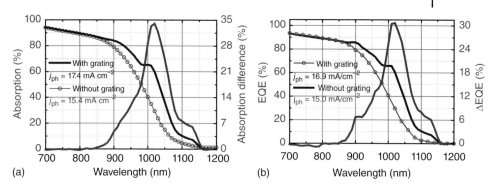

Figure 3.6 (a) Calculated absorptance of a silicon wafer solar cell with a thickness of 40 μm, AR coating and planar surface with rear side grating ($\Lambda = 990$ nm, $d = 160$ nm) (brown line). Also shown is the absorptance for a reference case without grating (blue circles) and the difference between the cases (red line). (b) Calculated EQEs and difference in EQE for the same samples.

spectrally and depth dependent absorptance profile within the solar cell has been calculated (not shown) and has been used to calculate the EQE using the commercial software Sentaurus TCAD [62] (Figure 3.6b). The calculated difference in short circuit current density induced by the grating is $\Delta j_{SC} = 1.85$ mA cm^{-2}. In these calculations, it has been assumed that the rear side grating can be equally well passivated, that is, has the same surface recombination as an untextured sample.

3.5 Realization

This section deals with realization methods of diffractive gratings for solar cells. In most general terms, the task is set to create a pattern in a functional material which is situated at the rear side of a solar cell structure. As given in more detail in the previous chapters, the aspired pattern consists of one- or two-dimensional gratings with lateral feature sizes of less than 1 μm. The fabrication of such features requires the use of advanced micro- and nanostructure technology. The foundation of this technology has its roots in the standardized processes developed for the semiconductor industry of the late twentieth century [63]. We note that this chapter is not intended to provide a comprehensive review of the state of the art of nanostructure technology, but a practical classification of tools that are suitable for the fabrication of grating structures on solar cells. For a deeper view into the practical aspects the interested reader is referred to Ref. [64]. In the following text we will review three exemplary techniques that had the largest impact on the fabrication of rear side structured solar cells in recent years.

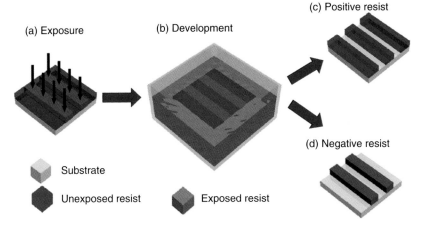

Figure 3.7 Schematic view of (a) exposure and (b) development of (c) a positive and (d) a negative resist.

3.5.1
Electron-Beam Lithography

In most general terms, lithography is a technique to transfer a generated pattern into a medium. The recording medium must be sensitive to a local interaction with an illuminating particle (or wave) source. The physical nature of these particles yields a straightforward classification into electron-, photon-, and ion-based lithography. Herein, lithography denotes the transfer of a pattern into a resist. The resist is mostly a polymer that changes its solubility in a developer solution upon local interaction with electrons, photons, or ions in comparison to the non-illuminated polymer (Figure 3.7). By this means, a spatially heterogeneous exposure leads to a solubility pattern that is memorized as a fixed structure in the resist. Two different kinds of resists are available: positive and negative. In a positive resist, the exposed areas will be dissolved in the subsequent development stage (Figure 3.7c), whereas in a negative resist, the exposed areas will remain intact after the development (Figure 3.7d).

In electron-beam lithography (EBL) [65], a beam of electrons is tightly focused to generate a pattern. The serial nature of the process is manifested in a scanning procedure across the surface to be patterned. Thus, EBL is most versatile at the point of initial design and preliminary experimental studies since it offers sub-wavelength resolution and almost complete pattern flexibility. These advantages outbalance the principal restriction of EBL application to limited areas, as long as no mass production is required. However, shifting this limit is routinely achieved nowadays using advanced EBL tool architectures such as variable shaped beam lithography [66, 67], which allows an efficient exposure of many points simultaneously. Since sub-10 nm scales can be achieved by EBL-based nanofabrication procedures [68, 69] it is still the first choice for the fabrication of micro- and nanostructured surfaces for a multitude of optical and electronic applications.

An EBL generated pattern is usually written in a resist with a distinct sensitivity to the applied electron dose. To date, one of the most commonly used EBL resists in research and development is still polymethyl methacrylate (PMMA), which has been invented already in 1968 [70]. Upon electron beam irradiation the long polymeric chains of PMMA are fragmented into shorter ones. Shorter chains have a higher solubility in the developer, typically methyl isobutyl ketone (MIBK).

In most cases, PMMA or other resists are dissimilar to the desired functional materials that are to bear the final pattern. Hence, the initial resist pattern must subsequently be transferred into the desired functional material(s). Nanopattern transfer techniques can be distinguished with respect to their either subtractive or additive interaction with the functional material, as it is schematically shown in Figure 3.8. The first case comprises etching techniques, while the most relevant version of the latter case is the lift-off method.

As for the first option, the subtractive interaction with a functional material, that is, its removal (Figure 3.8a–e), is mostly achieved by etching. A specific etching method is primarily chosen with respect to its material selectivity, etch

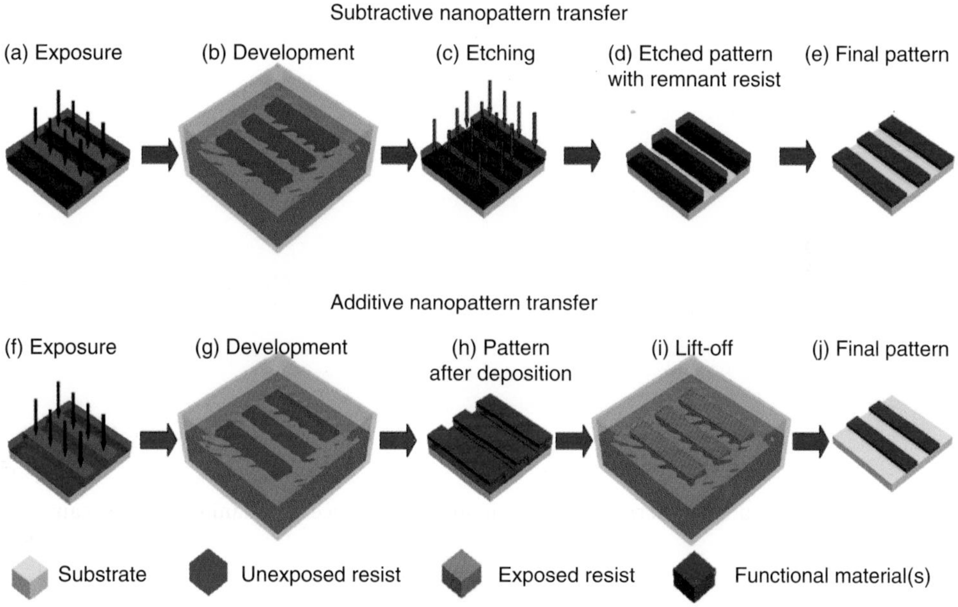

Figure 3.8 Schematic view of the transfer of a nanopattern written in a positive resist (a–e) by a subtractive transfer method such as etching and (f–j) by an additive transfer method such as lift-off. Note that when a positive resist is used, the final pattern made of the functional material(s) corresponds to the unexposed resist areas for the subtractive transfer method and to the exposed resist areas for the additive transfer method, respectively.

rate, and directionality (isotropic or anisotropic). In principle, one distinguishes between wet and dry etching. Wet etching methods make use of liquid acidic or alkaline solutions to dissolve the material. If the material itself shows no inherently anisotropic properties, these liquids etch rather isotropically and produce a lateral undercut, which limits the smallest feasible feature sizes. A lateral undercut is present if the effective angle between the sidewalls of a pattern and the supporting substrate is smaller than 90°. More favorable, dry etching methods allow tuning the directionality of the etching. Practically, a higher anisotropy of the applied etching method enables higher aspect ratio vertical structures and smaller undercuts, which in turn can lead to smaller lateral feature sizes. If the functional material to be structured (for example, Si, SiO_2, SnO, metals, etc.) exhibits a considerable etching rate in the presence of reactive etching gases (such as SF_6 or CH_4), reactive ion etching (RIE), reactive ion beam etching (RIBE) or inductively coupled plasma (ICP) etching are applicable. Otherwise, ion beam etching (IBE, also called *ion beam milling*) can be used as a purely physical process that utilizes accelerated inert (for example, argon) ions to ablate the target material [71].

Figure 3.9 displays an example of a grating structure that was etched into the rear side of a p-type mono-crystalline Si wafer (250 µm thickness, 4″ diameter, 100 surface orientation) for the purpose of light trapping [10] in a solar cell. In this case the silicon itself is the functional material. The grating was written by EBL into PMMA. The resist pattern was then transferred into the Si wafer by an ICP etching step. The grating parameters were determined by numerical optimization of the optical response of the solar sell. This led to a grating period of 320 nm (190 nm lines and 130 nm spaces) and a vertical depth of 190 nm. After the pattern transfer, further process steps for the completion of the device included thermal oxidation to grow a 105 nm-thick silica layer on both sides of the wafer, deposition of an

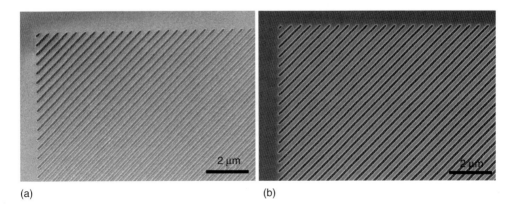

Figure 3.9 Normal view scanning electron micrographs of an exemplary rear side diffractive grating fabricated by EBL and etching at two intermediate fabrication steps. (a) Resist pattern with a pitch of 320 nm. (b) The same grating after ICP etch transfer into silicon and removal of the remnant resist.

additional silica buffer layer, and metallization with aluminum. The desired light trapping properties could be demonstrated both in the optical and the electronic properties of the device [72].

An additive transfer of a nanopattern to a substrate is achieved by the application of a directional physical vapor deposition technique to the binary resist pattern. The material is equally deposited onto the substrate in the cleared areas and onto the non-dissolved resist otherwise (see Figure 3.8f–j). By "lifting off" the remaining resist, including the deposited material on top of it, the inverse pattern is obtained: a thin film of the deposited material at the formerly cleared areas. If the resist is polymer-based, the lift-off step can be performed in, for example, acetone, and may be supported by ultrasonication. For this method, a sufficient adhesion of the thin film to the substrate surface is crucial. Therefore, an adhesion promoter like an ultrathin layer of titanium or chromium is usually applied. Owing to the lift-off step the aspect ratio of micro- and nano-structures generated by this method is limited to a width to height ratio of less than one. If the height of the deposited material is too large, it will pin down the resist pattern, which consequently cannot be lifted off any more. Simply increasing the resist height to avoid this diminishes the obtainable lateral resolution.

In the past, PMMA was proven to be the first choice for the lift-off process in countless contributions including demonstrations of extreme spatial resolution [73]. For reference, we describe a procedure that has been established at our lab to obtain metallic nanodisks with diameters down to 100 nm [74]. Those disks, for instance, can be employed to fabricate a plasmonic intermediate reflector inside a tandem solar cell. An 85 nm-thick layer of the PMMA-copolymer AR-P 610 from Allresist Berlin GmbH is prepared by spin coating and tempered at 210 °C for 10 min. Subsequently, the resist AR-P 671 from the same supplier is spin coated with the same thickness and baked at 180 °C for 30 min. This preparation results in two PMMA layers with two different electron selectivities in order to provide an undercut resist profile. Next, a 10 nm-thick gold layer is thermally evaporated to ensure a conducting surface during the subsequent electron beam exposure step. After the wet chemical removal of the gold conductance layer in an aqueous solution of potassium and potassium iodide, the samples are developed in a 1 : 1 MIBK:isopropanol solution for 30 s, rinsed with isopropanol and blown dry with nitrogen. After exposure and development, an adhesion promoting titanium film with a thickness of 3 nm and gold as the desired functional material are deposited onto the resist pattern by physical vapor deposition. For the herein described application, typical thicknesses of deposited films are of the order of 25–80 nm. Subsequently, the sample is immersed in acetone for more than 4 h and finally the lift-off is performed supported by ultrasonication. The resulting nanostructures comprise the evaporated gold thin film with the inversion of the exposed lateral pattern. Owing to the lift-off procedure we usually obtain non-perpendicular side walls with typical angles of 10–15° with respect to the substrate normal. In Figure 3.10, two examples of gold nanodisks on silicon and in different arrangements obtained with this lift-off procedure are shown.

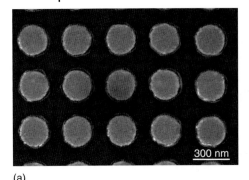

Figure 3.10 Normal view scanning electron micrographs of exemplary gold nanodisks on silicon wafers fabricated by EBL and the lift-off technique. (a) Gold nanodisks with a diameter of 300 nm arranged on a two-dimensional grid with lattice periods of 500 nm in each lateral direction. (b) Gold nanodisks with a diameter of 200 nm arranged in a hierarchical pattern for illustration purposes: while the disks form labels reading as "NANO," the labels themselves are arranged such that they visualize the word "MICRO."

3.5.2
Self-Organizing Photonic Crystals

Self-organizing processes like the assembly of block copolymers, colloids, or liquid crystalline materials offer another route to produce diffractive gratings. Especially, reflection layers based on colloidal photonic crystals (artificial opals) have been considered for use in silicon solar cells [11, 75, 76]. The 3D photonic structures consist of a close-packed array of polymer or silica colloids and can be prepared at low cost and on large areas. Sophisticated instruments, which are often necessary in lithographic methods, are not needed. However, it is more difficult to control defects, which evolve during the self-assembly of the colloids, and to modify the structure of the colloidal crystal [77, 78].

An example of an artificial opal prepared by self-assembly of monodisperse PMMA spheres is given in Figure 3.11. The colloids are arranged in a face-centered cubic packing (fcc), which is found to be the preferred crystal structure in these opals [80]. Furthermore, cracks, that are formed during drying of the colloidal crystal due to volume shrinkage are noticeable. Although cracks, colloid displacements, and other crystal defects are observable, the long range order of the crystal is conserved over the entire area. The orientation of the 111 plane of the fcc packing perpendicular to the substrate results in selective Bragg-peaks, whose position can be tuned by adjusting the diameter of the colloids used to prepare the crystal.[2]

---

2) The relation between the light reflected from a fcc-packed colloidal crystal, the lattice constant of the crystal and the angle of incidence can be described by a modified Bragg-equation: $\lambda_{\{1,1,1\}} = 2\sqrt{\frac{2}{3}}d\sqrt{n_{\text{eff}}^2 - \sin^2\theta}$ where d is the diameter of the colloids, θ is the angle of incidence to the 111 plane and n_{eff} is the effective refractive index of the colloidal crystal.

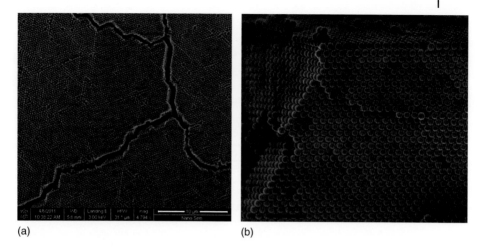

Figure 3.11 Two examples of artificial opals: (a) SEM image of the surface. The colloids form an fcc packing with the ⟨111⟩ plane oriented perpendicular to the substrate. The order extends over drying cracks, which are formed during the self-assembly of the colloids. (b) Side view of another opal showing the close-packed fcc structure [79]. Reproduced with permission from [79] © Wiley-VCH Verlag GmbH & Co. KGaA.

The defect density on the other hand determines the fraction of scattered light and, therefore, the transmittance of the opal.

The fabrication of an opaline film that can be applied as a reflection layer in silicon solar cells includes several steps: Firstly, the synthesis of monodisperse colloids with defined colloid size to control the reflection wavelength of the crystal and, secondly, the crystallization of these colloids. Furthermore, to facilitate the integration of the photonic structure into the solar cell, it is desirable that an electrical connection between the absorbing layer and the rear contacts of the cell through the opal can be achieved. This requires another fabrication step, since colloids suitable for the preparation of artificial opals usually are not made of conductive material. Polymer or silica colloids are accessible using emulsion polymerization or sol–gel (Stöber) processes. To prepare monodisperse spheres from different polymers, the surfactant-free emulsion polymerization is a convenient method. With this technique large quantities of polymer beads are available whose diameter is tunable between about 150 and 450 nm, which correspond to opals with a Bragg-reflectance peak between about 300 and 1000 nm. The colloid size is adjusted by the monomer to water ratio at constant reaction temperature and initiator concentration [81].

Colloidal crystals were produced on silicon, structured silicon wafers, and silicon solar cells using different methods like spray-coating, spin-coating, vertical crystallization and horizontal crystallization with and without coating knives [75, 82, 83]. While spray-coating and horizontal crystallization enables the preparation of large opaline films with areas of several hundred square centimeters the vertical crystallization seems to be the most appropriate method to fabricate

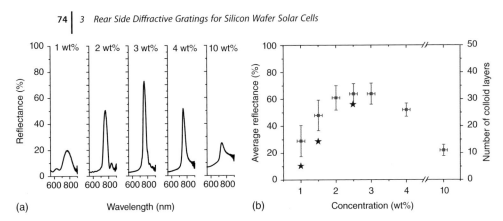

Figure 3.12 (a) Reflectance spectra of opaline films prepared by vertical crystallization of 340 nm PMMA spheres. The substrates were drawn out of aqueous suspensions containing different %weight of colloids with a speed of about 200 nm s^{-1}. (b) Averaged maximum reflectivity of opaline films plotted against the concentration of the colloidal suspensions.

high-quality opals [117]. Hereby, a substrate on which the opal is to be deposited, is either drawn out of an aqueous or ethanolic suspension of colloids or is left to stand in the suspension until the solvent has evaporated [84–86]. The assembly of colloids takes place in the meniscus at the interface between substrate, air, and solvent. The increased evaporation rate in the meniscus area induces a particle flow, which causes the self-assembly of the concentrated colloids into the fcc-structure. If the substrate is drawn out of the colloidal suspension constant crystallization conditions in the meniscus can be achieved. Hence, homogeneous opaline films with adjustable film thickness are accessible [86].

In Figure 3.12, reflection spectra of opaline films prepared with this technique are shown. The film thickness of these films was adjusted by varying the concentration of the colloidal suspension leading to thicker films for higher concentrated suspensions. Since the intensity of the Bragg-reflectance depends on the number of reflecting colloid layers it can be increased by fabricating thicker opals. Defects arising during the crystallization of the colloids limit the maximum reflectivity. It has been shown, however, that the self-assembly process can level out defects of the substrate. Opals grown on a randomly textured amorphous silicon substrate with a roughness up to twice the size of the colloids exhibited an ordered fcc packing after one to three monolayers of disordered spheres. The optical properties of the opal were comparable to the one prepared on flat substrates [87]. Thus, the colloids can be crystallized directly on different types of silicon solar cells.

One approach to realize electrical conductivity of the photonic structure is the preparation of inverted opals. Thereby, the interstitials of the colloidal crystal are filled with a conductive material and the colloids are removed afterwards. Conductive inverted opals from silicon, ZnO, and Al-doped ZnO were fabricated by chemical vapor deposition (CVD) and atomic layer deposition (ALD) using silica or polymer opals as templates [70, 88, 89]. Thus, photonic structures could be integrated directly in silicon solar cells [90, 91]. The combination of self-assembling monodisperse colloids and replication of the resulting opals with vapor deposition methods enables the fabrication of reflection layers with relative ease especially on large substrates. It is the advantage of the concept that defects in the photonic crystal (especially the cracks) reduce the reflectance only very slightly. However, by filling the drying cracks of the colloidal crystal with a conductive material the conductance of the photonic structure can be increased. Thus, they might even be advantageous for the electrical properties of the replica.

3.5.3
Fabrication of Rear Side Gratings via Interference Lithography and Nanoimprint Lithography

When it comes to the fabrication of periodic (but also aperiodic) patterns on large areas, one technology of choice is interference lithography (IL) [92]. In IL, a laser beam is split into two or more beams, which are expanded and superimposed on a photoresist coated sample (see Figure 3.13). The resulting interference pattern partially modifies the photoresist, so that after a subsequent development process a surface relief is formed. Various kinds of profile shapes arranged from 1D linear gratings to 3D photonic crystals can be realized on very large areas up to the square meter scale (depending on the complexity and the number of superimposed beams) [93]. Therefore, this technology is very well suited for photon management applications, in particular in PV, were a processing of large areas is required.

However, the process of IL itself is very elaborate and complex and, therefore, is not suited as production technology. The solution is to combine IL with replication processes such as nanoimprint lithography (NIL), to establish an

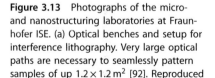

(a) (b)

Figure 3.13 Photographs of the micro- and nanostructuring laboratories at Fraunhofer ISE. (a) Optical benches and setup for interference lithography. Very large optical paths are necessary to seamlessly pattern samples of up 1.2 × 1.2 m² [92]. Reproduced with permission from [92] © SPIE. (b) Roller-nanoimprint lithography tool imprinting an etching mask on a 125 × 125 mm² [93]. Reproduced with permission from [93] © SPIE.

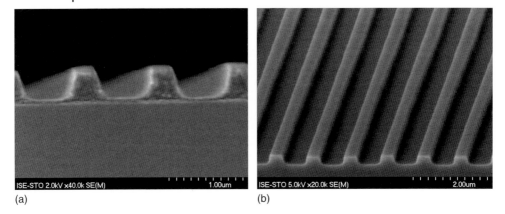

Figure 3.14 SEM micrographs of linear gratings. (a) Via soft-UV-NIL imprinted pattern on a monocrystalline silicon wafer. (b) Grating structure in silicon after plasma etching and resist removal [101]. Reproduced with permission from [101] © Elsevier.

economically attractive route for high definition patterning.[3] To this end, the patterned photoresist master is replicated using electroplating processes to generate so called shims (typically made of nickel). Then, a cast molding process is applied to replicate stamps for the NIL in a soft and UV-transparent material (for example, polydimethylsiloxane PDMS). These processes can be seen as preliminary steps before the actual repetitive texturing of silicon wafers for the formation of rear side gratings can take place.

The most promising type of NIL for fast large-area processing is known in literature as Soft-UV-NIL [94]. This nomenclature is based on the use of soft stamp materials to allow a conformal contact on large areas [95] and UV-curing resist materials to avoid the need of thermo-cycles as known in hot embossing processes [96]. To drive industrial feasibility of this process further, so called Roller-NIL processes were developed [97–99]. At Fraunhofer ISE we developed a tool for Roller-NIL especially designed for imprinting opaque, stiff, brittle, and potentially rough substrates to pattern etching masks for solar cell texturization in in-line processes [100]. This tool is also shown in Figure 3.13. In UV-NIL, a photosensitive resist is applied (for example, via spin coating), then the stamp is pressed into this resist and, while maintaining the pressure, the resist is cured. After the demolding of the stamp a patterned resist layer remains on the substrate. This patterned resist layer is then used as an etching mask in plasma etching processes. To achieve the best possible pattern transfer via plasma etching, it is essential that the residual layer thickness of the imprinted pattern is as small and as homogeneous as possible. Figure 3.14 shows scanning electron microscopy (SEM) micrographs of a linear grating first imprinted in a negative tone resist and then after plasma etching and resist removal. The period is 1 µm and the pattern

---

3) Note that master structures can also be generated with other techniques, for example, electron beam lithography.

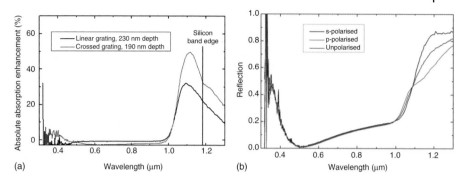

Figure 3.15 (a) Measured absolute absorption enhancements of a linear and a crossed grating. (b) Polarization dependent reflection measurement of the linear grating [101]. Reproduced with permission from [101] © Elsevier.

depth is around 300 nm for the imprinted and around 200 nm for the etched sample.

Using this processing scheme, linear and crossed gratings were realized on the rear of 200 µm thick monocrystalline silicon wafers [101]. The wafers had a flat front, which was coated with a silicon nitride anti-reflection coating (ARC). On the rear side, in which the grating was etched, a 450 nm-thick silicon oxide layer was deposited via PECVD (Plasma enhanced chemical vapor deposition). This dielectric layer is supposed to level the rear silicon oxide interface on which the aluminum is deposited by evaporation. Samples of different etching depths for the linear and the crossed grating as well as a flat reference were processed. Reflection measurements were used to quantify the absorption enhancement introduced by the gratings. In Figure 3.15a, measured spectral absorption enhancements of the best linear and crossed gratings related to the flat reference are shown.

These measurements verify that even in 200 µm-thick silicon wafers a considerable absorption enhancement can be achieved. The crossed gratings are superior to linear gratings, which is in good agreement with theoretical considerations [13]. The absorption enhancement above 1.2 µm wavelength however indicates that there is also considerable parasitic absorption in the rear reflector. This was found to be related to a modulation at the silicon oxide/aluminum interface, which can be seen in polarization-dependent measurements on linearly textured samples as also shown in Figure 3.15b. This parasitic absorption can be minimized by applying a dielectric buffer layer using a spin coating process [101]. To evaluate the potential benefit of these grating structures on solar cell performance, one has to distinguish between usable and parasitic absorption. As a preliminary step before solar cell fabrication, this can be evaluated using wave optical simulation methods.

Besides these optical aspects, while introducing a grating structure into the rear surface of a solar cell, care has to be taken concerning the electrical properties. One concept for effectively applying a photonic structure on the rear side of solar cells is to decouple the electrically active layers from the optically active layers. This can be achieved by passivating the flat rear surface (for example, by a thin aluminum oxide layer) and then applying a layer, in which the photonic structure is realized (for example, amorphous silicon). Very low values for surface recombination velocity S_{rear} of $18\,\mathrm{cm\,s^{-1}}$ were measured in house after the texturing processes. A similar concept was already proven to be effective on solar cell level [102]. There, also the flat rear was passivated and the photonic structure was realized by spin coating a solution of silicon oxide micro-spheres, which were subsequently inverted by silicon carbide.

3.6
Topographical Characterization

The advent of micro- and nanotechnology in recent decades went hand in hand with the establishment of various characterization techniques that enable material characterization to take place at smaller and smaller length scales. Concerning prototypical nanostructured solar cells, the assessment of their electronic and optical properties is usually complemented by investigations revealing information on their topographical characteristics. Out of the multitude of potential topographical characterization methods, we will give a brief introduction to atomic force microscopy (AFM), SEM, and focused ion beam (FIB) milling. The topographical information provided by these tools allows establishing a reproducible relationship between the desired functionality of complex nanostructured devices such as, for example, rear side diffractive gratings of a solar cell and their geometrical and structural peculiarities.

3.6.1
Atomic Force Microscopy

The most direct way of sampling the topography of a fabricated structure is the probing of the sample surface with a mechanical stylus using an AFM [103, 104]. Usually, the stylus is a sharp tip with a radius of curvature of less than 10 nm which is situated on a flexible cantilever. The movement of the cantilever is monitored by the deflection of a reflected laser beam while the sample is scanned by a piezoelectric stage with sub-nm resolution. There are two basic modes of AFM operation. In the contact mode the tip touches the surface and the tip deflection is used as feedback for the z-axis actuator. In the dynamic mode (or non-contact mode) the tip oscillates above the surface close to its resonance frequency and either the change of the oscillation amplitude or the oscillation phase upon interaction with the surface is used for feedback. In the contact mode the physical interpretation of the measured signal is more straightforward than in the dynamic mode while in the dynamic mode there is less wearing down of tip and sample and the obtainable spatial resolution is higher [104]. Commercial AFM systems reach a scan frequency of a few scan-lines per second. The acquisition of a complete AFM image (for example, 512 × 512 pixels), therefore, takes several minutes.

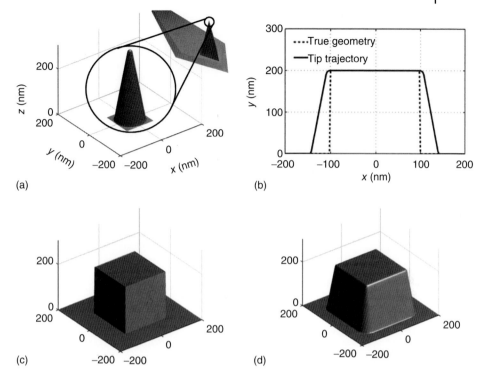

Figure 3.16 Illustration of the tip convolution effect of an AFM. (a) Assumed AFM probe with a tip radius of 10 nm and a cone semi-angle of 10°. (b) Offset tip trajectory at a rectangular grating groove. (c) True 3D geometry. (d) Simulated measurement after scanning with the tip shown in (a).

On a flat surface and under controlled environmental conditions (for example, in ultrahigh vacuum) it is possible to reach atomic resolution with an AFM. However, this does not hold for corrugated surfaces. In this case the tip–sample interaction will always take place at the point on the tip apex closest to the surface. Especially at narrow gaps and steep edges this point might be far away from the tip and, hence, the recorded surface will be distorted. This tip convolution effect is depicted in Figure 3.16. For the illustration the geometrical parameters of a standard AFM tip were taken (Figure 3.16a, tip radius is 10 nm, cone semi-angle is 10°) and the dilation of a box with 200 nm edge length was calculated with an algorithm as detailed in Ref. [105]. As can be seen from Figure 3.16b–d, there is a considerable deviation between the simulated scan result and the true geometry. If the tip geometry is known, a partial deconvolution of the AFM scan is possible. However, there might remain narrow and deep features where the tip possibly did not touch the sample surface. In those regions the true geometry remains uncertain. For a more accurate characterization of samples with steep and narrow features special high-aspect ratio probes with longer tips and smaller cone angles are available. Those tips can be operated only in non-contact mode and are much more fragile than conventional tips.

3.6.2
Scanning Electron Microscopy

SEM is an irreplaceable inspection technique for micro- and nano-fabrication. In an SEM, an electron beam is produced from a cathode by thermionic- or cold field emission and accelerated by an electric field to energies typically in the range of 1–25 keV. The electron beam is shaped by a series of electromagnetic lenses inside an electron column and focused into a vacuum chamber that contains the sample under investigation [106]. The size of the focal spot can be as small as 1 nm in modern instruments. An image of the sample is obtained by scanning the beam across the surface. The electrons interact with the sample within a penetration depth of up to several microns, depending on the sample and the detailed operation conditions. As a result, electrons are emitted from the sample primarily as either secondary or backscattered electrons. Further interaction products resulting from the interaction of the electron beam with the sample comprise Auger electrons, cathode-luminescence photons, and X-rays in reflection as well as elastically and inelastically scattered electrons in transmission. A comprehensive review on the interaction schemes and according detection principles in electron microscopy was published recently by Garcia de Abajo [107].

In an SEM, secondary electrons are produced as a result of interactions between the beam electrons and weakly bound electrons in the conduction band of the sample. They have comparably low energies (smaller than 50 eV), so only those formed within the first few nanometers of the sample have enough energy to escape and to be detected. Thus, the detection of secondary electrons will primarily give information on the topographical characteristics of the sample surface. Complementary, fast electrons arise from backscattering events of incoming electrons with the atomic cores inside the bulk of the sample. They will exhibit approximately the same energy as the primary electrons; however, the probability of the occurrence of a scattering event is proportional to the square root of the atomic number. Thus, heavy-core materials with high atomic numbers lead to a stronger backscattering signal than materials with low atomic numbers. This fact can be used to visualize a contrast between two materials with nanometer resolution. While the electron beam is scanned across the investigated surface, an SEM image is formed as the result of the user-defined, balanced detection of secondary, and backscattered electron emission from the sample surface at each single data point.

To illustrate the different imaging results one can exploit by an SEM, we rely on the silicon grating structure within the context of rear side structured solar cells. In Figure 3.17, we show cleaved edge of rear side diffractive gratings fabricated by EBL, ICP etching, thermal oxidation, and further thin-film deposition

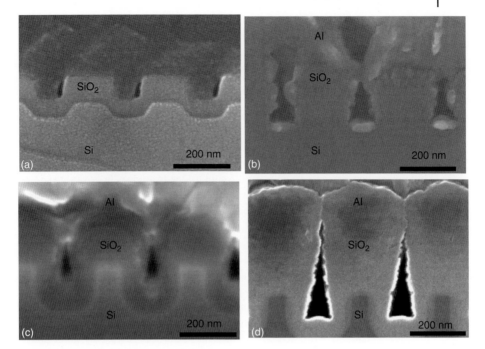

Figure 3.17 Cross sectional scanning electron micrographs of silicon gratings fabricated by EBL at different fabrication steps. (a) The same grating as in Figure 3.9 after a thermal oxidation step. (b–d) Variations of the grating after successive depositions of silica, leading to a total silica thickness of 130, 230, and 450 nm, respectively. Finally, the gratings were covered by a 2 µm thick layer of aluminum as rear reflector [72]. Reproduced with permission from [72] © EU PVSEC.

[72]. During the acquisition of the images in Figure 3.17a,c,d, mainly backscattered electrons were detected to enhance material contrast, which allowed a precise measurement of the respective layer thicknesses. By contrast, for the SEM image in Figure 3.17b, secondary electrons were the main source of information concerning surface topography rather than material contrast.

3.6.3
Focused Ion Beam Milling

In recent years, the use of FIBs has gained great importance in the field of micro- and nano-structuring. The applications thereby range from high resolution imaging over the preparation of samples for other inspection techniques to direct nano-patterning for prototyping [108]. In most commercial FIB systems a stream of Ga^+ ions is extracted from a liquid metal ion source (LMIS) and accelerated to energies of about 30 keV. The ion beam is then shaped by a set of electrostatic lenses and focused to a small spot of about 10 nm diameter or less. Upon impact on the sample, the ions cause a collision cascade and stop within a few tens of nanometers. As a result of the interaction, secondary electrons, sputtered secondary ions, and scattered primary ions are emitted, which can be analyzed to obtain an image while scanning the ion beam. The physical sputtering furthermore allows to purposefully remove material (ion beam milling), for example, for pattern engraving [108, 109]. Finally, the ion beam can also be used to decompose organic precursor gases supplied by a gas-injection system. This allows the local deposition of metals and dielectrics (for example, Pt, Au, W, or SiO_2) for additive patterning purposes (ion beam assisted deposition).

Nowadays, an FIB column is usually combined with an SEM column within a single instrument [108, 110]. In such a dual beam workstation the electron beam and the ion beam intersect at a common spot under an angle of 50° or more so that the SEM can be used for the *in situ* monitoring of the ion beam operation. This is also the most useful configuration for the inspection of buried nano-structures.

Without FIB machining the preparation of a cross-section for SEM inspection usually requires the cleaving of the sample. Thereby, the position of the cross-section can only be controlled coarsely and at the cleaved edge small structures or pieces of material might chip away. Differently, by FIB milling a sample cross-section can be prepared at any arbitrary position with nanometer precision and with a smooth surface. The basic principle of the procedure is illustrated in Figure 3.18a. First the ion beam is used to deposit a layer of protective material (for example, Pt) on top of the structure. This ensures a clean cutting edge and reduces ripples on the face of cross-section after milling. Then the ion beam is used to mill a trench which is large enough to allow an unobstructed view of the electron beam to the region of interest. Usually the trench has to be wider than the desired cross-section as material that is re-deposited at the edges during milling can block the view. This effect can be seen in Figure 3.18c. Finally, the electron beam is used to acquire an SEM image using an imaging mode with high material contrast.

A cross-section through a thin-film solar cell with a rear side grating that was prepared in the described fashion is shown in Figure 3.18b. In this case, the grating was one-dimensional and all necessary information could be obtained from a single cross-section. But, also, complicated three dimensional geometries can be analyzed by FIB milling. Therefore, consecutive layers of material are milled away after an SEM image has been acquired. This process is repeated several times until a sufficient number of cross-sectional images have been taken. The procedure is indicated by the black lines in Figure 3.18a. In this fashion a tomographic reconstruction of the sample can be obtained [111]. The voxel size of the obtained volume dataset is determined by the slice thickness along the z-direction and by the SEM resolution along the x- and y-direction. Typical values are 10 nm laterally and 15 nm along the z-axis.

After the image acquisition several image processing steps have to be performed. The image distortion along the y-axis due to the inclined viewing angle has to be removed and all images have to be registered in a global coordinate system. This can be done, for example, by least squares fitting [112] of a prominent common feature or a fiducial mark. The images are then cropped to the region of interest. Finally, the hardest part is the segmentation of the images into different

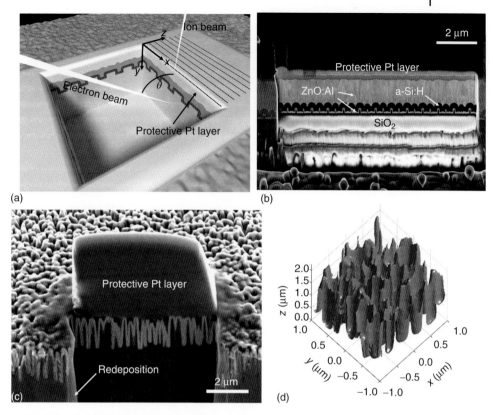

Figure 3.18 Focused ion beam tomography. (a) Basic principle and beam geometry. (b) Sliced rear side grating in an amorphous silicon thin-film solar cell. (c) Slice through a rough silicon surface. (d) Tomographic reconstruction of the geometry in (c).

structural features. For this purpose a vast variety of algorithms exists [113, 114] and often one has to try which one works best for each sample individually. If the material contrast is high and the noise level is low, simple methods like gray-level binning or region growing (for example, "flood filling") might be sufficient. Noisy images and features with discontinuous borders might require the application of more sophisticated methods like level set evolution [115]. If the boundary of the features of interest can be parameterized by an analytical function (for example, circles or ellipses) a generalized Hough transform might be useful [116].

An example of a tomographic reconstruction of a rough silicon surface with high aspect ratio features is shown in Figure 3.18c,d. In this particular case, the structure was coated conformally with a layer of ZnO by ALD before the FIB milling. This was necessary to obtain a sufficient contrast between the silicon needles and the voids of the etched pores. After that a simple region growing algorithm was sufficient to extract the silicon structures.

3.7
Summary

The concept of using diffractive gratings for light trapping shows high potential for silicon solar cells. The fundamental limit of absorption enhancement with gratings exceeds the Lambertian factor of $4n^2$ for ideal scattering and non-isotropic illumination. However, while scattering affects the entire spectrum, the effect of gratings so far has been restricted to a limited spectral region.

Optimizing gratings for a certain application is a complex task and needs to be done on a case to case basis. A theoretical optimization requires sophisticated simulation tools that take into account the electrical and optical impact of the grating texture on the solar cell. This requires wave-optical treatment of the grating itself in combination with semiconductor device simulation of the solar cell. In this chapter, we have discussed a variety of methods (RCWA, electro-optically coupled simulations, and the matrix method) for this purpose.

Manufacturing grating structures requires high-precision texturing techniques on a micrometer and sub-micrometer level. One of the challenges of the grating concept is tuning the texturing processes to deliver the required optical characteristics while maintaining a high electrical quality of the solar cell. Furthermore, the texturing process should be cheap and should be applicable to large areas to be industrially relevant. We have discussed electron-beam-, interference-, and nano-imprint lithography as well as self-organization of spherical beads colloids in this context.

A thorough characterization of the textured solar cell is required to assess the effect of the grating and optimize the processing. A large number of techniques are available for this purpose. In this chapter we concentrate on topographical methods namely AFM, SEM, and FIB milling.

To conclude the chapter, some recommendations as to the design of a diffractive structure for a solar cell is given. One aspect here concerns the geometry of the grating. All results here indicate that two-dimensional geometries are superior to one-dimensional (linear) ones, as the phase space is filled more efficiently. Concerning the period of the grating, a good choice seems to be to choose the period in a way that as much light as possible is diffracted into directions that are totally internally reflected. This can be achieved, for example, if the direction for first order of diffraction, for a certain design wavelength, coincides with the critical angle of total internal reflection. This criterion is met if the period equals the design wavelength, which appears to be a good rule of thumb. It should be noted, however, that some theoretical results indicate that the optimum is found for even larger periods. An explanation for this behavior is that, even though a part of the diffracted light leaves the cell through the escape cone, the average path length is still higher as more light is diffracted into the beneficial angular range.

References

1. Kray, D. and McIntosh, K. (2009) Analysis of ultrathin high-efficiency silicon solar cells. *Phys. Status Solidi A*, **206**, 1647–1654.

2. Shah, A. (ed) (2010) *Thin-Film Silicon Solar Cells*, EPFL Press.
3. Poortmans, J. and Arkhipov, V. (eds) (2006) *Thin Film Solar Cells*, John Wiley & Sons, Ltd.
4. Sheng, P., Bloch, A.N., and Stepleman, R.S. (1983) Wavelength-selective absorption enhancement in thin film solar cells. *Appl. Phys. Lett.*, **43** (6), 579–581.
5. Heine, C. and Morf, R.H. (1995) Submicrometer gratings for solar energy applications. *Appl. Opt.*, **34** (14), 2476–2482.
6. Kiess, H., Morf, R., Gale, M.T., and Widmer, A.E. (1990) Light trapping in solar cells using submicron gratings. Technical Digest of the International PVSEC-5, Kyoto, Japan, pp. 587–590.
7. Kiess, H., Morf, R., and Heine, C. (1997) in *Diffractive Optics for Industrial and Commercial Applications* (eds J. Turunen and F. Wyrowski), Akademie Verlag, Berlin, pp. 361–389.
8. Zaidi, S.H., Gee, J.M., and Ruby, D.S.(2000) Diffraction grating structures in solar cells. Proceedings of IEEE Specialists Conference on Photovoltaics, pp. 395–398.
9. Niggemann, M. *et al.* (2002) Trapping light in organic plastic solar cells with integrated diffraction grating. Proceedings of 17th European Photovoltaic Solar energy Conference.
10. Haase, C. and Stiebig, H. (2006) Light trapping in thin-film silicon solar cells with periodic structures. Conference Record of the 2006 IEEE 4th World Conference on Photovoltaic Energy Conversion, pp. 1509–1512.
11. Bermel, P., Luo, C., Zeng, L., Kimerling, L.C., and Joannopoulos, J.D. (2007) Improving thin-film crystalline silicon solar cell efficiencies with photonic crystals. *Opt. Express*, **15**, 16986–17000.
12. Kirchartz, T. (2009) in *Physics of Nanostructured Solar Cells* (ed. V. Badescu), Nova Science Publishers, pp. 1–40.
13. Mellor, A., Tobias, I., Marti, A., Mendes, M., and Luque, A. (2011) Upper limits to absorption enhancement in thick solar cells using diffraction gratings. *Prog. Photovoltaics Res. Appl.*, **19**, 676–678.
14. Yu, Z., Raman, A., and Fan, S. (2010) Fundamental limit of nanophotonic light trapping in solar cells. *Proc. Natl. Acad. Sci. U.S.A.*, **107**, 17491–17496.
15. Gjessing, J., Sudbo, A., and Marstein, E. (2011) Comparison of periodic light-trapping structures in thin crystalline silicon solar cells. *J. Appl. Phys.*, **110**, 033104.
16. Battaglia, C. *et al.* (2012) Light trapping in solar cells: can periodic beat random? *ACS Nano* **6** (3), 2790–2797, http://pubs.acs.org/doi/abs/10.1021/nn300287j.
17. Peters, M., Rüdiger, M., Hauser, H., Hermle, M., and Bläsi, B. (2011) Diffractive gratings for crystalline silicon solar cells—optimum parameters and loss mechanisms. *Progress in Photovoltaics: Research and Applications*, **20** (7), 862–873, doi: 10.1002/pip.1151.
18. Ferry, V. *et al.* (2010) Light trapping in ultrathin plasmonic solar cells. *Opt. Express*, **18**, A237–A245.
19. Chou, W. *et al.* (2011) Nanoimprinting-induced efficiency enhancement in organic solar cells. *Appl. Phys. Lett.*, **99**, 99–102.
20. Bläsi, B. *et al.* (2011) Photon management structures originated by interference lithography. Proceedings of Silicon PV, pp. 712–718.
21. Bell, J.M., Derrick, G.H., and McPhedran, R.C. (1982) Diffraction gratings in the quasi-static limit. *Opt. Acta*, **29**, 1475–1489.
22. Lalanne, P. (1996) Effective medium theory applied to photonic crystals composed of cubic or square cylinders. *Appl. Opt.*, **35**, 5369–5380.
23. Southwell, W.H. (1983) Gradient-index antireflection coatings. *Opt. Lett.*, **8**, 584–586.
24. Southwell, W.H. (1991) Pyramid-array surface-relief structures producing antireflection index matching on optical surfaces. *J. Opt. Soc. Am. A*, **8**, 549–553.

25. Bernhard, C.G. (1967) Structural and functional adaptation in a visual system. *Endeavour*, **26**, 79–84.
26. Schubert, M.F., Mont, F.W., Chhajed, S., Poxson, D.J., Kim, J.K., and Fred Schubert, E. (2008) Design of multilayer antireflection coatings made from co-sputtered and low-refractive-index materials by genetic algorithm. *Opt. Express*, **16**, 5290–5298.
27. Leem, J.W., Song, Y.M., and Yu, J.S. (2011) Broadband antireflective germanium surfaces based on subwavelength structures for photovoltaic cell applications. *Opt. Express*, **19**, 26308.
28. Chyan, J.Y., Hsu, W.C., and Yeh, J.A. (2009) Broadband antireflective poly-Si nanosponge for thin film solar cells. *Opt. Express*, **17**, 4646–4651.
29. Saeta, P.N., Vivian, V.E., Pacifici, D., Munday, J.N., and Atwater, H.A. (2009) How much can guided modes enhance absorption in thin solar cells? *Opt. Express*, **17**, 20975–20990.
30. Haug, F.-J., Söderström, T., Cubero, O., Terrazzoni-Daudrix, V., and Ballif, C. (2009) Influence of the ZnO buffer on the guided mode structure in Si/ZnO/Ag multilayers. *J. Appl. Phys.*, **106**, 044502.
31. Kroll, M., Fahr, S., Helgert, C., Rockstuhl, C., Lederer, F., and Pertsch, T. (2008) Employing dielectric diffractive structures in solar cells – a numerical study. *Phys. Status Solidi A*, **205**, 2777–2795.
32. Dewan, R., Marinkovic, M., Noriega, R., Phadke, S., Salleo, A., and Knipp, D. (2009) Light trapping in thin-film silicon solar cells with submicron surface texture. *Opt. Express*, **17**, 23058.
33. Biswas, R. and Xu, C. (2011) Nanocrystalline silicon solar cell architecture with absorption at the classical $4n^2$ limit. *Opt. Express*, **19**, A664–A672.
34. Callahan, D.M., Munday, J.N., and Atwater, H.A. (2012) Solar cell light trapping beyond the ray optic limit. *Nano Lett.*, **12**, 214–218.
35. Chao, C.-C., Wang, C.-M., and Chang, J.-Y. (2010) Spatial distribution of absorption in plasmonic thin film solar cells. *Opt. Express*, **18**, 11763–11771.
36. Madzharov, D., Dewan, R., and Knipp, D. (2011) Influence of front and back grating on light trapping in microcrystalline thin-film silicon solar cells. *Opt. Express*, **19**, A95–A107.
37. Naqvi, A., Söderström, K., Haug, F.-J., Paeder, V., Scharf, T., Herzig, H.P., and Ballif, C. (2011) Understanding of photocurrent enhancement in real thin film solar cells: towards optimal one-dimensional gratings. *Opt. Express*, **19**, 128–140.
38. Zanotto, S., Liscidini, M., and Andreani, L.C. (2010) Light trapping regimes in thin-film silicon solar cells with a photonic pattern. *Opt. Express*, **18**, 4260–4274.
39. Andersson, V., Tvingstedt, K., and Inganäs, O. (2008) Optical modeling of a folded organic solar cell. *J. Appl. Phys.*, **103**, 094520.
40. Rim, S.-B., Zhao, S., Scully, S.R., McGehee, M.D., and Peumans, P. (2007) An effective light trapping configuration for thin-film solar cells. *Appl. Phys. Lett.*, **91**, 243501.
41. Campbell, P. (1993) Enhancement of light absorption from randomizing and geometric textures. *J. Opt. Soc. Am. B*, **10**, 2410–2415.
42. Campbell, P. and Green, M.A. (1987) Light trapping properties of pyramidally textured surfaces. *J. Appl. Phys.*, **62**, 243–249.
43. Smith, A.W. and Rohatgi, A. (1993) Ray tracing analysis of the inverted pyramid texturing geometry for high efficiency silicon solar cells. *Sol. Energy Mater. Sol. Cells*, **29**, 37–49.
44. Smith, A.W. and Rohatgi, A. (1993) A new texturing geometry for producing high efficiency solar cells with no antireflection coatings. *Sol. Energy Mater. Sol. Cells*, **29**, 51–65.
45. Popov, E. and Bozhkov, B. (2001) Corrugated waveguides as resonance optical filters advantages and limitations. *J. Opt. Soc. Am. A*, **18**, 1758–1764.
46. Ulbrich, C. *et al.* (2008) Directional selectivity and ultra-light-trapping in solar cells. *Phys. Status Solidi A*, **205**, 2831–2843.

47. Jackson, J.D. (1999) *Classical Electrodynamics*, John Wiley & Sons, Inc.
48. Taflove, A. and Hagness, S.C. (2000) *Computational Electrodynamics: The Finite-Difference Time-Domain Method*, 2nd edn, Artech House.
49. Lalanne, P. (1997) Improved formulation of the coupled-wave method for two-dimensional gratings. *J. Opt. Soc. Am. A*, **14**, 1592–1598.
50. Li, L. (1997) New formulation of the Fourier modal method for crossed surface-relief gratings. *J. Opt. Soc. Am. A*, **14**, 2758–2767.
51. Stefanou, N., Yannopapas, V., and Modinos, A. (2000) MULTEM 2: a new version of the program for transmission and band-structure calculations of photonic crystals. *Comput. Phys. Commun.*, **132**, 189–196.
52. Sakoda, K. (2001) *Optical Properties of Photonic Crystals*, Springer.
53. Mie, G. (1908) Beiträge zur Optik trüber Medien, speziell kolloidaler Metallösungen. *Ann. Phys.*, **25**, 377–445.
54. Mellor, A., Tobías, I., Martí, A., and Luque, A. (2011) A numerical study of Bi-periodic binary diffraction gratings for solar cell applications. *Sol. Energy Mater. Sol. Cells*, **95**, 3527–3535.
55. Born, M. and Wolf, E. (1999) *Principles of Optics: Electromagnetic Theory of Propagation, Interference and Diffraction of Light*, 7th edn, Cambridge University Press, Cambridge.
56. IEC www.iec.ch (accessed 10 November 2014).
57. Würfel, P. (2005) *Physics of Solar Cells, From Principles to New Concepts*, Wiley-VCH Verlag GmbH.
58. Green, M. (1995) *Silicon Solar Cells – Advanced Principles and Practice*, Centre for Photovoltaic Devices and Systems, University of New South Wales.
59. Moharam, M.G. (1995) Formulation for stable and efficient implementation of the rigorous coupled-wave analysis of binary gratings. *J. Opt. Soc. Am. A*, **12**, 1068–1076.
60. Lalanne, P. and Jurek, M.P. (1998) Computation of the near-field pattern with the coupled-wave method for TM polarization. *J. Mod. Opt.*, **45**, 1357–1374.
61. Peters, M., Rüdiger, M., Bläsi, B., and Platzer, W. (2010) Electro-optical simulation of diffraction in solar cells. *Opt. Express*, **18**, A584–A593.
62. SENTAURUS (0000) Synopsys Inc., Mountain View, CA, http://www.synopsys.com/TOOLS/TCAD/Pages/default.aspx.
63. Campbell, S. (1996) *The Science and Engineering of Microelectronic Fabrication*, Oxford University Press, New York.
64. Bhushan, B. (2006) *Springer Handbook of Nanotechnology*, Springer, Heidelberg.
65. Pease, R. (1981) Electron beam lithography. *Contemp. Phys.*, **22**, 265–290.
66. Zeitner, U. and Kley, E.-B. (2006) Advanced lithography for micro-optics. *Proc. SPIE*, **6290**, 629009.
67. Hahmann, P., Bettin, L., Boettcher, M., Denker, U., Elster, T., Jahr, S., Kirschstein, U.-C., Kliem, K.-H., and Schnabel, B. (2007) High resolution variable-shaped beam direct write. *Microelectron. Eng.*, **84**, 774–778.
68. Broers, A., Hoole, A., and Ryan, J. (1996) Electron beam lithography–resolution limits. *Microelectron. Eng.*, **32**, 131–142.
69. Vieu, C., Carcenac, F., Pepin, A., Chen, Y., Mejias, M., Lebib, A., Manin-Ferlazzo, L., Couraud, L., and Launois, H. (2000) Electron beam lithography: resolution limits and applications. *Appl. Surf. Sci.*, **164**, 111–117.
70. Haller, I., Hatzakis, M., and Srinivasan, R. (1968) High-resolution positive resists for electron-beam exposure. *IBM J. Res. Dev.*, **12**, 251–256.
71. Koehler, J. and Koehler, M. (1999) *Etching in Microsystem Technology*, Wiley-VCH Verlag GmbH.
72. Voisin, P., Peters, M., Hauser, H., Helgert, C., Kley, E.-B., Pertsch, T., Bläsi, B., Hermle, M., and Glunz, S.(2009) Nanostructured back side silicon solar cells. Proceedings of the 24th European Photovoltaic Solar energy Conference, Hamburg, Germany, 2009, pp. 1997–2000.

73. Cumming, D., Thoms, S., Beaumont, S., and Weaver, J. (1996) Fabrication of 3 nm wires using 100 keV electron beam lithography and poly (methyl methacrylate) resist. *Appl. Phys. Lett.*, **68**, 322–324.
74. Fahr, S., Rockstuhl, C., and Lederer, F. (2010) Improving the efficiency of thin film tandem solar cells by plasmonic intermediate reflectors. *Photonics Nanostruct.*, **8** (4), 291–296.
75. Bielawny, A., Üpping, J., Miclea, P.T., Wehrspohn, R.B., Rockstuhl, C., Lederer, R., Peters, M., Steidl, L., Zentel, R., Lee, S.-M., Knez, M., Lambertz, A., and Carius, R. (2008) 3D photonic crystal intermediate reflector for micromorph thin-film tandem solar cell. *Phys. Status Solidi A*, **205**, 2796–2810.
76. O'Brien, P.G., Kherani, N.P., Chutinan, A., Ozin, G.A., John, S., and Zukotynski, S. (2008) Silicon photovoltaics using conducting photonic crystal back-reflectors. *Adv. Mater.*, **20**, 1577–1582.
77. Marlow, F., Muldarisnur, Sharifi, P., Brinkmann, R., and Mendive, C. (2009) Opals: status and prospects. *Angew. Chem. Int. Ed.*, **48**, 6212–6233.
78. Lange, B., Fleischhaker, F., and Zentel, R. (2007) Chemical approach to functional artificial opals. *Macromol. Rapid Commun.*, **28**, 1291–1311.
79. Egen, M., Braun, L., Zentel, R., Tännert, K., Frese, P., Reis, O., and Wulf, M. (2004) Artificial opals as effect pigments in clear-coatings. *Macromol. Mater. Eng.*, **289**, 158–163.
80. Woodcock, L.V. (1997) Entropy difference between the face-centered cubic and hexagonal close-packed crystal structures. *Nature*, **385**, 141–143.
81. Egen, M. and Zentel, R. (2004) Surfactant-free emulsion polymerization of various methacrylates: towards monodisperse colloids for polymer opals. *Macromol. Chem. Phys.*, **205**, 1479–1488.
82. Ye, J., Zentel, R., Arpiainen, S., Ahopelto, J., Jonsson, F., Romanov, S.G., and Sotomayor Torres, C.M. (2006) Integration of self-assembled three-dimensional photonic crystals onto structured silicon wafers. *Langmuir*, **22**, 7378–7383.
83. Jiang, P. and McFarland, M.J. (2004) Large-scale fabrication of wafer-size colloidal crystals, macroporous polymers and nanocomposites by spin-coating. *J. Am. Chem. Soc.*, **126**, 13778–13786.
84. Jiang, P., Bertone, J.F., Hwang, K.S., and Colvin, V.L. (1999) Single-crystal colloidal multilayers of controlled thickness. *Chem. Mater.*, **11**, 2132–2140.
85. Gu, Z.-Z., Fujishima, A., and Sato, O. (2002) Fabrication of high-quality opal films with controllable thickness. *Chem. Mater.*, **14**, 760–765.
86. Egen, M., Voss, R., Griesebock, B., Zentel, R., Romanov, S., and Torres, C.S. (2003) Heterostructures of polymer photonic crystal films. *Chem. Mater.*, **15**, 3786–3792.
87. Üpping, J., Salzer, R., Otto, M., Beckers, T., Steidl, L., Zentel, R., Carius, R., and Wehrspohn, R.B. (2011) Transparent conductive oxide photonic crystals on textured substrates. *Photonics Nanostruct. Fundam. Appl.*, **9**, 31–34.
88. Steidl, L., Frank, S., Weber, S.A., Panthöfer, M., Birkel, A., Koll, D., Berger, R., Tremel, W., and Zentel, R. (2011) Electrodeposition of ZnO nanorods on opaline replica as hierarchically structured systems. *J. Mater. Chem.*, **21**, 1079–1085.
89. Suezaki, T., O'Brien, P.G., Chen, J.I.L., Kherani, N.P., and Ozin, G.A. (2009) Tailoring the electrical properties of inverse silicon opals – A step towards optically amplified silicon solar cells. *Adv. Mater.*, **21**, 559–563.
90. Üpping, J., Bielawny, A., Wehrspohn, R.B., Beckers, T., Carius, R., Rau, U., Fahr, S., Rockstuhl, C., Lederer, F., Kroll, M., Pertsch, T., Steidl, L., and Zentel, R. (2011) Three-dimensional photonic crystal intermediate reflectors for enhanced light-trapping in tandem solar cells. *Adv. Mater.*, **23**, 3896–3900.
91. Suezaki, T., Chen, J.I.L., Hatayama, T., Fuyuki, T., and Ozin, G.A. (2010) Electrical properties of p-type and n-type doped inverse silicon opals – towards

optically amplified silicon solar cells. *Appl. Phys. Lett.*, **96**, 242102.

92. Gombert, A., Bläsi, B., Bühler, C., Nitz, P., Mick, J., Hoßfeld, W., and Niggemann, M. (2004) Some application cases and related manufacturing techniques for optically functional micro structures on large areas. *Opt. Eng.*, **43**, 2525–2533.

93. Bläsi, B., Hauser, H., Walk, C., Michl, B., Guttowski, A., Mellor, A., Benick, J., Peters, M., and Jüchter, S. (2012) Photon management structures for solar cells. Proceedings of SPIE Photonics Europe.

94. Plachetka, U., Bender, M., Fuchs, A., Vratzov, B., Glinsner, T., Lindner, F., and Kurz, H. (2004) Wafer scale patterning by soft UV-Nanoimprint Lithography. *Microelectron. Eng.*, **73–74**, 167–171.

95. Bietsch, A. and Michel, B. (2000) Conformal contact and pattern stability of stamps used for soft lithography. *J. Appl. Phys.*, **88** (7), 4310–4318.

96. Haisma, J., Verheijein, M., van den Heuvel, K., and van den Berg, J. (1996) Mold-assisted nanolithography: a process for reliable pattern replication. *J. Vac. Sci. Technol. B*, **14**, 4124.

97. Tan, H., Gilbertson, A., and Chou, S.Y. (1998) Roller nanoimprint lithography. *J. Vac. Sci. Technol. B*, **16** (6), 3926–3928.

98. Schift, H. (2006) Roll embossing and roller imprint, in *Science and New Technology in Nanoimprint*, (ed. Y. Hirai) Frontier Publishing Co., Ltd., Japan, ISBN: 4-902410-09-5.

99. Hyun Ahn, S. and Jay Guo, L. (2008) High-speed roll-to-roll nanoimprint lithography on flexible plastic substrates. *Adv. Mater.*, **9999**, 1–6.

100. Hauser, H., Michl, B., Schwarzkopf, S., Kübler, V., Müller, C., Hermle, M., and Bläsi, B. (2012) Honeycomb texturing of silicon via nanoimprint lithography for solar cell applications. *IEEE J. Photovoltaics.* **2** (2), 114–122, doi: 10.1109/JPHOTOV.2012.2184265

101. Hauser, H., Mellor, A., Guttowski, A., Wellens, C., Benick, J., Müller, C., Hermle, M., and Bläsi, B. (2012) Diffractive backside structures via nanoimprint lithography. to be published in. *Energy Procedia.* **27**, 337–342.

102. Berger, P., Hauser, H., Suwito, D., Janz, S., Peters, M., Bläsi, B., and Hermle, M. (2010) Realization and evaluation of diffractive systems on the back side of silicon solar cells. *Proc. SPIE*, **7725**, 772504. doi: 10.1117/12.854553

103. Binnig, G. and Quate, C.F. (1986) Atomic force microscope. *Phys. Rev. Lett.*, **56**, 930–933.

104. Giessibl, F.J. (2003) Advances in atomic force microscopy. *Rev. Mod. Phys.*, **75**, 949–983.

105. Villarrubia, J.S. (1997) Algorithm for scanned probe microscope image simulation, surface reconstruction, and tip estimation. *J. Res. Nat. Inst. Stand. Technol.*, **102**, 425–454.

106. Goldstein, J., Newbury, D., Joy, D., Lyman, C., Echlin, P., Lifshin, E., Sawyer, L., and Michael, J. (2003) *Scanning Electron Microscopy and X-ray Microanalysis*, 3rd edn, Springer, New York.

107. Garcia de Abajo, F. (2010) Optical excitations in electron microscopy. *Rev. Mod. Phys.*, **82**, 209.

108. Gierak, J. (2009) Focused ion beam technology and ultimate applications. *Semicond. Sci. Technol.*, **24**, 043001.

109. Volkert, C.A. and Minor, A.M. (2011) Focused ion beam microscopy and micromachining. *MRS Bull.*, **32**, 389–399.

110. Sudraud, P. (1988) Focused-ion-beam milling, scanning-electron microscopy, and focused-droplet deposition in a single microcircuit surgery tool. *J. Vac. Sci. Technol. B*, **6**, 234.

111. Uchic, M.D., Holzer, L., Inkson, B.J., Principe, E.L., and Munroe, P. (2011) Three dimensional microstructural characterization using focused ion beam tomography. *MRS Bull.*, **32**, 408–416.

112. Orchard, J. (2005) in *Image Analysis and Recognition*, Lecture Notes in Computer Science, vol. **3656** (eds M. Kamel and A. Campilho), Springer, Berlin, Heidelberg, pp. 116–124.

113. Haralick, R.M. and Shapiro, L.G. (1985) Image segmentation techniques. *Comput. Vision, Graphics, Image Process.*, **29**, 100–132.
114. Pal, N.R. and Pal, S.K. (1993) A review on image segmentation techniques. *Pattern Recognit.*, **26**, 1277–1294.
115. Chan, T.F. and Vese, L.A. (2001) Active contours without edges. *IEEE Trans. Image Process.*, **10**, 266–277.
116. Ballard, D. (1981) Generalizing the hough transform to detect arbitrary shapes. *Pattern Recognit.*, **13**, 111–122.
117. Kocher, G., Khunsin, W., Arpiainen, S., Romero-Vivas, J., Romanov, S.G., Ye, J., Lange, B., Jonsson, F., Zentel, R., Ahopelto, J., and Sotomayor Torres, C.M. (2007) Towards Si-based photonic circuits: integrating photonic crystals in silicon-on-insulator platforms. *Solid-State Electron.*, **51**, 333–336.

4
Randomly Textured Surfaces

Carsten Rockstuhl, Stephan Fahr, Falk Lederer, Karsten Bittkau, Thomas Beckers, Markus Ermes, and Reinhard Carius

Randomly textured surfaces constitute an indispensable means to enhance the efficiency of thin-film solar cells. Their purpose is to scatter the light into the active layer of the solar cell and to suppress reflection losses at the entrance facet. Their functionality has been discussed in the context of wafer-based solar cells, but the understanding of their functionality in thin-film solar cells poses challenges that have not been considered for a long time. Most importantly, the thickness of the light absorbing layer in thin-film solar cells amounts only to a few hundreds of nanometers, which is comparable to the wavelengths of interest and to the heights and lateral sizes of the typical surface features of the random texture. That match on all length scales forces a resonant scattering regime and prohibits the application of approximate theories, which can help disclose their optical properties. Such understanding, however, is the prerequisite for their optimization. With the example of a solar cell made from hydrogenated amorphous silicon, we summarize the insights obtained from experimental and theoretical work that fully respect such a resonant size domain. We discuss the scattering properties of random textures and their ability to enhance absorption. Unique to this discussion is the consideration of randomly textured surfaces as fabricated and as integrated already into commercial solar cells. This assures that the unprecedented insights are applicable to real-world devices. Furthermore, explicit guidelines for the fabrication of improved solar cells can be extracted from the optimization we perform.

4.1
Introduction

Photon management, as it is at the heart of this volume, suggests enhancing the efficiency of solar cell devices made of a light absorbing layer with a given thickness by integrating an assistive structure into it. The purpose of these structures is to steer the flow of light in solar cells such that more photons are absorbed, thereby enhancing the solar cell efficiency [1]. It is worth mentioning that the

Photon Management in Solar Cells, First Edition. Edited by Ralf B. Wehrspohn, Uwe Rau, and Andreas Gombert.
© 2015 Wiley-VCH Verlag GmbH & Co. KGaA. Published 2015 by Wiley-VCH Verlag GmbH & Co. KGaA.

enhancement of the efficiency with an acceptable increase in efforts is the primary challenge to be addressed to provide energy from sustainable sources on a reliable and competitive base in the near future [2]. The more efficient the solar cell and the cheaper the energy from non-fossil resources, the more attractive such power gets when compared to carbon-based sources. The ultimate goal is to achieve a sustainable energy source at minimal costs and while going in this direction, to reach grid parity is usually understood as being decisive. This suggests that the costs of solar cell energy is required to be identical to the grid price for the customer.

Photon management can be implemented by the following two different strategies. One strategy for single-junction solar cells is to integrate structures into the solar cell that modify the spectral composition of sun light into a more favorable distribution [3]. This can be done by down-converting high energy photons into low energy photons with an efficiency larger than unity [4]. This helps avoiding thermalization losses. It can be also done by up-converting a multiple number of photons with an energy below the optical band gap of the light absorbing layer into a single photon with an energy above [5]. The second strategy, on which we concentrate in this chapter, is to modify the propagation of the impinging photons such that their absorption probability is enhanced. Roughly spoken, this enhances the number of generated electron–hole pairs and thereby the short circuit current density. We note that in the radiative limit also the open circuit voltage is enhanced [6], but the effect is not considered here in detail. Various structural suggestions have been put forward, which aim at improved absorption. The majority of proposed structures possesses a deterministic geometry such as gratings [7–9], photonic crystals [10–12], stacks of layers [13, 14], or metallic nanoparticles [15–17]. Each of these structures exploits a particular regime in the light–matter interaction or combinations thereof, such as diffraction, refraction, scattering, or the excitation of resonances, to tailor the light propagation. As documented in various other chapters of this volume, the deterministic character has pros and cons. Most notably, in most cases the functionality is limited to a narrow spectral domain that is intimately linked to the geometry and/or the material properties of the assistive structure. This is in contrast to the requirement of the photon management to enhance the solar cell efficiency, if possible, across the entire spectral range for which the solar cell is absorptive; but, most notably, near the absorption edge. This spectral domain for thin-film solar cells made of hydrogenated amorphous silicon (a-Si:H) spans approximately 200 nm, that is, from 550 nm to 750 nm. While considering such a huge spectral domain, it gets appealing to rely on structures with a less deterministic geometry, that is, disordered structures.

Using randomly textured surfaces promises to overcome the deficiencies of the aforementioned approaches. Moreover, they are an integral part of many commercial solar cells where they prove their applicability. Randomly textured surfaces found their way into the market since they can be fabricated with marginal efforts by relying on bottom-up approaches at low costs. Their optical action is usually considered to suppress reflection losses and to scatter the light strongly into the solar cell [18]. But the documented applicability made it less important, for a long time, to understand the properties of such randomly textured surfaces in detail. The functionality has been deeply discussed when randomly textured surfaces have been integrated into wafer-based, rather thick, solar cells [19]. And this understanding has never been questioned since then; especially after being integrated into thin-film solar cells. However, most notably for such solar cells, the approximations that have been previously assumed to understand the optical action are no longer justified. More adapted methods have to be put in place to understand the optical action of such random textures and their ability to enhance the absorption; and, thereby, the efficiency of the solar cell. Such understanding, as always, is the prerequisite to optimize those random textures. Since a canonical limit suggesting an upper bound for the efficiency of solar cells, the Yablonovitch limit [20–22], has not been reached, a further optimization seems to be feasible. There is plenty of space for possible improvements and the major goal of this chapter is to outline the most recent achievements in this direction.

The purpose of this chapter is therefore to sensitize the reader to the peculiarities of randomly textured surfaces in the context of thin-film solar cells, to provide a basic understanding for their functionality that is obtained from simulations and experiments that take into account all details of the randomly textured surface, and the actual solar cell. Moreover, strategies are outlined that go beyond a mere description of actual surfaces but which target on optimizing them. The chapter is structured as follows. In the second section an overview of the research methodology in this field is given. The third section discusses in detail the optical properties of an isolated interface. We distinguish between properties as observable in the near- and in the far-field. The discussion on how these properties of an isolated surface translate into their ability to enhance the absorption in a single-junction solar cell is discussed in the fourth section. This section contains also various suggestions on how to optimize such textures by proposing entirely new surfaces that outperform existing ones. Overall, this chapter should motivate the integration of unprecedented random textures into solar cells which serve the purpose of photon management more efficiently than those already available.

4.2
Methodology

Randomly textured surfaces are in most cases an integral part of actual thin-film solar cells and can be fabricated by various means. A feasible approach is to dip a glass substrate coated with a transparent conductive oxide (TCO) of sufficient thickness for some tens of seconds into a diluted acid, for example, HCl. The aggressive nature of the acid causes material ablation and, depending on various process parameters, craters of different height and size are fabricated [23]. Another approach is the use of as-grown textures, where the surface morphology depends on the growth conditions of the TCO [24]. Pioneering work to understand such random textures has been made in the 1980s by E. Yablonovitch and co-workers [20, 21]. Most notably, they described how such macroscopic random textures shift the absorption edge of wafer-based thick solar cells toward longer wavelengths. The investigations were made using a ray-optical treatment and statistical arguments to discuss the impact of the random texture in an actual device. By assuming that the surface causes a complete randomization of the direction in which a light-ray propagates and while neglecting absorption, it was shown that the effective path length can be enhanced by a factor of $4n^2$ when compared to an unstructured solar cell where n is the refractive index of the light absorbing layer. The resulting limit is called the Yablonovitch limit and it constitutes a benchmark any photon management has to compete with.

Nonetheless, the assumptions cease to hold if thin-film solar cells are considered. The thin active layer of the solar cell questions a perfect randomization and the size of the geometrical features of the randomly textured surface in every dimension prohibits the application of ray-optics. Instead, the light propagation through such cells is required to be discussed using wave-optical methods. After a description of typical solar cells usually considered, we outline selected wave-optical methods that are used in the context of photon management and describe which properties can be extracted. We also outline a suitable experimental technique, that is, a near-field scanning optical microscope NSOM that can be used to map the optical near-fields with sub-wavelength resolution close to the interface.

4.2.1
Structure of a Referential Solar Cell and Description of Available Substrates

A sketch of the considered thin-film solar cell is shown on the top of Figure 4.1. The starting point for building the solar cell is a glass substrate on which a TCO film is deposited. The surface profile of this TCO layer is subject to the considerations. The randomly textured surface is fabricated into that TCO layer. Information on the topology of this interface can be obtained by measuring it by an atomic force microscope. This interface is subsequently covered by an a-Si:H p-i-n layer stack. The thickness of this layer is subject to modifications but it is usually chosen between 250 nm and 350 nm. Although a thicker active layer would stipulate a better absorption, the short diffusion length of the generated electron–hole pairs in the active layer, which is caused by a high defect density and further reduced due to light induced degradation [25], prohibits the use of thicker layers. The solar cell, if fabricated as a single-junction solar cell, is terminated with a back reflector that can be either a metal or a sequence of layers from a TCO and a metal. The latter might be favorable with respect to the electrical properties of the solar cell and reduces plasmonic losses in the back reflector [26]. If the a-Si:H solar cell is the top cell of a tandem solar cell, that is, if combined with another active layer made of microcrystalline silicon (μc-Si:H) that can be usually chosen to be much thicker, a dielectric intermediate layer (IL) is used to optically improve the tandem solar cell. Since the μc-Si:H solar cells can be made so thick that the currents generated from the top and the bottom cell are identical, the attention can be limited to the top solar cell alone while assuming an infinite extended dielectric half space as the terminating material of the solar cell. In the majority of works the sequence of layers is assumed to be conformally deposited on the TCO interface, that is, it is assumed that the sequence of subsequent layers possess the same topography. Motivated by experimental observations, we restrict our considerations to such a scenario and do not discuss previous work where a non-conformal layer sequence has been considered.

The exploitation of self-organization strategies to fabricate the randomly textured surfaces has been promoting heuristic approaches for optimizing the textures in the past. This has been usually put in place by varying systematically individual parameters of the fabrication process and observing their impact on the solar cell efficiency. Such approach led inevitably to different structures that are nowadays considered as useful but which possess disparate topological features. Selected examples are shown in the bottom of Figure 4.1. They will be considered in the following discussions. The commercially available substrate of Asahi (Asahi-U) is characterized by a fine structure with an autocorrelation length of the critical features in the order of 160 nm, a root-mean-square (rms) value for the height of 35 nm and a maximum peak-to-valley height in the order of 250 nm. In contrast, the Jülich substrate [27] has much larger critical features (autocorrelation length of 1400 nm) and also has a much larger height modulation (rms value of 160 nm and a peak-to-valley height of 1000 nm). A third representative substrate which we consider is provided by colleagues from the EPF Lausanne [28]. It has properties inbetween the other samples (autocorrelation length of 140 nm and 81 nm and 550 nm as rms value and peak-to-valley height, respectively). It has to be stressed that we do not intend to discuss all possible substrates that have been documented

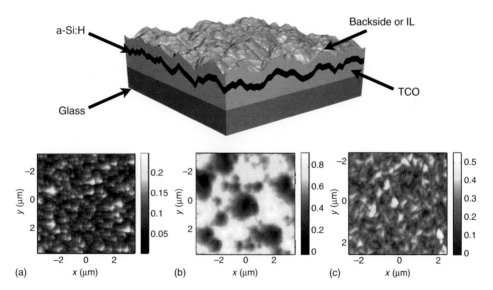

Figure 4.1 On the top the basic configuration of the considered thin-film solar cell comprising a random texture is shown. At the bottom selected topographies of representative surfaces are displayed (height scale is also in micrometer). (a) Commercial substrate from Asahi, Japan (b) substrate from FZ Jülich (Germany) and (c) substrate from EPF Lausanne, Switzerland. Reproduced with permission from [36]. Copyright 2010, OSA.

in the past. The chosen ones were selected for representative purposes only since they posses distinct and unique properties.

We finally wish to stress that all further considerations only address the optical properties of the surfaces. A final decision for or against a certain texture necessarily has to take into account also their electric properties. However, a combined discussion of optical and electric properties for textures is beyond the scope of this section. Moreover, many of the electrical properties characterizing the solar cell are not known with the necessary accuracy as to consider them reliable in an analysis.

4.2.2
Rigorous Methods

As could be seen in Figure 4.1, the critical features of the surfaces are entirely comparable to the wavelength in the light absorbing layer. Therefore, in order to describe the light propagation in the solar cells, Maxwell's equations should be solved directly. They read in time domain as

$$\nabla \times \mathbf{E}(\mathbf{r},t) = -\frac{\partial \mathbf{B}(\mathbf{r},t)}{\partial t}, \qquad \nabla \cdot \mathbf{D}(\mathbf{r},t) = \rho(\mathbf{r},t),$$

$$\nabla \times \mathbf{H}(\mathbf{r},t) = \frac{\partial \mathbf{D}(\mathbf{r},t)}{\partial t} + \mathbf{j}(\mathbf{r},t), \qquad \nabla \cdot \mathbf{B}(\mathbf{r},t) = 0,$$

with $\mathbf{E}$, $\mathbf{H}$, $\mathbf{D}$, $\mathbf{B}$, $\mathbf{j}$, and ρ as the electric and magnetic field, the electric displacement, the magnetic induction, and the current and charge density, respectively. They all depend on space and time. While considering the spectral domain of interest to coincide with the visible, the usual simplification as done in the optics can be imposed, that is, no external charges or conduction current, and all materials are non-magnetic. Maxwell's equations are only complete while defining constitutive relations between electric field and electric displacement. In time domain, local constitutive relations are written in terms of a response function that is convoluted with the electric field from the past. In frequency domain, this leads to a multiplication of the frequency dependent electric field amplitude with a dispersive permittivity $\varepsilon(\mathbf{r},\omega)$. All geometrical information concerning the solar cell is included in that spatially dependent permittivity. In most optical simulations, these intrinsic properties of the materials involved can be adopted from values tabulated already in literature. However, attention should be paid to use the correct material properties since they depend often, especially for a-Si:H, sensitively on the fabrication conditions. Moreover, the measurement of the absorption coefficient close to the optical band-gap constitutes an experimental challenge and techniques such as photothermal deflection spectroscopy are suggested to be of use.

Solving Maxwell's equations for a given geometry and a given incident wave field is the primary task to be performed. In the past, various methods have been used for this purpose. Among others, methods developed for micro-optical problems were applied that require the unit cells to be periodically arranged. Examples thereof are the Fourier modal method, the rigorous coupled wave analysis, or the Chandezon method. The requirement of a periodic structure sounds contradictory to the stochastic nature of the samples, but the unit cells can be usually made sufficiently large to suppress any spurious impact of the artificially introduced periodicity. Such methods were often used to discuss simplified solar cells in which the dimensionality is reduced, that is, the solar cell was assumed to be infinitely extended into one dimension. This reduces the three-dimensional Maxwell's equations to two sets of two-dimensional equations. Such simplifications were made to obtain insights into the basic properties of random textures since simulations can be done in a short time without requiring super computing facilities.

In order to simulate actual solar cell devices that fully take into account a large scale textured surface (in the order of 10 μm × 10 μm) methods such as the finite-element method, the finite integration method, or the finite-difference time-domain (FDTD) method have been used to solve Maxwell's equations on rigorous grounds. Especially the last method turned out to be a useful tool and has been widely applied. The FDTD method solves Maxwell's equations directly in time domain by approximating the differential operators with difference operators [29]. Discretizing the fields on a Yee grid, where the electric and magnetic field are discretized half a unit cell apart in space and time, assures that the fields remain to be free of divergence; hence, only the curl equations need to be considered. The finite computational domain can be truncated either by imposing periodic or perfectly conducting boundaries. Furthermore, using perfectly matched layers [30] allows to simulate an open space. With all these methods, the electric and magnetic field can be calculated everywhere in space for an individual configuration, that is, a geometry of the solar cell and a given illumination scenario. The spatially dependent field is then the primary source of information and any further analysis can rely on it.

4.2.3
Scalar Methods

In order to avoid the use of massive supercomputing facilities, scalar optical methods are frequently exploited to simulate light scattering properties. Instead of directly solving Maxwell's equations, scalar quantities are derived by applying several approximations. In the approach of Carniglia [31] which was later modified by Krč *et al.* [32], the surface topography was reduced to the root mean square (rms) roughness. The variety of textures with different shapes but identical rms roughnesses were distinguished by a wavelength dependent correction function that has to be determined by adjusting simulation results to the experiment. Such an adjustment is done for the scattering into air. The strongest and also most questionable assumption is that for the scattering into the absorbing layer, the identical correction function can be used.

A more detailed approach considers the local phase shift while traversing the roughness zone of the interface. In the work of Dominé *et al.* [33], the exact surface topography is taken into account and the basic property is a two-dimensional scalar field. The assumption is that light is traversing vertically through the roughness zone without directly intersecting with light from different locations. The far-field light scattering properties of textures, that are commonly used as a front contact in thin-film solar cells, can be well described by this model. The results are compared to experiment and rigorous methods, both for scattering into air and silicon showing a good agreement.

4.2.4
Properties of Interest

Once the fields are calculated everywhere in space, three quantities of interest can be extracted: the spatial distribution of the scattered light in the solar cell or just behind an isolated interface, the angular distribution of the scattered light into the far-field, and the local absorption. The first property is not needed to be discussed but it will be shown how such information can provide insights into the functionality of the solar cell.

The way the light is scattered by a random texture, which is assumed to separate two infinitely extended half spaces, can be calculated by using methods from Fourier optics, that is, in particular the angular spectrum method. To be applicable, it requires, however, a periodic arrangement of the unit cell. Again, with methods such as FDTD, huge unit cells can be considered, rendering the conclusion to be independent of the absolute size of the super cell. The angular domain into which the light is scattered can then be calculated by Fourier transforming the complex field oscillating at a fixed frequency in a plane behind the interface. Assuming the interface to be extended in the x- and y-direction and considering the z-axis as the principal propagation direction, this Fourier transformation reads as

$$\tilde{\mathbf{E}}(k_x, k_y) = \left(\frac{1}{2\pi}\right)^2 \int\int_{-\infty}^{\infty} \mathbf{E}(\mathbf{r}) e^{-i(k_x x + k_y y)} dx dy. \tag{4.1}$$

The share of energy that is diffracted in transmission (t) or reflection (r) into a given direction can be calculated by considering the z- component of the time averaged Poynting vector for each diffraction order normalized to the energy of the incident field, that is,

$$\eta_r = \Re\left[\frac{k_r(k_x, k_y, \omega)}{k_i}\right] |\tilde{\mathbf{E}}_r(k_x, k_y)|^2, \tag{4.2}$$

$$\eta_t = \Re\left[\frac{k_t(k_x, k_y, \omega)}{k_i}\right] |\tilde{\mathbf{E}}_t(k_x, k_y)|^2 \tag{4.3}$$

with $k_{r/t}$ being the z-component of the wavevector in the reflected or transmitted domain, that is, $k_{r/t} = \sqrt{\frac{\omega^2}{c_0^2}\varepsilon(\omega)_{r/t} - k_x^2 - k_y^2}$, k_i is the z-component of the incident wave vector in the incident half-space and $\tilde{\mathbf{E}}_{r/t}(k_x, k_y)$ is the reflected or transmitted amplitude of the plane wave propagating into a specific direction. To further reduce the amount of information, an angularly resolved scattering (ARS) function can be introduced. There, the energy as scattered into an annular angular cone is integrated and taken as a measure of how much energy is deflected into a certain direction. The technique is often used to discuss the light scattering in transmission, but the method can be extended to discuss the angular distribution of scattered light in reflection. The ARS is often shown as a function of the angle, ranging from $-\pi$ to π with 0 being the forward direction. A further condensation of the scattering properties in transmission can be done by introducing the haze. The haze is the amount of energy diffusely scattered in a forward direction normalized to total transmitted energy. Although such reduction of the scattering response to eventually just a single number is often useful to compress the huge amount of information, it clearly constitutes a loss of information. Care has to be taken while evaluating the functionality of an optical texture solely on the basis of such a piece of compressed information.

The absorption in a solar cell is best evaluated by computing the divergence of the time averaged Poynting vector in each spatial coordinate, that is,

$$\langle \mathrm{div}\mathbf{S}(\mathbf{r},\omega)\rangle = -\frac{1}{2}\mathfrak{J}[\omega\mathbf{P}(\mathbf{r},\omega)\mathbf{E}^{*}(\mathbf{r},\omega)] \tag{4.4}$$

with $\mathbf{P}(\mathbf{r},\omega)$ being the electric polarization defined as the product between the dispersive permittivity in a given spatial point and the electric field. Integrating across the spatial domain of interest and normalizing to the incident power allows to calculate the global absorption of a given device for a given illumination. In solar cells where only the active layer is considered to be absorptive, the difference of the incident energy to the sum of the reflected and transmitted energy equally serves as a measure for the absorptance. A further quantity that is often useful is the absorption enhancement. It requires to normalize the absorbed power to the power absorbed in the same solar cell without the assistive structure.

If these properties need to be known across an entire spectral domain, the simulations should be repeated for each wavelength of interest. With such procedure the spectrally resolved absorptance or scattering properties can be extracted. Weighting the absorptance by the solar spectrum as perceived on earth, that is, the AM1.5g spectrum, provides a measure for the number of absorbed photons. Assuming that each absorbed photon generates one electron–hole pair and integrating the number of electron–hole pairs, allow to compute the short circuit current density for a given solar cell. That is, in most cases, the quantity that ultimately matters and which should be used to quantify or evaluate the photon management.

4.2.5
Near-Field Scanning Optical Microscopy

Since FDTD simulations provide the electromagnetic field distribution of a given thin-film structure, the experimental access to this property can be used to validate the numerical methods and vice versa. Moreover, by combining optical simulations and experimental techniques, methods can be derived to obtain information about local scattering properties and local light trapping efficiency. Near-field scanning optical microscopy (NSOM, also known as SNOM) is used to measure the local light intensity distribution. NSOM is one kind of scanning probe microscopy. Two different principles exist. First, *NSOM with aperture*, where a tapered optical fiber tip illuminates the sample or collects the transmitted or reflected light. Second, *apertureless NSOM*, where a metallic tip on a cantilever is illuminated with a focused laser beam and the back-scattered light is detected. The latter takes advantage of the near-field effects of a metallic tip, which significantly enhance the scattering efficiency of the surface. The former has a much higher flexibility in the optical configuration since the tip can be used as a light source as well as a detector. In both cases, the tip scans across the surface by piezoelectric systems.

The optical simulation by FDTD yields the electric field distribution inside the whole computing domain. In the experiment, direct access is only possible above the surface of the sample, typically in air. For a direct comparison, NSOM with aperture in collection mode, where the sample is illuminated from the far-field by a laser and the tip is used as the detector, is the most appropriate technique. Here, the optical resolution of the system is determined by the geometrical size of the aperture and approaches 50 to 80 nm. A shear-force feedback system is used to avoid contact of the tip with the surface and to keep the distance between them constant. For larger distances, the feedback loop is switched off and the tip is scanned at constant heights.

4.3
Properties of an Isolated Interface

The optical effect of an isolated random texture might be considered from two perspectives. At first, the distribution of light in the near-field can be considered. This allows to access directly the optical properties of the randomly textured surface and enables the study of the impact of certain features on the field distribution. Moreover, the study of the field behind a randomly textured surface is an important means since it allows to compare results from simulations. Such comparison provides confidence that the properties of the texture are properly reflected in the simulations so that further analysis can be done where the absorption enhancement can be reliably calculated. Secondly, the ability of the texture to deflect the light from its normal propagation direction can be considered to understand the functionality of the random texture. Clearly, the deflection causes an enhancement of the optical path in the solar cell, which is beneficial. Then, the local properties of the random texture are no more discernible and only the overall action is accessed. Insights from both approaches are summarized in the following.

4.3.1
Near-Field Properties

Both, FDTD simulations and NSOM experiment, provide the local light intensity distribution very close to the surface. Figure 4.2 shows the topography and the

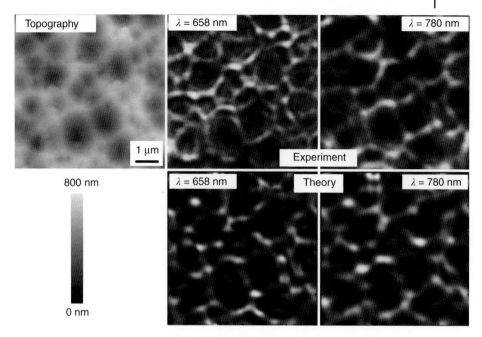

Figure 4.2 Topography of a randomly textured ZnO:Al layer and local light distribution very close to the surface at wavelengths of $\lambda = 658$ nm and 780 nm (top). In comparison, results from FDTD simulations are shown on the bottom.

experimental results for wavelengths of $\lambda = 658$ nm and 780 nm on the top and results obtained via FDTD on the bottom [34].

The agreement between experiment and simulation is very good, which proves that the assumptions the model is based on are reasonable and that the simulations provide reliable information. A strong light localization along the rims of the craters is found. This is attributed to light focusing and the lightning rod effect [35]. The fine structure in the local light intensities inside the craters results from evanescent waves that are excited by total internal reflection in the ZnO:Al layer. The intensity of these evanescent waves decreases exponentially with increasing distance from the surface and are, therefore, only visible at distances below half the wavelength.

A large absorption enhancement can be achieved by trapping the light inside the absorbing layer, which has in the case of an a-Si:H a larger refractive index than its surrounding. Usually this can be achieved for propagation vectors in a-Si:H which correspond to evanescent waves in the adjacent medium. Hence, the investigation of the intensity distribution in the optical near-field is crucial for the understanding and optimization of surface textures.

To study the propagation of light in air, cross-sectional images are generated [35]. In the experiment, the NSOM tip is retracted and scanned across the surface at a constant height. The cross-sectional image is calculated from the measured light intensities at different heights across a selected scanning line. In simulation, these images can be directly extracted from the calculation domain.

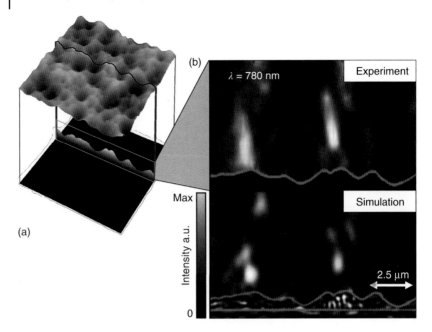

Figure 4.3 The generation of cross-sectional images of local light intensity from a NSOM experiment is shown in (a). One scanning line is selected and the measured light intensities above this line is plotted for different heights forming the image in (b). This result is compared to FDTD simulation at the same line. The corresponding wavelength is 780 nm.

The results are shown for one scanning line at a wavelength of 780 nm in Figure 4.3. Again, the experimental and theoretical results agree well. Above the highest points of the surface structure, light localization is found being extended in height. These localizations appear as light spots and jet-like structures. This can be explained by the intersection of differently scattered light that emerges from different locations at the surface. Very close to the surface, hot-spot-like localizations can be observed, particularly in the FDTD results. These structures are accompanied by evanescent waves.

The simulations have a great advantage wherein light intensities can also be studied inside the layers which is also seen in Figure 4.3. Here, complex interference patterns inside the ZnO:Al layer are found, which result from standing waves not only in the growth direction of the layer but also perpendicular to it; suggesting the possibility of the excitation of guided modes in the layer.

4.3.2
Far-Field Properties

The quantities which will be used to characterize the scattering properties were introduced above. To glimpse the optical properties, the haze is shown for different textures in Figure 4.4(a) while illuminating the interface by a plane wave

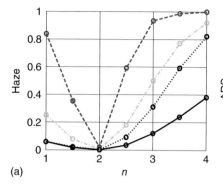 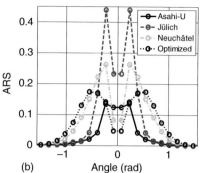

Figure 4.4 Here the impact of various textures is considered that separate two infinite half-spaces. (a) shows the haze of the selected substrates as a function of the index of a second half space. (b) shows ARS for the transmitted light into the second half space corresponding to Si. Reproduced with permission from [36]. Copyright 2010, OSA.

at a wavelength of 633 nm [36]. In addition, the haze of an optimized profile is included. This optimized profile is discussed in section 4.4.2. The haze is shown for scattering at an interface between TCO (refractive index of 1.915) and a second half space whose refractive index is subject to modifications being unique for simulation whereas in experiments such modification is impossible and the study of the scattering properties at a high-index material interface constitutes a major challenge. Usually, high index materials are strongly absorptive and the characterization of scattering is exceedingly difficult. But just these studies are important because the interface between the TCO and the a-Si:H exhibits just this characteristic. The haze in the simulation attains a minimum for an index of the second half space close to two. If the indices of both media match, the interface is physically present but has no optical effect. This result leads to an important conclusion. The topology of the interface is of secondary importance. Of importance are particularly the optical properties which are dictated by the product of a typical geometrical length scale and the refractive index difference of the materials at both sides of the interface. Only if this product is in the order of the free space wavelength at which the photon management is important, the texture operates efficiently [37].

The haze, as seen in Figure 4.4(a), is smallest for the Asahi-U substrate and largest for the substrate from Jülich. The large height of the latter substrate introduces a sufficient phase contrast that scatters the light. As a rule of thumb it has to be argued that a simple binary structure with a lateral size larger than the wavelength and a filling factor of 0.5 scatters the light in a forward direction only off-axis, if the optical path difference in one half of the unit cell amounts to π when compared to the other half. Therefore, a sufficient modulation depth is necessary to scatter the light. And the strong scattering regime is only reached for a sufficiently large index contrast. A smaller index contrast requires a larger modulation depth for compensation.

But the haze is not the only parameter of importance. The texture should not just scatter the light, but it should scatter the light into a sufficiently large angular domain. The quantity that permits to evaluate this aspect is the ARS function. The ARS function for the considered textures is shown in Figure 4.4 (b) while assuming an interface between TCO and a material having approximately the real part of the refractive index of a-Si:H, that is, $n = 4$. It can be seen, that although scattering is strongest at the substrate from Jülich, the angular cone into which the light is scattered is rather narrow and most of the light is confined to an angular domain close to the optical axis. That is reminiscent of the large correlation length associated to small spatial frequencies. The substrate from EPFL performs better since much light is scattered into a large angular domain. That is possible because the critical features of the textures are much smaller and the spatial frequencies are larger.

4.4
Single-Junction Solar Cell

To know the scattering properties from the random textures, either from experiments or from simulations, is only half the information. Further to such analysis of the scattering properties it is essential to understand how they translate into the ability of the texture to promote the light absorption in the active layer of the solar cell. To understand why a certain texture is more beneficial for that purpose when compared to others is a necessary prerequisite for an optimization. To understand this relation, it has to be stressed that it is not necessary to consider the entire spectral domain, which would complicate the analysis and make it much more demanding. The significant amount of scattered light in the solar cell strongly suppresses the interference effects of light inside the light absorbing layer. In particular, close to the optical band gap they would cause Fabry–Perot transmission resonances at particular wavelengths. This spectral behavior prevents the extrapolation of the absorption characteristic at an individual wavelength to the entire spectral domain. However, such resonance phenomena are not observed in the absorption profiles of most thin-film solar cells. The random texture suppresses the interference and its only impact is to shift the absorption edge toward longer wavelengths. Therefore, the performance of photon management can be extracted from the analysis of the absorption enhancement at a single wavelength, that is, preferably a wavelength where a referential cell absorbs half the incident energy. In the following we provide such an analysis of the textures considered before. We also discuss strategies to optimize the textures such that they outperform existing substrates.

4.4.1
Absorption Enhancement

To quantify the performance of random textures for absorption enhancement, a cell as shown in Figure 4.1, can be considered. It is possible to investigate the absorption enhancement for a solar cell made of a-Si:H alone or by considering

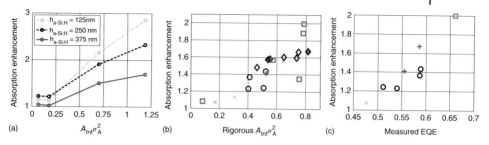

Figure 4.5 (a) Absorption enhancement for different substrates and different thickness of the a-Si:H films. The substrates are shown according to their merit criterion, where the order is Asahi-U ($A_{int}\sigma_A^2 \approx 0.1$), the substrate from Jülich ($A_{int}\sigma_A^2 \approx 0.2$), the substrate from EPFL ($A_{int}\sigma_A^2 \approx 0.68$) and the optimized substrate ($A_{int}\sigma_A^2 \approx 1.2$). The functional dependency for a larger number of substrates can be seen in (b), where the merit criterion is related to the absorption enhancement in a 250 nm thick a-Si:H layer. (c) shows the correlation of the measured EQE with the theoretically predicted absorption enhancement. The different icons in (b) and (c) correspond to substrates that have been described in different publications. Details can be found in Ref. [38]. (a) reproduced with permission from [36]. Copyright 2010, OSA. (b) and (c) reproduced with permission from [38]. Copyright 2011, AIP Publishing LLC.

the absorption enhancement in a conceptual cell that resembles only the top cell of a tandem cell. Then, the material in the transmitted region has to be a dielectric conducting material, that is, TCO, to match the properties of an intermediate layer as used in a realistic solar cell.

For randomly textured surfaces whose scattering properties were discussed in the previous section, the absorption enhancement in an a-Si:H solar cell is shown in Figure 4.5(a) for three different thicknesses for the active layer [36]. The absorption enhancement is shown as a function of a texture-dependent merit criterion, that is, four different textures have been considered for this graph. The merit criterion has been established to correlate the absorption enhancement to the scattering properties. It is defined as the product between the total amount of scattered light A_{int}, that is, the haze, multiplied by the square of the Gaussian width σ_A that approximates in a least-square sense the ARS function at larger angles [36]. That merit criterion is very useful since it reduces a complicated response from an involved structure to a single scalar number. Of course, that reduction comes at the expense of this criterion not being rigorous. It conceptually unifies the idea that the texture should primarily scatter as much light as possible into a broad angular domain. Both properties enter the merit criterion; although weighted in a different manner. With the desire to establish a linear dependency for different wavelengths and for different thicknesses, the square of the Gaussian width of the ARS had to be taken. Therefore, an important conclusion to be drawn is that scattering is important but it is even more important to scatter into a large angular domain.

That also leads to the superiority of the substrate from EPFL ($A_{int}\sigma_A^2 \approx 0.68$) concerning the ability to enhance the absorption, as can be seen in Figure 4.5(a). The caption explains which texture belongs to which data point. The absorption enhancement is larger when compared to other interfaces considered; independent of the thickness of the active layer. To verify the applicability of the merit criterion to a larger number of substrates, their absorption enhancement is shown in Figure 4.5(b) for an individual thickness of the active layer of 250 nm [38]. The half-space where the light is transmitted is assumed to be a dielectric intermediate layer. Although statistical deviations always exist, it can be seen that the scattering properties of each individual interface can be related to the absorption enhancement. The predictive power of the merit criterion deteriorates for textures that either scatter the light excessively strong or where the absorption enhancement is excessively large. Then, the approximation made while reducing the information is not justified anymore, for example, the ARS function might be only insufficiently approximated by a Gaussian function.

The last important topic to be mentioned is the comparison of the simulated absorption enhancement to a measured external quantum efficiency (EQE). To be able to link the results from a theoretical analysis to a measured quantity is an important contribution to motivate further research and to justify the means. By assuming a perfect collection efficiency, that is, each absorbed photon generates a electron–hole-pair that contributes to the cell current, the absorption enhancement can be related to the EQE. We consider here an EQE as measured by colleagues at EPFL for selected interfaces after they were integrated into a tandem solar cell, that is, all the assumptions imposed on the simulations are matched by the experimental devices. The EQE is only the EQE of the top cell. Results are also shown in Figure 4.5(c) [38]. Clearly, the measured EQE is related to the estimated absorption enhancement. This also suggests, by going one step back, that estimating the scattering properties with the merit criterion as discussed, serves eventually as a good measure to judge the suitability of a certain interface for the purpose it has been developed for.

In a short summary of the descriptive part of our work, we wish to state that in order to be beneficial, the surface texture should have a sufficiently large height modulation and it should have critical features in the order of a few hundreds of nanometers only [37]. Structures with larger feature sizes might be able to scatter the light; but unfortunately only in a narrow angular cone around the optical axis. After such analysis the natural question arises whether these textures can be modified to work better. This is an essential question most notably for experimentalists working in the field. Alternatively, one may also ask, what an optimized randomly textured surface should look like. The following subsection will give an answer to these questions.

4.4.2
Design of Optimized Randomly Textured Interfaces

As discussed above, the quest for a suitable optimization strategy that allows to propose new textures that outperform existing ones remains. For the moment, we might even accept to detach ourselves from any constraints imposed by reality and we may ask ourselves, what would be the optimal texture? Of course, experimental

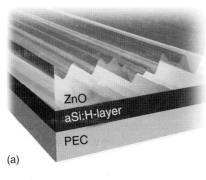 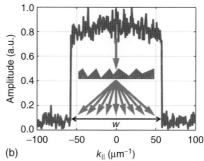

Figure 4.6 (a) Geometry of the simplified solar cell under considerations. (b) Angular distribution of the scattered light by an optimized surface. The angular domain into which the light should be scattered is denoted by an angular width w. The inset explains the geometry in further detail. The surface was designed such that it scatters a normally incident plane wave equally distributed into an angular domain of width w.

constraints have to be taken into account eventually but the definition of a target function by theory/numerics may force the proposition of experimentalists to rely on unorthodox methods to actually achieve these surfaces.

The consideration is restricted at first, for the sake of simplicity, to a solar cell that is infinitely extended into one dimension, that is, the structure is two-dimensional. Furthermore, the solar cell is reduced to a back reflector made of a perfect electric conductor (PEC), which is covered by an a-Si:H layer of 350 nm thickness and a TCO layer exhibiting a textured interface with air. This simplified geometry allows the optimization of the TCO texture using some approximations [39]. The geometry is shown in Figure 4.6(a). The absorptance in the solar cell, and so the evaluation of the actual photon management, is performed rigorously.

Summarizing the insights from the preceding sections, it turns out that the interface should scatter the light strongly into a sufficiently large angular cone. The absolute size of the cone needs to be determined and, even more important, the random texture that scatters the light into the desired cone. The latter task can be performed while using a scalar theory, Fourier optics, and the thin-element approximation. The latter approximation neglects diffraction of the incident light while it traverses the structure. The field behind the textured interface is given by the amplitude of the incident illumination (which is constant for a plane wave) and a spatially varying phase. The magnitude of the phase depends on the optical path length which a conceptional ray experiences while traversing the structure. Once the desired spatially dependent phase-only transmission function is known, it can be translated into a geometrical profile while considering a design wavelength [39].

The optimization of the phase-only transmission function that allows to scatter the light into a predefined angular domain can be obtained while exploiting a technique known as the iterative Fourier transform algorithm (IFTA) [40]. The optimization iterates between the real space directly behind the structure and the angular domain by Fourier transforming the respective fields back and fourth. The constraint imposed in real space is a constant amplitude of the

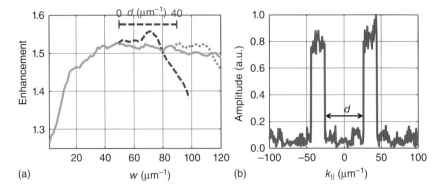

Figure 4.7 (a) Enhancement of the short circuit current density of a one-dimensional solar cell comprising optimized textures that scatter the light equally distributed into an angular cone of width w (gray solid line). (b) Suppressing a central cone of width d (see inset) into which the light is scattered can further enhance the short circuit current density, as can be seen from the dashed and the dotted lines in (a).

transmission function while the constraint imposed in the reciprocal space is the desired amplitude distribution for the scattered light. The phase in each domain constitutes the free parameters that is found usually after a small number of iterations. In general, the optimization starts with some random phase profile for the transmission function.

As can be seen exemplarily is Figure 4.6(b), a surface can be designed such that it scatters the light equally distributed into an angular cone of a size characterized by w. The enhancement of the short circuit current density rigorously calculated for the cell described above having the different optimized surface textures is shown by the gray solid line in Figure 4.7(a). Here unpolarized light was used for illumination and a design wavelength of 550 nm was chosen to translate the phase-only transmission function into a geometrical profile. It can be clearly seen that for an increasing size of the scattering cone, the short circuit current density gets enhanced. Hence, the light that is diffracted into a larger angular cone by the textured surface clearly experiences an enhanced optical path in the light absorbing layer. This translates into a more efficient solar cell. Beyond an angular cone of a specific size, the enhancement of the short circuit current density saturates. This indicates that the light is sufficiently redistributed inside the solar cell and that all the supported modes are excited with some finite energy. A further redistribution does not cause any change. The opposite actually holds. By further increasing the angular domain into which the light is scattered, an increasing amount of spatial frequencies excites only evanescent modes since the critical period associated with the angular frequency is too small. This is detrimental since the light will progressively experience an effective medium only and it will not resolve the spatial details of the scattering surface. This causes a degradation and, hence, an upper bound for a beneficial angular cone can be identified.

The structure can be even further improved by departing from a cell with an optimal outer width w for the scattering cone and progressively suppressing the light scattered in forward direction, that is, a cone of width d as shown in Figure 4.7(b). Naturally, this part of the scattered light experiences the lowest amount of optical path length enhancement since it propagates almost perpendicularly to the optical axis. Steering preferentially the light into larger angular domains might even further improve the properties. That this is indeed possible, can be equally concluded from the dashed line in Figure 4.7(a). Nonetheless, an excessive suppression of this central angular cone into which the light is scattered is not advised as can be seen from the degradation of the short circuit current density enhancement for larger d. The strong suppression appears since the scattering profile resembles a periodic structure in the limiting case of an equal size for the inner and the outer cone. Then, photon management is not improved across the entire solar spectrum but only in a narrow spectral domain that acts sensitively on a grating with a single period. Therefore, an optimal balance between the inner and the outer width of the angular cone has to be found and can be extracted from the figure.

A similar strategy can be put in place to optimize a texture for a regular solar cell. As has been shown before, what is important for the scattering interface is to strongly scatter the light and to scatter the light into a rather large angular cone. A merit criterion was introduced above that can serve as a simple measure to estimate how beneficial the integration of that texture into a solar cell can be. The merit criterion can be also calculated by relying on the approximations used above, that is, scalar and thin element approximation, and the IFTA algorithm can be used to determine scattering surfaces that maximize the criterion or to make it at least larger when compared to existing surfaces. Results from a rigorous analysis where the scattering properties of such optimized texture and their ability to enhance the absorption in a referential thin-film solar cell have been already shown in Figures 4.4 and 4.5. As can be seen in these figures, the haze of the optimized interface is less than that of the substrate from EPFL, but the scattering takes place into a much larger angular cone. This causes the merit criterion as evaluated on rigorous grounds to be larger than those of available substrates. This immediately translates into the ability of the interface to enhance the absorption. While discussing the values as shown in Figure 4.5 it has to be kept in mind that the absolute absorptance is not unbound but a finite quantity, that is, absorptance larger than unity is not possible. This necessarily causes a saturation of the absorption enhancement and limits its maximum achievable value. This occurs already for the thickest thin-film considered in Figure 4.5. It remains to be noted that the optimized texture outperforms existing substrates.

The question may arise as to why such interfaces have not been fabricated yet. Although technological challenges may exist to achieve their fabrication, nanofabrication methods suitable for the implementation are explored at the moment that might be of use to realize thin-film solar cells possessing such optimized interfaces, for example, nano-imprint lithography [41]. Fabricating a master structure with high effort that contains the optimized surface texture and its translation into a TCO might be a reasonable strategy to be explored in the near future. It finally remains to be mentioned that the method of optimization as outlined in this section is not the only strategy that has been explored in the past. Other optimization algorithms, for example, genetic algorithms [42] or methods based on simulated annealing can be equally used for that purpose and are well documented in the literature.

4.5
Intermediate Layer in Tandem Solar Cells

In the previous section, we described strategies to optimize an a-Si:H Single-junction solar cells. Another approach to enhance the efficiency of thin-film solar cells is based on the concept of multi-junction solar cells. In particular, tandem solar cells with an a-Si:H top cell and a μc-Si:H bottom cell have the great advantage that they can be prepared by the same process technology and that the different optical band gaps of the absorbing layers enhance the total amount of absorbed light [43, 44].

One important issue of such serial connected tandem solar cells is the current matching of both individual cells, since the current has to flow through both cells. Therefore, the total current density is given by the minimal current density of the two cells. Due to the effects of light induced degradation, the thickness of the a-Si:H top cell is much stronger limited than the thickness of the bottom cell [45]. The larger optical band gap of a-Si:H as compared to μc-Si:H leads to a higher voltage at the top cell. Both aspects demonstrate the need of a redistribution of light absorption between the two cells. The improvement of light absorption in the a-Si:H top cell leads to an improvement of the entire tandem cell.

One promising way to achieve this is the implementation of intermediate layers (IL) between the top and the bottom cell [43, 45, 46]. The basic functionality is the spectrally dependent reflection of light in the spectral range, where the absorption of both cells is comparable. For longer wavelength, where the absorption in a-Si:H is quite low, the intermediate layer has to transmit the light [47].

For the optimization of the IL, we assumed an a-Si:H top cell with a thickness of 250 nm below the textured ZnO layer. The refractive index of the IL is representatively assumed to be $n_{IL} = 2$. The thickness of this layer is subject to variation [48]. For three different scenarios, the absorption enhancement in the top cell is shown in Figure 4.8 for two different wavelengths (780 nm and 658 nm). The absorption enhancement is given by the integrated absorption in the entire top cell with IL normalized to the absorption without the IL. The forthcoming layer stack below the IL differs in the three scenarios. In Figure 4.8(a), the device is finalized by an air half space, whereas an additional 40 nm μc-Si:H layer is assumed between the IL and air in Figure 4.8(b) and a semi-infinite μc-Si:H half space is placed below the IL in Figure 4.8(c). (Figure 4.8d) shows the topography of the surface that is taken into account for the simulations and Figure 4.8(e) illustrates the layer stack and the direction of light illumination.

For the IL alone, a decrease of absorption in the top cell is found as shown in Figure 4.8(a). This is due to the fact, that the refractive index of the IL is between that of a-Si:H and air which leads to an anti-reflection effect. For the thin μc-Si:H

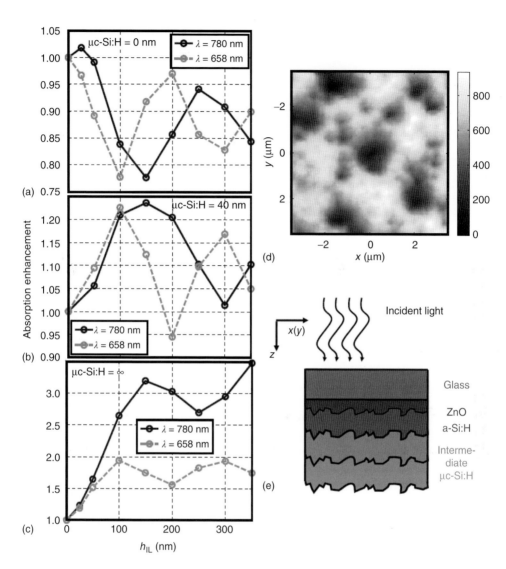

Figure 4.8 Absorption enhancement in the a-Si:H layer of a tandem solar cell as a function of the thickness of the intermediate layer, h_{IL}, for two relevant wavelengths (solid line denotes $\lambda = 780$ nm and dashed line denotes $\lambda = 658$ nm). (a) Cell consisting of a 250 nm a-Si:H layer below a randomly textured ZnO:Al film deposited on a glass substrate. (b) An additional 40 nm layer of μc-Si:H was sandwiched between the intermediate layer and air. (c) The μc-Si:H layer has infinite thickness. (d) Surface profile in nanometers of a fabricated solar cell retrieved with an atomic force microscope. (e) Schematic of the layer stack. Reproduced with permission from [48]. Copyright 2009, AIP Publishing LLC.

layer as well as the µc-Si:H half space, an improved absorption is found in the top cell. In all cases, Fabry–Perot oscillations are found with periods that are larger than that for flat interfaces, suggesting that the incidence light is deflected and has a smaller wave vector component in longitudinal direction. This shows the impact of textured interfaces and demonstrates that the IL has to be optimized for each given surface texture [49].

4.6 Conclusions

To concisely sum up, in this chapter we have been discussing the means and the approaches to discuss and to explore the optical properties of randomly textured interfaces that are integrated into single- or multi-junction thin-film solar cells. Research to disclose these optical properties is timely and rewarding since these textures are already an integral part of many solar cells commercially available. There, they have been proven as being extremely useful to suppress undesired back-reflections at the entrance facets, and thereby to enhance the in-coupling of light into the solar cell, as well as to strongly scatter the light, and thereby to enhance the optical path of the light in the active layer of the solar cell which enhances the probability of absorption. To disclose these properties, specifically in thin-film solar cells, is extremely challenging since the characteristic features of randomly textured solar cells are comparable to the optical wavelength. This inhibits the application of simplifying the concepts to understand the optical actions of such textures and requires to solve Maxwell's equations with no further approximations. Moreover, the thickness of the active layer is often smaller than the modulation height of the randomly textured surface, which equally challenges the understanding on how such textures affect the propagation of light in the solar cell. Therefore, rigorous numerical tools and experimental techniques that can predict the distribution of the electromagnetic fields in the near-field have to be put in place in order to disclose the mechanisms and the effects that occur in actual solar cells.

The work we have been presenting can be roughly divided into two parts. The first was devoted to the descriptive analysis of systems already in place. For these solar cells experimental results regarding the actual efficiency of solar cells that contain the different randomly textured surfaces exist. The predictions of the functionality of those textures from the analysis described above can then be compared to the action of an ultimate device. In our work we hopefully documented that many different textures exist. But for a good texture to be efficient the light has to be scattered at first by the texture pretty strongly. But it is of even more importance that the light is scattered into a sufficiently large angular cone. This suggests that the scattering of light into a narrow angular cone is of marginal use since the achievable enhancement of the optical path remains to be incremental. Therefore, emphasis should be put on achieving textures that scatter the light into a large angular cone which requires sufficient fine and strong surface modulation.

With that in mind we went further and have been proposing optimized randomly textured surface. They have been proven to outperform existing textures in their ability to enhance the absorption of light in the solar cell. The structures that have been outlined are not yet realized. But their performance should be a strong motivation for experimentalists to identify strategies to fabricate such structures and to integrate them into solar cell devices. Even admitting that the structures we have been discussing so far might not represent the optimum of whatever can be achieved, we have at least documented in this chapter the entire methodology that can be put in place to work toward that goal. With that we hope that we have stimulated further research in the field of photon management for solar cells, and beyond.

Acknowledgments

We are indebted to numerous colleagues, coworkers, and project partners working in the field of photon management for the numerous discussions and their advices on many practical aspects in this field of research. We would explicitly like to thank Franz-Joseph Haug and his colleagues from EPFL, Switzerland, for providing us with detailed information on the topology of their substrates and on the performance of solar cells made thereof. We would like to acknowledge the partial financial support of this work by the Deutsche Forschungsgemeinschaft (Nanosun and Nanopho), the Federal Ministry of Education and Research (Nanovolt and Infravolt) and the Thuringian State Government (SolLux). Some computations utilized the IBM p690 cluster JUMP of the Forschungszentrum in Jülich, Germany.

References

1. Mallick, S.B., Sergeant, N.P., Agrawal, M., Lee, J.Y., and Peumans, P. (2011) Coherent light trapping in thin-film photovoltaics. *MRS Bull.*, **36** (6), 453–460, doi: 10.1557/mrs.2011.113.
2. Aberle, A.G. (2009) Thin-film solar cells. *Thin Solid Films*, **517** (17), 4706–4710, doi: 10.1016/j.tsf.2009.03.056.
3. Strümpel, C., McCann, M., Beaucarne, G., Arkhipov, V., Slaoui, A., Švrček, V., del Canizo, C., and Tobias, I. (2007) Modifying the solar spectrum to enhance silicon solar cell efficiency-an overview of available materials. *Sol. Energy Mater. Sol. Cells*, **91** (4), 238–249, doi: 10.1016/j.solmat.2006.09.003.
4. Trupke, T., Green, M.A., and Würfel, P. (2002) Improving solar cell efficiencies by down-conversion of high-energy photons. *J. Appl. Phys.*, **92** (3), 1668–1674.
5. Trupke, T., Green, M.A., and Würfel, P. (2002) Improving solar cell efficiencies by up-conversion of sub-band-gap light. *J. Appl. Phys.*, **92** (7), 4117–4122, doi: 10.1063/1.1505677.
6. Nelson, J. (2003) *The Physics of Solar Cells*, Imperial College Press.
7. Dewan, R. and Knipp, D. (2009) Light trapping in thin-film silicon solar cells with integrated diffraction grating. *J. Appl. Phys.*, **106** (7), 074–901, doi: 10.1063/1.3232236.
8. Kroll, M., Fahr, S., Helgert, C., Rockstuhl, C., Lederer, F., and Pertsch,

T. (2008) Employing dielectric diffractive structures in solar cells - a numerical study. *Phys. Status Solidi A*, **205** (12), 2777–2795, doi: 10.1002/pssa.200880453.

9. Wang, W., Wu, S., Reinhardt, K., Lu, Y., and Chen, S. (2010) Broadband light absorption enhancement in thin-film silicon solar cells. *Nano Lett.*, **10** (6), 2012–2018, doi: 10.1021/nl904057p.

10. Duché, D., Escoubas, L., Simon, J.J., Torchio, P., Vervisch, W., and Flory, F. (2008) Slow Bloch modes for enhancing the absorption of light in thin films for photovoltaic cells. *Appl. Phys. Lett.*, **92** (19), 193–310, doi: 10.1063/1.2929747.

11. Bermel, P., Luo, C., Zeng, L., Kimerling, L.C., and Joannopoulos, J.D. (2007) Improving thin-film crystalline silicon solar cell efficiencies with photonic crystals. *Opt. Express*, **15** (25), 16 986–17 000.

12. Zeng, L., Yi, Y., Hong, C., Liu, J., Feng, N., Duan, X., Kimerling, L.C., and Alamariu, B.A. (2006) Efficiency enhancement in Si solar cells by textured photonic crystal back reflector. *Appl. Phys. Lett.*, **89** (11), 111–111, doi: 10.1063/1.2349845.

13. Chen, F.C., Wu, J.L., and Hung, Y. (2010) Spatial redistribution of the optical field intensity in inverted polymer solar cells. *Appl. Phys. Lett.*, **96** (19), 193–304, doi: 10.1063/1.3430060.

14. Schubert, M.F., Mont, F.W., Chhajed, S., Poxson, D.J., Kim, J.K., and Schubert, E.F. (2008) Design of multilayer antireflection coatings made from co-sputtered and low-refractive-index materials by genetic algorithm. *Opt. Express*, **16** (8), 5290–5298.

15. Rockstuhl, C., Fahr, S., and Lederer, F. (2008) Absorption enhancement in solar cells by localized plasmon polaritons. *J. Appl. Phys.*, **104** (12), 123–102, doi: 10.1063/1.3037239.

16. Pala, R.A., White, J., Barnard, E., Liu, J., and Brongersma, M.L. (2009) Design of plasmonic thin-film solar cells with broadband absorption enhancements. *Adv. Mater.*, **23** (10), 3504–3509, doi: 10.1002/adma.200900331.

17. Atwater, H.A. and Polman, P. (2010) Plasmonics for improved photovoltaic devices. *Nat. Mater.*, **9** (3), 205–213, doi: 10.1038/nmat2629.

18. Müller, J., Rech, B., Springer, J., and Vanecek, M. (2004) TCO and light trapping in silicon thin film solar cells. *Sol. Energy*, **77** (6), 917–930, doi: 10.1016/j.solener.2004.03.015.

19. Campbell, P. and Green, M.A. (1987) Light trapping properties of pyramidally textured surfaces. *J. Appl. Phys.*, **62** (1), 243–249, doi: 10.1063/1.339189.

20. Yablonovitch, E. (1982) Statistical ray optics. *J. Opt. Soc. Am. A (1917–1983)*, **72** (7), 899–907.

21. Yablonovitch, E. and Cody, G.D. (1982) Intensity enhancement in textured optical sheets for solar cells. *IEEE Trans. Electron Devices*, **29** (2), 300–305, doi: 10.1109/T-ED.1982.20700.

22. Green, M.A. (2002) Lambertian light trapping in textured solar cells and light-emitting diodes: analytical solutions. *Prog. Photovoltaics*, **10** (4), 235–241, doi: 10.1002/pip.404.

23. Kluth, O., Rech, B., Houben, L., Wieder, S., Schöpe, G., Beneking, C., Wagner, H., Löffl, A., and Schock, H.W. (1999) Texture etched ZnO:Al coated glass substrates for silicon based thin film solar cells. *Thin Solid Films*, **351** (1-2), 247–253, doi: 10.1016/S0040-60909900085-1.

24. Nicolay, S., Despeisse, M., Haug, F.J., and Ballif, B. (2011) Control of LPCVD ZnO growth modes for improved light trapping in thin film silicon solar cells. *Sol. Energy Mater. Sol. Cells*, **95** (3), 1031–1034, doi: 10.1016/j.solmat.2010.11.005.

25. Staebler, D.L. and Wronski, C.R. (1977) Reversible conductivity changes in discharge-produced amorphous Si. *Appl. Phys. Lett.*, **31** (4), 292–294, doi: 10.1063/1.89674.

26. Haug, F.J., Söderström, T., Cubero, O., Terrazzoni-Daudrix, V., and Ballif, C. (2009) Influence of the ZnO buffer on the guided mode structure in Si/ZnO/Ag multilayers. *J. Appl. Phys.*, **106** (3), 044–502, doi: 10.1063/1.3203937.

27. Berginski, M., Hüpkes, J., Schulte, M., Schöpe, G., Stiebig, H., Rech, B., and Wuttig, M. (2007) The effect of front ZnO:Al surface texture and optical

transparency on efficient light trapping in silicon thin-film solar cells. *J. Appl. Phys.*, **101** (7), 074–903, doi: 10.1063/1.2715554.
28. Steinhauser, J., Faÿ, S., Oliveira, N., Vallat-Sauvain, E., and Ballif, C. (2007) Transition between grain boundary and intragrain scattering transport mechanisms in boron-doped zinc oxide thin films. *Appl. Phys. Lett.*, **90** (14), 142–107, doi: 10.1063/1.2719158.
29. Taflove, A. and Hagness, S.C. (2000) *Computational Electrodynamics: The Finite-Difference Time-Domain Method*, 2nd edn, Artech House.
30. Berenger, J.P. (1994) A perfectly matched layer for the absorption of electromagnetic waves. *J. Comput. Phys.*, **114**, 185–200, doi: 10.1006/jcph.1994.1159.
31. Carniglia, C. (1979) Scalar scattering theory for multilayer optical coatings. *Opt. Eng.*, **18**, 104–115.
32. Krč, J., Zeman, M., Kluth, O., Smole, F., and Topič, M. (2003) Effect of surface roughness of zno: Al films on light scattering in hydrogenated amorphous silicon solar cells. *Thin Solid Films*, **426** (1), 296–304.
33. Dominé, D., Haug, F., Battaglia, C., and Ballif, C. (2010) Modeling of light scattering from micro-and nanotextured surfaces. *J. Appl. Phys.*, **107** (4), 044–504.
34. Bittkau, K., Beckers, T., Fahr, S., Rockstuhl, C., Lederer, F., and Carius, R. (2008) Nanoscale investigation of light-trapping in a-Si:H solar cell structures with randomly textured interfaces. *Phys. Status Solidi A*, **205** (12), 2766–2776, doi: 10.1002/pssa.200880454.
35. Rockstuhl, C., Lederer, F., Bittkau, K., and Carius, R. (2007) Light localization at randomly textured surfaces for solar-cell applications. *Appl. Phys. Lett.*, **91** (17), 171–104, doi: 10.1063/1.2800374.
36. Rockstuhl, C., Fahr, S., Bittkau, K., Beckers, T., Carius, R., Haug, F.J., Söderström, T., Ballif, C., and Lederer, F. (2010) Comparison and optimization of randomly textured surfaces in thin-film solar cells. *Opt. Express*, **18** (S3), A335–A341, doi: 10.1364/OE.18.00A335.
37. Fahr, S., Kirchartz, T., Rockstuhl, C., and Lederer, F. (2011) Approaching the Lambertian limit in randomly textured thin-film solar cells. *Opt. Express*, **19** (S4), A865–A874, doi: 10.1364/OE.19.00A865.
38. Rockstuhl, C., Fahr, S., Lederer, F., Haug, F.J., Söderström, T., Nicolay, S., Despeisse, M., and Ballif, C. (2011) Light absorption in textured thin film silicon solar cells: a simple scalar scattering approach versus rigorous simulation. *Appl. Phys. Lett.*, **98** (5), 051–102, doi: 10.1063/1.3549175.
39. Fahr, S., Rockstuhl, C., and Lederer, F. (2008) Engineering the randomness for enhanced absorption in solar cells. *Appl. Phys. Lett.*, **92** (17), 171–114, doi: 10.1063/1.2919094.
40. Gerchberg, R.W. and Saxton, W.O. (1972) A practical algorithm for the determination of the phase from image and diffraction plane pictures. *Optik*, **35**, 237–246.
41. Battaglia, C., Escarré, J., Söderström, K., Erni, L., Ding, L., Bugnon, G., Billet, A., Boccard, M., Barraud, L., de Wolf, S., Haug, F., Despeisse, M., and Ballif, C. (2011) Nanoimprint lithography for high-efficiency thin-film silicon solar cells. *Nano Lett.*, **11** (2), 661–665, doi: 10.1021/nl1037787.
42. Lin, A. and Phillips, J. (2008) Optimization of random diffraction gratings in thin-film solar cells using genetic algorithms. *Sol. Energy Mater. Sol. Cells*, **92** (12), 1689–1696, doi: 10.1016/j.solmat.2008.07.021.
43. Fischer, D., Dubail, S., Anna Selvan, J.A., Vaucher, N.P., Platz, R., Hof, C., Kroll, U., Meier, J., Torres, P., Keppner, H., Wyrsch, N., Goetz, M., Shah, A., and Ufert, K.D. (1996) The micromorph solar cell: extending a-Si:H technology towards thin film crystalline silicon. Conference Record of the 25th IEEE Photovoltaic Specialists Conference, Washington, DC, pp. 1053–1056, doi: 10.1109/PVSC.1996.564311.
44. Meier, J., Spitznagel, J., Kroll, U., Bucher, C., Faÿ, S., Moriarty, T., and Shah, A. (2004) Potential of amorphous and microcrystalline silicon solar cells. *Thin Solid Films*, **451–452**, 518–524, doi: 10.1016/j.tsf.2003.11.014.

45. Söderström, T., Haug, F.J., Terrazzoni-Daudrix, V., and Ballif, C. (2010) Flexible micromorph tandem a-Si/µc-Si solar cells. *J. Appl. Phys.*, **107** (1), 014–507, doi: 10.1063/1.3275860.
46. Söderström, T., Haug, F.J., Niquille, X., Terrazzoni, V., and Ballif, C. (2009) Asymmetric intermediate reflector for tandem micromorph thin film silicon solar cells. *Appl. Phys. Lett.*, **94** (6), 063–501, doi: 10.1063/1.3079414.
47. Bielawny, A., Üpping, J., and Wehrspohn, R.B. (2009) Spectral properties of intermediate reflectors in micromorph tandem cells. *Sol. Energy Mater. Sol. Cells*, **93** (11), 1909–1912, doi: 10.1016/j.solmat.2009.07.012.
48. Rockstuhl, C., Lederer, F., Bittkau, K., Beckers, T., and Carius, R. (2009) The impact of intermediate reflectors on light absorption in tandem solar cells with randomly textured surfaces. *Appl. Phys. Lett.*, **94** (21), 211–101, doi: 10.1063/1.3142421.
49. Fahr, S., Rockstuhl, C., and Lederer, F. (2010) The interplay of intermediate reflectors and randomly textured surfaces in tandem solar cells. *Appl. Phys. Lett.*, **97** (17), 173–510, doi: 10.1063/1.3509414.

5
Black Silicon Photovoltaics

Kevin Füchsel, Matthias Kroll, Martin Otto, Martin Steglich, Astrid Bingel, Thomas Käsebier, Ralf B. Wehrspohn, Ernst-Bernhard Kley, Thomas Pertsch, and Andreas Tünnermann

5.1
Introduction

Over the past years many research groups investigated lithographical [1, 2] and wet chemical etching technologies [3–6] to achieve a textured silicon surface with dimensions in the range of a few microns. Figure 5.1 shows a typical potassium hydroxide (KOH) etched silicon surface. As a result of this process, the reflectance of the silicon surface is reduced by multiple reflections at the textured front interface [7]. This effect is demonstrated in Figure 5.2. The first implementation of such surface structures was the COMSAT "non-reflective solar cell," published in 1975, which achieved an efficiency of 15.5% [8].

To avoid time consuming and costly chemical or lithographical processes many possible substituting techniques were investigated. Additional nanostructured antireflection (AR) coatings based on indium tin oxide nanowhiskers [9], porous silicon [123], as well as nanostructured silicon interfaces were discussed and demonstrated experimentally. Especially the latter case seems to be a promising approach to realize silicon interfaces with outstanding AR and light trapping properties by using cost-efficient processes.

In the following review chapter the name "black silicon" refers to all randomly structured silicon interfaces with lateral feature sizes in the submicron range and aspect ratios (structure height/lateral feature size) larger than one. Those surfaces exhibit a significantly reduced optical reflectance in the visible and near infrared spectral range and have a homogeneous black visual appearance. As the characteristic dimensions of those structures are in the range of the incident wavelength an adequate optical description in most cases requires wave optical methods.

5.1.1
Fabrication Methods

There are a variety of possibilities to achieve a black silicon structure at the silicon interface. From a historical point of view the development is a result of integrated

Photon Management in Solar Cells, First Edition. Edited by Ralf B. Wehrspohn, Uwe Rau, and Andreas Gombert.
© 2015 Wiley-VCH Verlag GmbH & Co. KGaA. Published 2015 by Wiley-VCH Verlag GmbH & Co. KGaA.

118 | 5 Black Silicon Photovoltaics

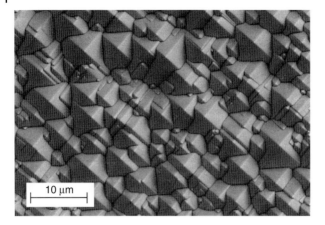

Figure 5.1 SEM image of a KOH textured silicon surface.

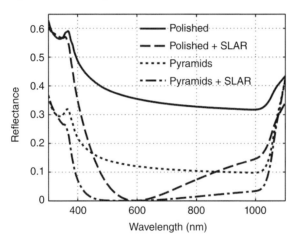

Figure 5.2 Calculated hemispherical reflectance of KOH etched pyramids on silicon with and without a single layer Si_3N_4 antireflection coating (SLAR).

circuit technologies, especially the structuring of silicon by plasma and dry etching processes [10–12]. One of the first works describing the formation of black silicon was published in 1979 by Schwartz and Schaible [13]. They reported the occasional appearance of nanostructured areas during plasma etching in a chlorine (Cl_2) plasma as an unwanted side effect. In the same year IBM patented the fabrication of black silicon by reactive ion etching (RIE) in sulfur hexafluoride (SF_6) and Cl_2 atmosphere with the focus on solar cell fabrication [14].

Five years later, Rothenberg and Kelly [15] reported the realization of black silicon-like structures by laser technologies. Owing to advances in the availability of laser sources, the interest in nanostructured silicon interfaces fabricated by ultra-short laser pulses strongly increased during the past years. It seems that it is possible to find a process window for many laser systems, starting from krypton fluoride (KrF) excimer laser at 248 nm [15], femtosecond laser systems at 390 nm [16] and 620 nm [17], pulsed Ti-sapphire lasers at 780 and 800 nm, as well as Nd:YAG systems at a wavelength of 1064 nm. Furthermore, the structuring can be achieved with femtoseconds, picoseconds, as well as nanosecond pulses [18–20] and a variety of ambient conditions [19].

The fabrication of nanostructured silicon by the usage of electrochemical and chemical techniques is already known since 1955 and 1960. Uhlir described the electrolytic shaping and achieved black or reddish surfaces [21]. Archer reported the chemical etching with a solution of hydrofluoric acid (HF) and nitric acid (HNO_3) [22]. During the following years the reaction kinetics of the so-called porous silicon were investigated and improved by many groups, for example, [23–26]. Owing to the luminescence of the nanoporous silicon in the visible spectrum, the research was mainly focused on the realization of light emitting devices [27]. A metal-assisted etching process in HF:HNO_3:H_2O solutions was published 1997 by Dimova-Malinovska *et al.* [28] and was further investigated during the past years, see for example, Li and Bohn [29], Koynov *et al.* [30, 31], Huang *et al.* [32, 33], or Kumar *et al.* [34–36].

5.1.2
Reactive Ion Etching

Especially RIE processes seem to be promising candidates for the structuring of solar cells and the implementation of an AR surface. In most cases the processes have a good reproducibility, excellent process control, and upscaling possibilities. In 1995, Jansen *et al.* [37] presented the so-called "black silicon method" for inductively coupled plasma (ICP) etching of silicon in a sulfur hexafluoride (SF_6)–fluoroform (CHF_3)–oxygen (O_2) atmosphere. However, the formation of black silicon is also possible in SF_6/O_2–atmosphere without CHF_3. In principle, fluorine radicals F^* are necessary for the formation of volatile species by etching the silicon substrate, such as silicon tetrafluoride SiF_4 [37, 38]. The plasma-generated oxygen radicals O^* form an atomic thin passivation layer of silicon oxyfluoride SiO_xF_y, which protects the surface against the etching with fluoride radicals. Sulfur fluoride ions SF_x^+ realize the physical sputtering of the SiO_xF_y layer and generate new etching points. Figure 5.3 shows the processes schematically. All processes together determine the geometry of the silicon interface.

The etching process does not start immediately after plasma ignition as the native oxide has to be removed by sputtering first. After that the structure formation might be further delayed by the time to build up a sufficiently high SF_x^+ concentration before some microparticles will precipitate the etch pit formation due to a trenching effect. After this initial delay, the etching depth increases linearly with time. Figure 5.4 shows the silicon surface after 2, 3, 5, and 10 min etching time in a SF_6/O_2 atmosphere.

The geometry of the black silicon structure depends strongly on the process parameters such as gas flows and process pressure. The influence of the oxygen

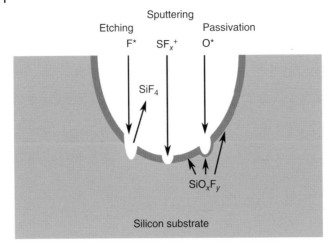

Figure 5.3 Schematic diagram of the ICP etching process with SF_6/O_2-plasma.

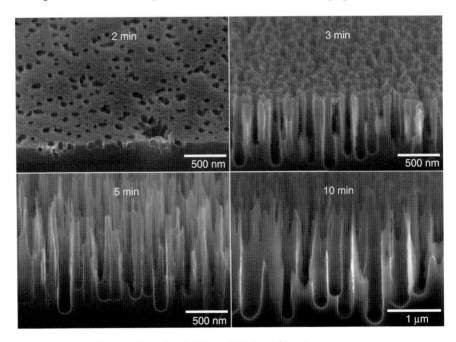

Figure 5.4 Silicon surface after 2, 3, 5, and 10 min etching time.

flow is illustrated in Figure 5.5. Increasing the flow raises the concentration of oxygen radicals. This leads to an enhanced passivation mechanism. Hence, the structure height decreases and even sidewalls can be obtained. With higher SF_6 flow the etching process increases and deep as well as more frayed structures appear. This effect is shown in Figure 5.6. Owing to the more isotropic etching, the silicon needles are partially opened at a sulfur hexafluoride flow of 129 sccm.

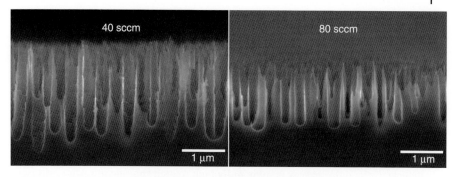

Figure 5.5 Black silicon surfaces prepared under different oxygen flows.

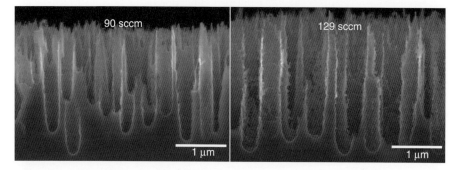

Figure 5.6 Black silicon surfaces prepared under different sulfur hexafluoride flows.

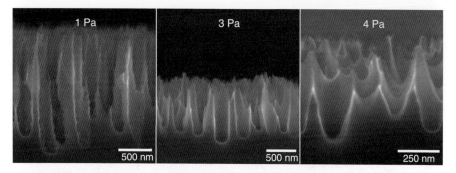

Figure 5.7 Black silicon surfaces prepared under different working pressures.

The steepness of the structures is influenced by the process pressure. To achieve pyramid-like structures with clearly positive sidewall slopes, the working pressure has to be increased. The resulting surface geometries for different values of the pressure are shown in Figure 5.7. With decreasing working pressure the mean free path of the particles increases. As a result the number of etching species from the plasma arriving at the silicon surface rises. Furthermore, the etching process is more directional.

5 Black Silicon Photovoltaics

As a result of the chemical and flow limited nature of the ICP black silicon process, the influence of ICP power, and RF BIAS on the structural properties is marginal.

5.1.3
Laser Processing

The structure evolution of black silicon under repeated pulsed laser irradiation can be understood as an interference effect. The tail of the incident pulse will interfere with the fraction of its pulse front that is scattered at present surface defects. If the pulse energy is sufficiently high to melt the silicon surface, this will create an inhomogeneous melt depth across the sample, shaped like a standing capillary wave (Figure 5.8a). Since the resolidification of the silicon surface occurs fast enough, this wave structure will freeze and form a characteristic ripple pattern, the so-called "Laser Induced Periodic Surface Structure" (LIPSS) [39]. During the irradiation with more and more pulses the ripple pattern changes into an amorphous, two-dimensional network of small beads (Figure 5.8b). These beads now concentrate the incident light into the valleys between them, giving an enhanced ablation here and leading to the final microstructure. The latter depends strongly on the applied laser parameters and ambient conditions, so

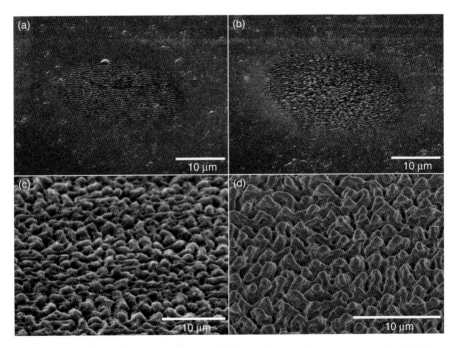

Figure 5.8 Black silicon evolution from LIPSS after 5 laser pulses (a) over amorphous network of small beads after 10 pulses (b) to final structures after 400 pulses (c/d). Structures (a–c) were fabricated in N_2, structure (d) in SF_6 atmosphere of 500 mbar.

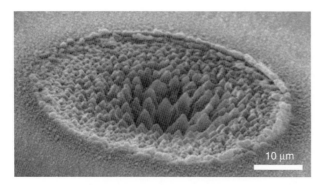

Figure 5.9 Surface morphology after irradiation of a stationary sample with 750 pulses (atmosphere of 500 mbar SF_6, laser wavelength 775 nm). The Gaussian intensity distribution of the pulse is reflected in the distribution of spike sizes and distances.

that a variety of structures with different characteristic lengths can be obtained, ranging from blisters to sharp, conical spikes (Figure 5.8c,d). In the following, the dependencies are discussed in principle.

Since the period of the LIPSS, forming in the initial stage of structure evolution, is proportional to the wavelength of the incident laser pulses, also the final microstructure (mean distance and height of structure features) scales coarsely with the wavelength [40, 41]. Similarly, an increase in laser fluence leads to a raised spike/blister separation and height (Figure 5.9) [42].

Additionally, as a result of repeated melting of the silicon surface, the ambient atmosphere under which the laser structuring is performed has a great influence on the developing microstructure. Her *et al.* [19] have shown that the silicon melt is highly susceptible to chemical reactions with a variety of background gases. Using Cl_2 or SF_6 as ambient gases they obtained conical microstructures with markedly sharp tips (Figure 5.8d). This can be understood as result of chemical reactions between silicon and volatile SF_6 fragments (e.g., SF_4, dissociated through high intensity laser pulse) or liquid silicon with Cl_2, respectively. Structuring in the presence of O_2 or N_2 leads to blistered structures and causes significant oxidation or nitridation of the microstructured surface.

Recently, the great interest in black silicon fabricated by femtosecond laser pulses arose because of an observation of Wu *et al.* in 2001 [43]. After structuring of silicon in the presence of SF_6, they noticed an additional, strong light absorptance in the sub-bandgap spectral range of silicon (>1100 nm; Figure 5.10). As could be shown later, this is a consequence of a strong non-equilibrium doping ("hyperdoping") of silicon with sulfur as a side-effect of the structuring process [44]. Lately, Sheehy *et al.* [45] and Smith *et al.* [46] demonstrated hyperdoping during laser structuring using solid powder and thin-film sources, eliminating the need for gaseous precursors and hence expanding the range of possible hyperdopants. Since strong doping to densities far above the limit of solid solubility changes the optical and electronical properties of the doped areas, this could allow for novel opto-electronic devices, especially for the spectral range above 1100 nm

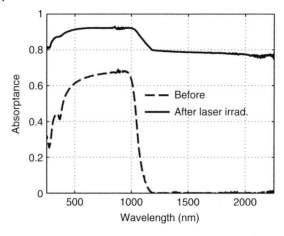

Figure 5.10 Absorptance spectra of a silicon wafer before (dashed line) and after (solid line) laser structuring in the presence of SF_6 (partial pressure 500 mbar, laser wavelength 775 nm, fluence 15 kJ m^{-2}, 400 laser pulses per area).

where the absorptance of silicon vanishes quickly. An exciting example is the photodetector by Carey [41] fabricated on a black silicon substrate hyperdoped with sulfur, which exhibits a responsivity of $10-100$ A W^{-1} for $\lambda = 600-1100$ nm and even a non-zero responsivity of several 10 mA W^{-1} up to 1600 nm.

5.1.4
Chemical and Electrochemical Etching

The formation of porous silicon in a solution of HF has been known for more than half a century. Galvanostatic etching of porous silicon under anodic bias in a HF containing electrolyte has first been reported by Uhlir in 1956 [21]. In the same year Fuller and Ditzenberger [47] described a method for purely chemical stain etching of the p doped region of silicon pn-junctions with a solution of HF and HNO_3. Stain films produced by this method were later investigated by Archer [22].

The interest in anodic etching of silicon increased strongly after Canham had demonstrated the fabrication of microporous silicon structures on p-doped substrates that exhibited photoluminescence (PL) in the visible spectral range [48]. The observed shift of the band-gap was attributed to a quantum confinement effect [48, 49]. Later a blue-shift of the PL from structures produced by anodic etching was also demonstrated for n-doped silicon and amorphous silicon [50, 51]. The reaction kinetics of the electrochemical etching (ECE) process were first described by Lehmann and Gösele [49]. They also pointed out that the presence of electron holes at the silicon surface is necessary for the reaction. An extended model for the etching mechanism was later given by Allongue et al. [52]. Review articles on the topic were written by Cullis et al. [53], and by Torres–Costa and Martín–Palma [54].

Currently, the most widely accepted model of the etching reaction assumes the initial passivation of the silicon surface with hydrogen followed by the breaking of the weak Si–H bonding and substitution with Si–OH by taking two holes into the process [54]. The replacing of OH$^-$ groups by F$^-$ ions leads to a strong polarization on the silicon atoms. Hence, HF molecules break the weakened Si–Si bonds. After the tearing of the silicon atom, the process starts again until it is interrupted. The morphology of the resulting porous silicon depends strongly on the electrolyte composition, the crystal orientation and the doping of the substrate [54]. On n-type substrates the injection of free holes by photogeneration was also found to be an important parameter [55].

The purely chemical etching of silicon in HF based solution relies, in principle, on the same chemical reactions as the anodic etching. The role of the electric current for hole injection, however, is replaced by an oxidizing agent, for example, HNO_3 or H_2O_2 [24]. A drawback of the chemical etching is that, depending on the oxidant, either the etching rate is slow or a delayed onset of the reaction occurs. A major improvement to this method therefore was made in 1997 when Dimova-Malinovska et al. [28] used a thin aluminum film to provide a fast start of the chemical etching of silicon in a HF:HNO_3:H_2O = 1 : 3 : 5 solution. Based on this finding, various oxidants and various noble metals, for example, in the form of nanoparticles, were used as a catalyst to achieve a fast etching reaction in later works. Those variants have become known as metal assisted chemical etching (MACE). A detailed review of the method was prepared by Huang et al. [33]. According to this article, the etching reacting can be understood as follows: The oxidant is preferentially reduced at the noble metal surface. The resulting holes then diffuse through the metal and are injected into the silicon. In the following, a reaction between the injected holes, HF, and silicon leads to the formation of water soluble H_2SiF_6. Because the hole density is highest at the metal–silicon interface, the etching rate reaches a maximum at the contact points between both materials. Figure 5.11 shows a single etching event with a gold particle; fabrication

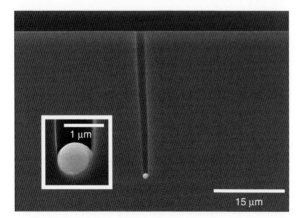

Figure 5.11 Single etching event in silicon with a gold particle. By courtesy of Michio Matsumura, Osaka University, Japan.

Figure 5.12 Silicon nanowire fabricated by HF/AgNO$_3$ solution. Reproduced with permission from Ref. [34].

details are published by Lee et al. [56, 57]. The most commonly used noble metals are Ag, Au, Pt, and Pd [33]. In most cases physical vapor deposition (PVD), such as thermal and electron beam evaporation [30, 31, 58] or sputtering [32] are used to deposit the metal. The use of spin coating was reported in Ref. [59]. A possible one-step method for this process is the usage of metal containing solutions, such as HF/AgNO$_3$ [60, 35, 61] or HF/KAuCl$_4$ [62]. Depending on the used technique it is possible to fabricate silicon nanowires with a length of more than 1 μm and lateral dimensions in the range of a 100 nm. Such a nanowire array is shown in Figure 5.12.

5.2
Optical Properties and Light Trapping Possibilities

5.2.1
Overview

As shown in the previous chapter, there is a wide variety of fabrication methods to modify the surface of silicon. All those methods have in common that they can reduce the reflectance of the silicon–air interface significantly. However, the structures that lead to this result can be vastly different. In the following, we give a brief overview of the results achieved with different fabrication methods.

The fabrication of black silicon by the treatment of the silicon surface with femtosecond laser pulses in an atmosphere of SF$_6$ was reported by different groups [43, 63]. The resulting features are quite large (several microns wide and more than 10 μm deep) and the lowest reported reflectance of those structures is less than 3%. Sarnet and coworkers [64] achieved black silicon structures by laser treatment even without the presence of SF$_6$. The geometrical characteristics are similar in both cases, but the features are reported to be more rounded without SF_6. Owing to the laser damage and the diffusion of impurities, laser treated black silicon usually exhibits a strong sub-bandgap absorption in the IR. As black silicon structures produced by femtosecond laser processes seem to be commonly much larger than the wavelength of the incident light, the AR effect of those structures can be understood in a geometric optical picture as a succession of multiple reflection and scattering events. The operation principle of such structures is thus quite similar to classical alkaline etched structures [7].

As mentioned before, black silicon structures with feature sizes in the sub-micron domain can be produced by MACE. Depending on the process parameters, this can produce either a dense nanowire array [65] or a porous silicon surface [30, 66]. While the nanowire arrays are usually several microns deep, the porous surface structures are much shallower with a depth of less than half a micron. However, in both cases the achievable average reflectance seems to be comparably low with reported values in the range of about 2–5%.

Black silicon structures produced by RIE with Cl_2 [67] or SF_6 and O_2 [68–70] also exhibit lateral features much smaller than 1 μm. The specific shape of the features depends strongly on the process parameters ranging from shallow sawtooth-like structures to long and steep needles (as described in the Section 5.1.1). The reported values of the average reflectance are in the range of 2–5%. Structures very similar to RIE black silicon were obtained by Sai and coworkers [71] by etching with an focused atom beam in a SF_6 atmosphere through a mask of porous Al_2O_3.

The AR properties of sub-micron black silicon structures cannot be explained in terms of geometrical optics. Instead they act as a gradient index AR coating [72]. If the lateral features of a structured interface are spaced considerably closer than the wavelength of the incident light, individual details are not resolved and the whole structure behaves like an effective medium with an average refractive index that increases with the material volume fraction. Provided that the thickness of this graded index layer is at least half a wavelength, such an interface exhibits almost zero reflectance [73]. Taking the absorbing spectral range of silicon, those requirements translate into characteristic lateral dimensions of less than 300 nm and a depth of about 500 nm. Those are indeed the typical dimensions of sub-micron black silicon structures with good AR properties. Similar values were also found by Boden and Bagnall [74]. They studied deterministic silicon moth-eye AR structures and found that a cone array with a period of about 200 nm and a depth of about 400 nm exhibits very good broadband AR properties.

Most publications on black silicon surfaces focus on optimizing the AR properties of the structures. The fact that black silicon might also be tuned to increase the absorption near the silicon absorption edge is often ignored. This light trapping can enhance absorption in the bulk silicon considerably, especially if one considers very thin substrates. In the following, we will have a closer look on various black silicon structures that were fabricated by RIE of silicon in an ICP with SF_6 and O_2 to elucidate this fact.

5.2.2
ICP-RIE Black Silicon

The fabrication of black silicon by an ICP-RIE process with SF_6 and O_2 etching gases offers multiple parameters to tune the geometry of the resulting structures. By adjusting the gas flows, the process pressure and the etching time, the lateral feature size and spacing, the structure depth and the sidewall angles of the black silicon structures can be varied in a wide range.

We used those tuning possibilities to fabricate four different black silicon structures on 400–450 µm thick silicon wafers. Cross-sections of those samples are shown in the top of Figure 5.13. At high process pressure the black silicon structures exhibit shallow sidewalls, a small lateral feature size and a low depth (type A and B). Reducing the process pressure leads to steeper sidewalls with needle-like features and both a larger feature size and depth (type C and D). The reflectance spectra in Figure 5.13 reveal that all structures have a residual hemispherical reflectance of less than 5% from the UV to the NIR. Owing to its small lateral size, structure A behaves like a prototypical gradient index AR coating; however, its low depth limits the performance in the NIR. The larger structures, especially type C and D, are not small enough to be understood as a pure effective medium. In those structures scattering and waveguiding are also expected to play an important role. Surprisingly, the

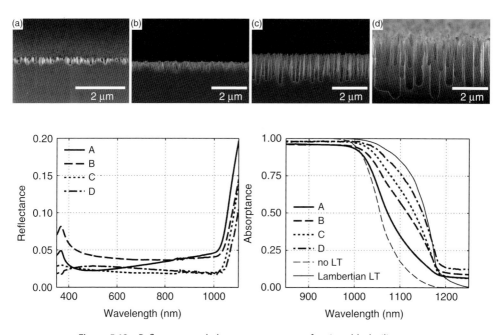

Figure 5.13 Reflectance and absorptance spectra of various black silicon structures produced by an ICP-RIE process. The wafer thickness was about 400 µm in all cases.

AR performance of the larger and steeper structures is even better than the performance of the gradient index structures, especially in the UV. The residual hemispherical reflectance of those is in the range of 2–3% throughout the spectrum.

The absorptance spectra in Figure 5.13 reveal that all investigated black silicon types show an increased absorption at the silicon absorption edge. This light trapping effect stems from the high silicon refractive index. The wavelength of light in silicon is at least 3.5 times smaller than the vacuum wavelength. Therefore, even small features might exceed the sub-wavelength limit. Additionally, due to the randomness of the structures, some special frequency features below cut-off are always present, which can promote forward scattering into propagating waves into the bulk silicon.

For comparison, two theoretical limits are included in the absorptance plot of Figure 5.13. The thin dashed line corresponds to the absorption of a planar silicon wafer with a reflectionless front interface but without any light trapping. The thin solid line corresponds to the light trapping due to ideal Lambertian forward scattering at the front interface. This limit is discussed extensively by Yablonovitch [75] and will be termed *Lambertian light trapping* in the following. The light trapping properties of the investigated black silicon structures cover the whole range between those limits and depend strongly on the feature size. To get a quantitative measure for the lateral size, we obtained top-view SEM images of all structures. From those we calculated the radial autocovariance function (ACV) and extracted the correlation length (taken as the first minimum of the ACV). Both the SEM images and the ACVs are shown in Figure 5.14. We find that the near band edge absorption increases monotonically with the correlation length of the structures. The small type A black silicon exhibits only a very weak light trapping and almost resembles the behavior of the planar wafer. This becomes intuitively clear if one considers that the correlation length of this structure is much smaller than the wavelength at the silicon absorption edge, even when scaled with the silicon refractive index. Therefore, hardly any scattering is expected in the NIR spectral domain. The type D black silicon on the contrary has a correlation length which is larger than the scaled wavelength. The performance of this structure is close to the Lambertian limit.

Of course one must be careful while reducing the dependence of the optical performance to a single geometrical figure of merit. Not only the correlation length but also the steepness of the features is considerably different between the structures of type A and type D. This might also have a strong influence on the scattering efficiency. However, from our findings we can at least deduce that a minimum correlation length in the order of the scaled wavelength is necessary for an efficient light trapping at the absorption edge. Up to now we have not found a structure where the light trapping performance decreases with increasing correlation length. Therefore, the question of the optimum structure remains unresolved.

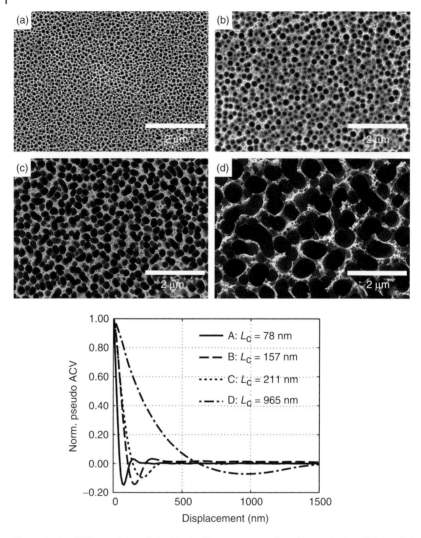

Figure 5.14 SEM top views of the black silicon structures from Figure 5.13 and the radial pseudo autocovariance functions that have been calculated from those images.

5.2.3
Influence of Dielectric Coatings

In the previous sections, the optical properties of black silicon were always discussed on the basis of structures with a bare silicon–air interface. However, a real world device usually requires the coating of the interface with some dielectric material either for the purpose of interface passivation or for the functionalization of the device (e.g., heterojunction formation). To take this circumstance into account we coated black silicon of type B (in the following termed *"shallow BS"*) and type C (in the following termed "*deep BS*") with two different coating methods and measured the optical reflectance and transmittance spectra.

First we coated the black silicon samples with Al_2O_3 by thermal atomic layer deposition (ALD), which is a highly conformal coating method. Absorptance spectra for various film thicknesses are shown in Figure 5.15. Additionally, we calculated the mean reflectance in the spectral range from 300 to 1000 nm and as a measure for the light trapping the mean absorptance in the range from 1000 to 1175 nm. The dependence of absorptance and mean reflectance on film thickness is shown in Figure 5.16. We find that thin layers up to 100 nm thickness have no negative effect on the AR properties of both structures. Thicker layers however cause an increased reflectance. The transition occurs when the coating layer is thick enough to fill the air voids between the black silicon features completely. This causes a smoothing of the interface and reduces the effective refractive index

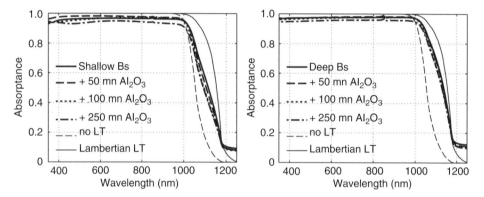

Figure 5.15 Influence of a conformal Al_2O_3 coating onto the optical properties of black silicon. Shallow BS corresponds to type B and deep BS corresponds to type C in Figure 5.13.

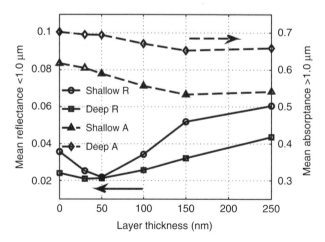

Figure 5.16 Dependence of the mean reflectance and the light trapping on the thickness of a conformal Al_2O_3 coating.

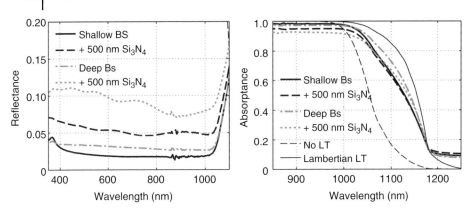

Figure 5.17 Influence of a thick non-conformal Si_3N_4 coating on the optical properties of the shallow (type B) and deep (type C) black silicon.

gradient. The light trapping decreases monotonically up to a layer thickness of about 150 nm. At thicker layers it stays constant. We assume this effect is due to the lowered refractive index contrast between the black silicon features and the background. This in turn reduces the scattering efficiency. Once all voids are filled, the contrast does not change anymore and the light trapping does not decrease any further. However, even in the worst case, the reduction of the light trapping performance is almost negligible.

In a second experiment we deposited a nominally 500 nm thick Si_3N_4 layer by sputtering. This coating method is highly non-conformal. The layer growth mostly takes place at the tips of the black silicon features while only little material is deposited onto the bottom of the pores of the black silicon. The effect is more pronounced the steeper and the deeper the structures are. Reflectance spectra of the black silicon samples, before and after sputtering, are shown in Figure 5.17. Both structures exhibit strongly increased reflection losses, with the deep black silicon being affected most. This is due to the formation of a thick Si_3N_4 layer with a smooth interface on top of the black silicon structure, which causes a strong Fresnel reflection. Except for the increased reflection losses, there is no additional negative effect on the light trapping properties of both samples. Thus we assume there is only little material deposited directly onto the black silicon features and most of the pores remain void. The scattering properties are therefore unaffected.

5.2.4
Influence of the Substrate Thickness and Limiting Efficiency

We have already demonstrated that ICP-RIE black silicon might exhibit an increased absorption in the NIR. In the following, we will address the question how this light trapping effect translates into an increased photocurrent if the thickness of the substrate is reduced. This is desirable for solar cell applications as it both saves silicon feedstock and reduces the non-radiative bulk recombination which, at least for high quality material, scales linearly with the wafer thickness

[76]. As an experimental variation of the wafer thickness in finely graded steps is quite complicated, we will rely on purely numerical results.

Numerical simulations of the black silicon absorption were performed using a synthetic model of the deep black silicon structure that was created on the basis of various cross-sectional and top-view SEM images. It has the same correlation length, depth, and sidewall angles as the real structure. A comparison of the synthetic model with the original black silicon structure is shown in the top part of Figure 5.18. The optical simulation was divided into two parts. In the spectral range from 350 to 800 nm all light will be absorbed in the substrate even if it is only 50 µm thin. In this spectral range the optical response is determined solely by the reflectance of the black silicon interface which was simulated with the finite difference time domain (FDTD) method [77]. Above 800 nm the propagation of the scattered light within the substrate has to be taken into account explicitly.

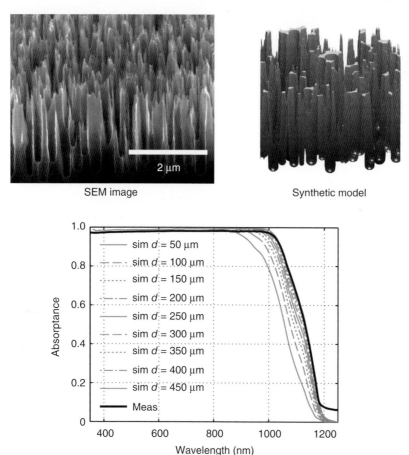

Figure 5.18 Comparison of the measured absorption spectrum with numerically calculated data for from a synthetic geometry model of the deep BS (type C). The distortion of the SEM image due to the inclined viewing angle has been corrected.

In this domain, the light scattering at the interface is simulated coherently with the Fourier modal method (FMM) [78] while the substrate is treated incoherently by a scattering matrix formalism [79] based on Mueller matrices and Stokes Vectors [80]. Details of the algorithm can be found elsewhere [81]. A comparison of the simulation with measured data is shown in the bottom of Figure 5.18. At a substrate thickness of 450 μm the measurement and the simulation are in excellent agreement.

The calculated absorption spectra were weighted by the AM1.5G solar photon flux and integrated. The resulting short circuit current density dependent on the substrate thickness is presented in Figure 5.19a. For comparison, the short circuit current densities for the case of no light trapping, Lambertian light trapping, and random pyramids with a single layer Si_3N_4 AR coating were calculated as well. Both the deep black silicon and the random pyramids exhibit nearly the same performance. At large substrate thickness, the additional light trapping only slightly outbalances the residual reflections of the real world structures. Therefore, the current gain compared to the reflectionless interface with no light trapping is less than 0.5 mA cm^{-2}. However, at lower substrate thicknesses the light trapping advantage becomes more pronounced and at 50 μm substrate thickness the current gain is almost 2 mA cm^{-2}.

From the short circuit current densities we calculated the limiting solar cell efficiency for a high quality solar cell with the so-called thin base approximation [76, 82]. This model assumes that recombination in the solar cell emitter is negligible and that the Fermi level splitting in the base of the solar cell is constant and equal to the applied voltage. With this assumption the steady state carrier concentrations and, hence, the impurity and Auger recombination rates can be calculated analytically. The recombination rates are converted into current densities by multiplying with the substrate thickness. Those currents increase

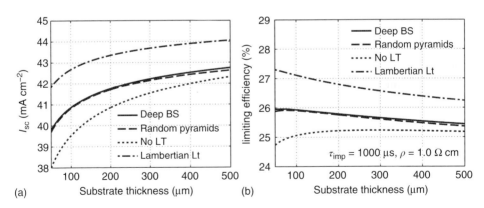

Figure 5.19 Comparison of the maximum short circuit current density (a) and the limiting efficiency (b) of different front interfaces: Deep ICP-RIE black silicon similar to structure C, random pyramids with a Si_3N_4 AR coating, a reflectionless interface without light trapping and a reflectionless interface with light trapping due to ideal Lambertian scattering.

with the applied voltage and counteract the current due to photo generation. From the resulting current–voltage dependence all solar cell parameters can be calculated. As the emitter and the surface recombination are neglected completely, the resulting efficiencies have to be understood as an upper bound for the theoretically achievable, and, not the technologically or economically reasonable.

The resulting limiting efficiencies for a p-doped substrate with a specific resistance of 1 Ω cm and a bulk lifetime of 1 ms are shown in Figure 5.19b. As it can be seen, the efficiencies of the structures with light trapping increase with decreasing substrate thickness while the efficiency without light trapping decreases. The reason for this behavior is that the light trapping sustains the photo current while the recombination losses decrease linearly with the substrate thickness. Without light trapping, however, the photo current decreases faster than the recombination losses and, therefore, the efficiency drops. The results reveal that both the deep black silicon as well as the random pyramids would enable efficient solar cells down to substrate thicknesses of 50 μm and probably even less. However, a direct comparison with the Lambertian light trapping also reveals that even larger gains are possible by further optimization of the light trapping.

5.3 Surface Passivation of Black Silicon

As discussed in the previous sections, various types of black silicon show excellent optical properties in comparison to common alkaline etched structures with AR coating [81]. However, the silicon surface is greatly enhanced and surface-near defects might be introduced by the etching process. As a result, a bare black silicon surface exhibits a strongly enhanced surface recombination rate for excess carriers.

As a rule of thumb, surface recombination will limit the efficiency of a solar cell if the surface recombination velocity gets in the order of the substrate thickness W divided by the minority carrier bulk lifetime τ_{bulk} [76],

$$S \approx \frac{W}{\tau_{bulk}}.$$

This means that already for bulk lifetimes in the order of a couple of tens of microseconds, moderate surface recombination velocities in the order of $S \approx 10^2$ cm s^{-1} would limit the solar cell efficiency. Hence, in order to take advantage of the optical benefits of black silicon, an efficient passivation of surface defects is crucial. However, this task is very challenging, as due to its extreme geometry black silicon is not easily treated with standard tools and processes established in the solar cell industry.

5.3.1
Requirements for Black Silicon Passivation

If a nanostructured surface shall be passivated effectively, a number of constraints are to be met. First, the material must be ideally suited to passivate a Si surface of preferably any type and doping level. Second, the rough structures must be covered highly conformal to avoid recombination at "hot spots" like inclusions, voids or pin-holes. Third, the process temperature should be kept rather low to avoid excessive diffusion of eventually remaining contaminants from the black silicon etching step. And finally, the coating should not significantly change the optical response of the black silicon surface. The latter point is usually fulfilled if conformality is achieved as it was already shown in the previous section. In general, an additional criterion will be that the amount of damage caused by the black silicon etching process itself has to be minimized. But since this requirement is not demanded by the passivation layer, it will be discussed later.

As a general rule, a good passivation of a Si surface (and the affected underlying bulk) requires a low density of defect states (traps) at the Si interface (D_{it}). This phenomenon is called *"chemical passivation."* In addition, a "field effect passivation" induced by a (preferably high) number of fixed charges ($\pm Q_f$) in the passivating layer is desired. The charge pushes one type of charge carriers away from the surface (preferably minorities) and thus, lowers the capture cross-sections, or recombination probability. Especially for nanostructured surfaces such as black silicon with large surface area and possibly damaged near-surface bulk, both passivation components should be present at a maximum level.

5.3.2
Possible Passivation Schemes

For the electronic passivation of Si surfaces, several materials and deposition techniques are available. In industrial solar cell production (mainly for processing p-type c-Si with phosphorus diffused n-type emitters), the front surface passivation is presently accomplished by deposition of a hydrogenated amorphous silicon nitride (SiN_x) layer. This material provides an acceptable level of passivation and, as unnecessary for black silicon, has the optimum refractive index for a single layer AR coating. SiN_x is usually deposited by plasma enhanced chemical vapor deposition (PECVD) at moderate to high temperatures (400–700 °C). The positive fixed charges in the film are attributed to so-called "K-centers" (Si atoms bound to 3 N atoms) and moderate charge densities of the order of $Q_f \sim +3 \times 10^{12}$ cm^{-2} are obtained [83]. However, PECVD is somewhat limited to the deposition on low aspect ratio structures. ICP-RIE structured black silicon may not easily be covered conformally by PECVD SiN_x as can be seen on the SEM images in Figure 5.20. While the shallow structure is covered quite well and some passivation was achieved, on the deep structure the PECVD process did not reach into the deep pores, which lead to poor passivation quality. Thus, PECVD deposited SiN_x is not the best choice for black silicon passivation, as

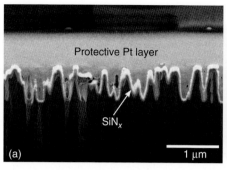
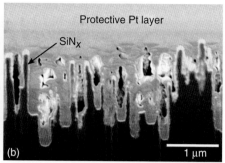

Figure 5.20 Cross-sectional images of a shallow (a) and a deep black silicon structure (b) with aspect ratios of AR ~3 and 10, respectively. Both were deposited with ~50 nm PECVD SiN$_x$. Whereas the shallow structure is well covered conformally, the deep structure exhibits almost no film at the pore bottoms.

conformality is not achieved on all structures, and the deposition temperature may be quite high.

In the same way, PECVD deposited SiO$_2$ films may be used for surface passivation. SiO$_2$ exhibits a lower density of recombination active states at the Si–SiO$_2$ interface than SiN$_x$ but also possesses less fixed charges ($Q_f \approx +1 \times 10^{12}$ cm^{-2}) [83]. Owing to its low refractive index, SiO$_2$ does not perform optically as well as SiN$_x$ on alkaline textures. Since both dielectrics carry positive fixed charges they are mainly suitable for passivation of n-type Si or lowly doped p-type substrates. For this type of material, the same disadvantages as for the discussed SiN$_x$ layers are expected.

Higher quality SiO$_2$ films may be obtained by clean thermal oxidation (wet in H$_2$O- or dry in O$_2$-atmosphere), which mainly leads to remarkably low defect densities of $D_{it} \approx 10^9 – 10^{10}$ cm^{-2}. Dry thermal oxidation was applied, for example, for the passivation of thin ($\approx$30 µm) macroporous Si membranes [84]. This demonstrates the capability of the method to produce highly conformal passivation layers. In the work of Yuan et al., a 20 nm thick thermal SiO$_2$ passivation layer led to the best black silicon solar cell result reported so far. They used density graded black silicon with a diffused phosphorus emitter, which was thermally oxidized and obtained an efficiency of 16.8% [85]. The rear side was "passivated" by an aluminum back surface field (Al-BSF). However, for sufficient thermal oxidation the samples have to be heated for some hours at high temperatures ($\approx$950–1100 °C) which are even higher than in PECVD processing. This may lead to accelerated diffusion of chemical impurities (e.g., Au or Ag, etc.) remaining from the black silicon etching step, and causes undesired low EQE (External Quantum Efficiency) response in the blue spectral region. At the same time, the lifetime may be compromised due to thermal degradation of the bulk. For an optimum SiO$_2$ passivation, an additional forming-gas annealing step (FGA) or annealing with a "sacrificial" Al layer [86] is needed. SiO$_2$ may be a good choice for the passivation of complex Si structures with high aspect ratios. However, due to the high temperature processing, the surfaces have to be very clean and free of defects.

Passivation by intrinsic hydrogenated amorphous Si (a-Si:H) deposited at moderate temperatures (300–400 °C) is another possibility. Even though, the material is actually a semiconductor, a relatively high level of passivation may be achieved. Very low surface recombination velocities (well below 2 cm s^{-1} in the as-deposited state) have been reported for 1.5 Ω cm p-doped Si [87]. In combination with suitable dopants, the layer may replace the classic emitter in a so-called "heterojunction with intrinsic thin layer (HIT)" solar cell [88]. High efficiencies are obtained by simultaneous junction formation and surface passivation. The material was used to passivate plasma structured black silicon yielding surface recombination velocities of $S \approx 50$ cm s^{-1} [89]. The HIT solar cell concept was also demonstrated to work on the previously mentioned macroporous black silicon membranes [90] where an efficiency of 7.2% was achieved. Hence, a-Si offers some passivation quality even on rather rough surfaces. However, since it is deposited by PECVD, the performance is somewhat limited if the surface aspect ratio becomes too large.

Finally, a material with outstanding passivation capabilities is Al_2O_3. The passivation of Si with Al_2O_3 deposited by pyrolysis was first reported by Hezel and Jaeger [91]. However, their results found no attention in the PV community for almost two decades. In 2006, the very effective surface passivation of Al_2O_3 was (re-) discovered by Agostinelli et al. and Hoex et al. [92, 93]. Both deposited Al_2O_3 by thermal- or plasma-enhanced (PE) ALD. ALD is a specific form of the chemical vapor deposition (CVD) [94], wherein the precursor (trimethylaluminum (TMA) for Al_2O_3) and oxidant (H_2O, O_3, or O_2 for PEALD) reactions are separated either in time or in space [95]. The Al_2O_3 process was described in great detail by Puurunen [94]. During one ALD deposition cycle, at maximum, one monolayer of the desired material is deposited on the surface. This gives subnanometer thickness control and leads to a highly conformal and homogeneous film growth. The growth per cycle mainly depends on the used precursors and deposition temperature. The reaction of TMA and H_2O to Al_2O_3 films is remarkably robust and occurs in a wide "ALD window" from room temperature up to $T_{dep} \approx 300$ °C. Best passivation results are usually obtained for films grown at $T_{dep} \approx 200$ °C. This makes the material compatible with the requirements of conformal growth and low temperature processing.

Recently, Al_2O_3 films were extensively studied and meanwhile first industrial ALD-tools are being developed and installed [96]. However, very good passivation results are obtained only if the Al_2O_3 layers, as in the case of SiO_2, receive a post deposition anneal (PDA). In contrast to annealing pure SiO_2, where a FGA is needed, the ambient does not play a role for Al_2O_3 annealing. Al_2O_3 is an insulator that induces a rather high density of negative fixed charges ($Q_f \approx 5 \cdot 10^{12} - 1 \cdot 10^{13}$ cm^{-2}) at an interface with Si and/or SiO_2. Even if deposited on Si–H terminated surfaces, at least one monolayer of SiO_x is formed during deposition. Thus, the benefits of strong field effect passivation with a rather low interface defect density of states ($\approx 10^{11}$ eV^{-1} cm^{-2}) are combined in this material. The good chemical passivation of dangling bonds is related to strong coordination of Si and excess oxygen during annealing [97] and additional selective hydrogenation of the near-subsurface bulk [98]. In general, the material is capable of passivating all kinds of doped c-Si including almost all doping levels [99–101]. A rather comprehensive review on the passivation properties of ALD deposited Al_2O_3 films may be found in Ref. [83]. In summary, ALD is an excellent method for depositing perfectly conformal films on (almost) any substrate with outstanding homogeneity and thickness control. In combination with its excellent passivation quality on Si surfaces, ALD deposited Al_2O_3 is an ideal candidate for the passivation of black silicon. All the mandatory requirements as defined above are integrated in one specific material/process combination. Thermal ALD deposited Al_2O_3 was successfully applied for the passivation of ICP-RIE structured black silicon surfaces by Otto et al., who report surface recombination velocity values of $S < 13 \text{ cm s}^{-1}$ for the nano-structured samples, whereas on a polished reference wafer values of $S < 12 \text{ cm s}^{-1}$ were obtained [102].

5.3.3
Passivation of Black Silicon Surfaces

5.3.3.1 Surface Damage and Sample Cleaning

After black silicon etching some kind of cleaning step has to be performed. Otherwise an effective passivation cannot be obtained. Usually, wet chemical cleaning is applied to remove residues from the surface. Contaminants may be, for example, shallowly implanted ions from the ICP-RIE process, atoms forced into the surface by laser structuring, or metallic remains after MACE with Au or Ag catalytic particles.

Hence, the enlarged surface area is not the only source of elevated recombination in such materials. In addition to extrinsic defects, also intrinsic defects may be introduced, for example, Si-vacancies originating from UV irradiation and high energetic ions during plasma etching or high laser light intensities that may even lead to amorphization [103]. Even during wet chemical processing, intrinsic defects may be formed that likely diffuse into the (near-surface) bulk. Of course, it is difficult to distinguish the different sources of recombination. However, H-atoms from a passivation layer as well as a strong field effect might help working around the issue of near-surface damage if the repulsing field keeps minority carriers far enough from the Si surface.

In ICP etched black silicon samples passivated by PECVD-SiN_x a damage removal etching step (DRE) could increase the lifetime by up to one order of magnitude (from ≈ 3 to $\approx 30\,\mu s$). The samples were etched with an acidic solution of HNO_3:HF (50:1) for 8 min. On the other hand, the effective lifetime of the reference sample without black silicon was still another order of magnitude larger than in the case of the DRE treated sample. Furthermore, a DRE usually has a negative impact on the black silicon optics. Zaidi et al. [69] also reported a certain improvement of cell performance upon DRE by KOH for a similar ICP process. The only way to essentially inhibit these defects is to adjust the nanostructuring process to a regime where minimal damage is introduced right from the beginning. Schaefer and Lüdemann reported on the optimization of an ICP-RIE etching step, aiming to dry-clean unstructured surfaces before solar cell processing [104]. This idea was adopted by Otto et al. [102], who found consistently that the damage during the first few minutes of ICP-RIE processing (e.g., induced by plasma ignition) may be reduced by long etching times. According to the model for ICP-RIE damage on planar Si presented by Schaefer, a film of etching residues may cover the surface [104]. About the first 10 nm may be heavily damaged, and the first 50 nm might be "modified." Thus, for effective surface passivation the etching parameters, such as etching time, process pressure, plasma power, self bias, etchant concentrations, and ratios might all play an important role.

Prior to the passivation of black silicon surfaces with conformal ALD deposited Al_2O_3 a standard RCA [105] cleaning procedure is recommended [102]. This way, only a few nanometers of material are removed, such that the structure is conserved and residues from the black silicon etching are removed including some of the heavily damaged zone. In the following, an example for the successful passivation of black silicon will be given.

5.3.3.2 Effective Passivation of ICP-RIE Black Silicon

To accurately determine the charge carrier lifetime and effective surface recombination velocity, the sample must be prepared identically on both sample sides. In the following example, bifacial black silicon etching was accomplished by ICP-RIE processing with the parameter variation indicated in Table 5.1. The structures differ in process pressure and etching time, resulting in two different geometries (shallow and deep) as described in the previous sections. The third sample is a so-called "intermediate" structure possessing approximately the shallow geometry, and only differing from the shallow structure in a longer etching time $t_2 > t_1$. All samples were cleaned in a standard RCA procedure starting with SC1 (5 : 1 : 1 solution of $H_2O:NH_4OH:H_2O_2$) followed by an HF dip and a SC2 (5 : 1 : 1 solution of $H_2O:HCl:H_2O_2$). The cleaning steps SC1 and SC2 were carried out at a temperature of 80 °C for 10 min. After each step the samples were rinsed in deionized water. After cleaning, the samples were dried and transferred into the ALD reactor, where a 100 nm thick layer of Al_2O_3 was deposited bifacially. Finally, the samples were annealed in a furnace at 400 °C for 30 min. Cross-sectional SEM images of the samples are shown in Figure 5.21. The 100 nm thick Al_2O_3 layer is wrapped around the black silicon needles conformally, even in the 1.7 μm deep pores of the deep structure. Hence, the requirement of conformal deposition is fulfilled. The discussion in the optics section showed that the optical response of the samples is hardly changed for thin Al_2O_3 films on black silicon surfaces.

There are many spectroscopy techniques to measure the minority charge carrier lifetime in semiconductors. A good description for commonly used methods can be found in Ref. [106]. The lifetimes of the samples in this example were determined by the quasi steady state photo conductance (QSSPC) method. The data presented in Figure 5.22 show the calculated surface recombination velocity of all three black silicon structures and a polished reference wafer that is dependent on the injection level after annealing. The surface recombination velocity can be

Table 5.1 ICP Process parameters for the three black silicon lifetime samples ($t_1 < t_2$; $p_1 < p_2$) and the resulting surface recombination velocities S_{eff} for the described passivation method by ALD deposited Al_2O_3.

Structure	Etching time	Process pressure	τ_{eff} (ms)	S_{eff} (cm s^{-1})
Polished reference	—	—	1.63	12
Shallow black silicon	t_1	p_1	0.50	45
Intermediate black silicon	t_2	p_1	1.48	13
Deep black silicon	t_1	p_2	0.24	80

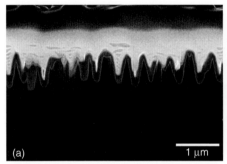 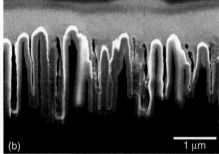

Figure 5.21 Cross-section of ICP-RIE structured black silicon samples covered with a perfectly conformal layer of Al_2O_3. The SEM images show an "intermediate" (a) and a "deep" (b) black silicon structure. The cross-sections were prepared by a focused ion beam (FIB).

calculated for a material with a high bulk lifetime by neglecting the bulk recombination (e.g., assuming that $\tau_{\text{bulk}} = \infty$). With this an upper bound for the surface recombination is given by

$$S \leq \frac{W}{2\tau_{\text{eff}}}$$

where the factor of 2 accounts for recombination on either side of the (identically structured and passivated) sample. A detailed derivation for this formula was given by Rein [106].

We compare the surface recombination velocity at an injection level of 5×10^{15} cm^{-3}. The polished sample exhibits the lowest recombination rate (i.e., longest lifetime). On the intermediate sample we achieve nearly the same value, indicating a similarly good passivation quality. This is a remarkable achievement, since the surface area is strongly increased and the discussed damage might have been introduced by ICP-RIE structuring. The shallow structure, possessing the geometrical same characteristics, exhibits a more than three times higher surface recombination velocity. Thus, the additional recombination can only be attributed to a higher level of impurities, for example, damage in the near-surface region. We assume that those have been reduced in the intermediate sample due to the longer etching time. It should be taken into account that the samples were

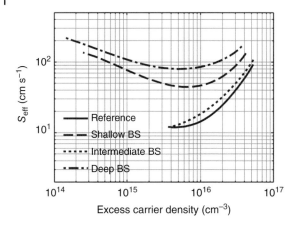

Figure 5.22 Surface recombination velocity of three different black silicon samples and a polished reference.

not identically doped, which also causes slightly different recombination rates. But the effect is not limiting the bulk lifetime, which was assumed to be infinite. The deep structure's surface recombination velocity, again, is a factor of two higher than for the shallow structure. This can be explained by the lower process pressure. Lower pressure means a longer mean free path of ions traveling in the plasma. Thus, plasma species impinge on the surface with higher energies. Also the higher aspect ratio of the deep surface structures originates from the higher physical etching component due to lower plasma pressure. From this observation we conclude that minimizing the damage induced by ICP-RIE processing is of the same importance as the passivation scheme itself.

5.4
Black Silicon Solar Cells

Owing to the multiplicity of black silicon fabrication methods as well as homo- and heterojunction solar cell concepts, the following paragraph should give a short review of published black silicon solar cells.

The realization of an efficient solar cell by laser treatment seems very challenging as the method causes an amorphization and a heavy doping of the silicon surface. Nayak *et al.* used a Ti:sapphire laser system for the fabrication of nanostructured silicon. By using three cleaning steps, (i) NaOH solution, (ii) isotropic etching in HNO_3:CH_3COOH:HF (30:10:4), and (iii) IMEC-clean [107], they were able to achieve an efficiency of 14.2% [63].

Kumar *et al.* realized a silicon nanowire solar cell by etching with HF/HNO3 solution. Afterwards a n+− p−p+ junction was fabricated by conventional wafer-based technologies. The resulting efficiency was in the range of 13.7% [34]. Yuan *et al.* [85] published a black silicon concept without an additional AR layer that demonstrated an efficiency of 16.8%. In this work a $HAuCl_4$:HF:H_2O_2:H_2O

solution was used for the etching process. By the implementation of multi-scale structures, that is, pyramid structures plus black silicon, the efficiency of this concept increased further to 17.1% [108].

Brendel et al. used porous silicon produced by ECE as a separation layer for a transfer process adapted from Yonehara et al. [109] and Tayanaka [110]. With this they were able to realize a monocrystalline silicon solar cell of only 25 µm thickness with an efficiency of 15.4% [111]. The resulting device however had a classical KOH front texture for light trapping. The porous silicon was only used for the separation of the cell from the substrate and was removed afterwards. A 43 µm thick solar cell made with the same process even reached an efficiency of 19.1% [112].

A very broad study on the prospects of the use of porous silicon as an AR coating for silicon solar cells was carried out by Bilyalov et al. [113]. They optimized the formation of porous silicon by chemical stain etching as well as by ECE to replace the standard emitter formation process for low cost multicrystalline silicon solar cells. They reported maximum efficiencies of 14.1 and 13.2% for chemical stain etching and ECE, respectively. Notably they were also able to demonstrate that the porous silicon layer alone has a sufficiently high passivation quality to make additional passivation layers superfluous for solar cells on low quality multicrystalline bulk material. Another multicrystalline Si solar cell that employed electrochemically etched porous silicon as front AR coating with an efficiency of 13.3% was demonstrated by Lipiński et al. [114]. Kim and coworkers [115] used ECE to form a 100 nm thin porous AR layer on top of a KOH texture on a crystalline Si solar cell. With this they were able to achieve an efficiency of 13.6% .

Jia et al. [116] fabricated a HIT silicon nanowires solar cell produced by MACE with an efficiency of 7.29%. A hybrid system consisting of a n-type Si nanowire array and a spin-coated conductive polymer was published by Garnett el al. [117]. The resulting device was only 4.5 µm thick and reached an efficiency of 6.1%. A hybrid Si nanowire solar cell on a thick Si wafer was presented by Shen et al. [118] and reached an efficiency of 9.7%. In this case the polymer spiro-OMeTAD was used to generate the heterojunction.

Another concept for a cost-effective fabrication of heterojunction solar cells on black silicon was presented by the authors [119, 120]. In this case a sub-nm thin insulator and a layer of a transparent conductive oxide were coated onto black silicon fabricated by RIE-ICP to form a semiconductor–insulator–semiconductor (SIS) system. The nanostructure was the same as the black silicon in the right picture of Figure 5.7. Owing to the poor quality and homogeneity of the thin insulator layer at the nanostructured surface, the open-circuit voltage was in the range of 427 mV and the efficiency was about 7.9% [120].

Finally, solar cells on RIE black silicon with a classical diffused emitter were demonstrated by Yoo et al. Thereby, efficiencies of 15.1 and 16.1% were achieved on RIE etched crystalline and multicrystalline silicon, respectively [70, 121].

A summary of the efficiencies that were reported for various types of black silicon solar cells is presented in Table 5.2.

In summary, the presented fabrication methods are able to realize acceptable solar cells. As a result of the excellent light trapping, high short-circuit currents

Table 5.2 Reported black silicon solar cells efficiencies and performance parameters under global AM1.5 spectra.

Black silicon process	Substrate	J_{sc} (mA/cm²)	V_{OC} (mV)	FF (%)	η (%)	References
fs-laser	c-Si	39.2	507	71.4	14.2	[63][a]
MACE	c-Si	37	544	68	13.7	[34][a],[b]
	c-Si	28	531	36	5.3	[34][a]
MACE	c-Si	27	476	56.2	7.29	[116][a]
MACE	c-Si	34.6	618	77.4	16.6	[57][a]
MACE	c-Si	34.1	612	80.6	16.8	[85][a]
MACE	c-Si	35.6	615	78.2	17.1	[108][c]
MACE	Thin c-Si	21.2	≈500	—	6.1[d]	[117][a],[e]
MACE	c-Si	31.3	527	58.8	9.7	[118][a]
ECE	Thin c-Si	37.8	650	77.6	19.1	[112][c],[f]
ECE	mc-Si	29.6	585	77	13.3	[114][a]
ECE	mc-Si	30.4	583	72.2	13.2	[113][a]
Chem. etch.	mc-Si	29.1	609	79.7	14.1	[113][a]
RIE	c-Si	36	427	51.6	7.94	[120][a]
RIE	c-Si	32.5	610	77	15.1	[70][a]
RIE	mc-Si	33.5	619	77.7	16.1	[121][a]

a) Laboratory measurement.
b) Black silicon only in the active region.
c) Independently confirmed.
d) Estimated from LBIC measurements.
e) 4.5 μm absorber.
f) KOH front texture, porous silicon used only for layer transfer.

are a common property of almost all cell concepts. The main disadvantage is the low open-circuit voltage compared to state-of-the-art technologies because of high surface recombination. However, it has been demonstrated that low surface recombination values and long carrier lifetimes can be achieved with novel passivation schemes.

On the roadmap for wafer-based crystalline silicon solar cells, the reduction of the wafer thickness is an important step to reduce material cost [122]. Hence, the implementation of nanostructured light trapping surfaces is a fundamental issue. Future technologies should therefore concentrate on the passivation problem and the development of cost-efficient solutions to achieve high efficient solar cells on black silicon substrates.

References

1. Zhao, J., Wang, A., Altermatt, P.P., Wenham, S., and Green, M.A. (1996) 24% efficient PERL silicon solar cell: recent improvements in high efficiency silicon cell research. *Sol. Energy Mater. Sol. Cells*, **41–42**, 87–99. doi: 10.1016/0927-0248(95)00117-4
2. Bläsi, B., Hauser, H., Höhn, O., Kübler, V., Peters, M., and Wolf, A.J. (2011) Photon management structures

originated by interference lithography. *Energy Procedia*, **8**, 712–718. doi: 10.1016/j.egypro.2011.06.206
3. Lee, D.B. (1969) Anisotropic etching of silicon. *J. Appl. Phys.*, **40** (11), 4569–4574. doi: 10.1063/1.1657233
4. Sopori, B.L. and Pryor, R.A. (1983) Design of antireflection coatings for textured silicon solar cells. *Solar Cells*, **8** (3), 249–261. doi: 10.1016/0379-6787(83)90064-9
5. Baraona, C.R. and Brandhorst, H.W. (1975) V-grooved silicon solar cells. 11th IEEE Photovoltaic Specialists Conference, Phoenix, AZ.
6. King, D.L. and Buck, M.E. (1991) Experimental optimization of an anisotropic etching process for random texturization of silicon solar cells. Proceedings of the 22nd IEEE Photovoltaic Specialists Conference, Las Vegas, NV, pp. 303–308. doi: 10.1109/PVSC.1991.169228.
7. Baker-Finch, S.C. and McIntosh, K.R. (2011) Reflection of normally incident light from silicon solar cells with pyramidal texture. *Prog. Photovoltaics*, **19** (4), 406–416. doi: 10.1002/pip.1050
8. Allison, J.F., Arndt, R., and Meulenberg, A. (1975) A comparison of the COMSAT violet and non-reflective solar cell. *COMSAT Tech. Rev.*, **5**, 211–224.
9. Chang, C.-H., Peichen, Y., Hsu, M.-H., Tseng, P.-C., Chang, W.-L., Sun, W.-C., Hsu, W.-C., Hsu, S.-H., and Chang, Y.-C.G. (2011) Combined micro- and nano-scale surface textures for enhanced near-infrared light harvesting in silicon photovoltaics. *Nanotechnology*, **22** (9), 095201. doi: 10.1088/0957-4484/22/9/095201
10. Endo, N. and Kurogi, Y. (1980) 1-µm MOS process using anisotropic dry etching. *IEEE J. Solid-State Circuits*, **15** (4), 411–416. doi: 10.1109/JSSC.1980.1051414
11. Perry, A.J. and Boswell, R.W. (1989) Fast anisotropic etching of silicon in an inductively coupled plasma reactor. *Appl. Phys. Lett.*, **55** (2), 148–150. doi: 10.1063/1.102127
12. D'Emic, C.P., Chan, K.K., and Blum, J. (1992) Deep trench plasma etching of single crystal silicon using SF6/O2 gas mixtures. *J. Vac. Sci. Technol., B*, **10** (3), 1105–1110. doi: 10.1116/1.586085
13. Schwartz, G.C. and Schaible, P.M. (1979) Reactive ion etching of silicon. *J. Vac. Sci. Technol.*, **16** (2), 410. doi: 10.1116/1.569962
14. Hansen, T.A. Johnson, C. Jr., and Wilbarg, R.R. (1980) Method for fabricating non-reflective semiconductor surfaces by anisotropic reactive ion etching. Patent Number 4,229,233, Issued October 21, 1980.
15. Rothenberg, J.E. and Kelly, R. (1984) Laser sputtering. Part II. The mechanism of the sputtering of Al2O3. *Nucl. Instrum. Methods Phys. Res., Sect. B*, **1** (2–3), 291–300. doi: 10.1016/0168-583X(84)90083-1
16. Chichkov, B.N., Momma, C., Nolte, S., von Alvensleben, F., and Tünnermann, A. (1996) Femtosecond, picosecond and nanosecond laser ablation of solids. *Appl. Phys. A*, **63** (2), 109–115. doi: 10.1007/s003390050359
17. Kautek, W. (1994) Femtosecond pulse laser ablation of metallic, semiconducting, ceramic, and biological materials. *Proc. SPIE*, **2207**, 600–611. doi: 10.1117/12.184768
18. Medvid, A., Onufrijevs, P., Dauksta, E., and Kyslyi, V. (2011) 'Black silicon' formation by Nd:YAG laser radiation. *Adv. Mater. Res.*, **222**, 44–47. doi: 10.4028/www.scientific.net/AMR.222.44
19. Her, T.-H., Finlay, R.J., Wu, C., Deliwala, S., and Mazur, E. (1998) Microstructuring of silicon with femtosecond laser pulses. *Appl. Phys. Lett.*, **73** (12), 1673. doi: 10.1063/1.122241
20. Zhu, X.N., Zhu, H.L., Liu, D.W., Huang, Y.G., Wang, X.Y., Yu, H.J., Wang, S., Lin, X.C., and De Han, P. (2011) Picosecond laser microstructuring for black silicon solar cells. *Adv. Mater. Res.*, **418–420**, 217–221. doi: 10.4028/www.scientific.net/AMR.418-420.217
21. Uhlir, A. Jr., (1956) Electrolytic shaping of germanium and silicon. *Bell Syst. Tech. J.*, **35** (2), 333–347.
22. Archer, R.J. (1960) Stain films on silicon. *J. Phys. Chem. Solid*, **14**, 104–110. doi: 10.1016/0022-3697(60)90215-8

23. Unagami, T. and Seki, M. (1978) Structure of porous silicon layer and heat-treatment effect. *J. Electrochem. Soc.*, **125**, 1339–1344.
24. Fathauer, R.W., George, T., Ksendzov, A., and Vasquez, R.P. (1992) Visible luminescence from silicon wafers subjected to stain etches. *Appl. Phys. Lett.*, **60** (8), 995–997. doi: 10.1063/1.106485
25. Smith, R.L. and Collins, S.D. (1992) Porous silicon formation mechanisms. *J. Appl. Phys.*, **71** (8), R1–R22. doi: 10.1063/1.350839
26. Hamilton, B. (1995) Porous silicon. *Semicond. Sci. Technol.*, **10** (9), 1187–1207. doi: 10.1088/0268-1242/10/9/001
27. Buriak, J.M. (2006) High surface area silicon materials: fundamentals and new technology. *Philos. Trans. R. Soc. Lond. A*, **364** (1838), 217–225. doi: 10.1098/rsta.2005.1681
28. Dimova-Malinovska, D.M., Sendova-Vassileva, N.T., and Kamenova, M. (1997) Preparation of thin porous silicon layers by stain etching. *Thin Solid Films*, **297** (1–2), 9–12. doi: 10.1016/S0040-6090(96)09434-5
29. Li, X. and Bohn, P.W. (2000) Metal-assisted chemical etching in HF/H2O2 produces porous silicon. *Appl. Phys. Lett.*, **77** (16), 2572–2574. doi: 10.1063/1.1319191
30. Koynov, S., Brandt, M.S., and Stutzmann, M. (2006) Black nonreflecting silicon surfaces for solar cells. *Appl. Phys. Lett.*, **88** (20), 203107. doi: 10.1063/1.2204573
31. Koynov, S., Brandt, M.S., and Stutzmann, M. (2007) Black multi-crystalline silicon solar cells. *Phys. Status Solidi RRL*, **1** (2), R53–R55. doi: 10.1002/pssr.200600064
32. Huang, Z., Zhang, X., Reiche, M., Liu, L., Lee, W., Shimizu, T., Senz, S., and Gösele, U. (2008) Extended arrays of vertically aligned sub-10 Nm diameter [100] Si nanowires by metal-assisted chemical etching. *Nano Lett.*, **8** (9), 3046–3051. doi: 10.1021/nl802324y
33. Huang, Z., Geyer, N., Werner, P., de Boor, J., and Gösele, U. (2011) Metal-assisted chemical etching of silicon: a review. *Adv. Mater.*, **23** (2), 285–308. doi: 10.1002/adma.201001784
34. Kumar, D., Srivastava, S.K., Singh, P.K., Husain, M., and Kumar, V. (2011) Fabrication of silicon nanowire arrays based solar cell with improved performance. *Sol. Energy Mater. Sol. Cells*, **95** (1), 215–218. doi: 10.1016/j.solmat.2010.04.024
35. Srivastava, S.K., Kumar, D., Singh, P.K., Kar, M., Kumar, V., and Husain, M. (2010) Excellent antireflection properties of vertical silicon nanowire arrays. *Sol. Energy Mater. Sol. Cells*, **94** (9), 1506–1511. doi: 10.1016/j.solmat.2010.02.033
36. Kumar, D., Srivastava, S.K., Singh, P.K., Sood, K.N., Singh, V.N., Dilawar, N., and Husain, M. (2009) Room temperature growth of wafer-scale silicon nanowire arrays and their Raman characteristics. *J. Nanopart. Res.*, **12** (6), 2267–2276. doi: 10.1007/s11051-009-9795-7
37. Jansen, H., de Boer, M., Legtenberg, R., and Elwenspoek, M. (1995) The black silicon method: a universal method for determining the parameter setting of a fluorine-based reactive ion etcher in deep silicon trench etching with profile control. *J. Micromech. Microeng.*, **5** (2), 115. doi: 10.1088/0960-1317/5/2/015
38. Winters, H.F. (1983) Surface processes in plasma-assisted etching environments. *J. Vac. Sci. Technol., B*, **1** (2), 469–480. doi: 10.1116/1.582629
39. van Driel, H.M., Sipe, J.E., and Young, J.F. (1985) Laser-induced coherent modulation of solid and liquid surfaces. *J. Lumin.*, **30** (1–4), 446–471. doi: 10.1016/0022-2313(85)90071-7
40. Tull, B.R., Carey, J.E., Mazur, E., McDonald, J.P., and Yalisove, S.M. (2006) Silicon surface morphologies after femtosecond laser irradiation. *MRS Bull.*, **31** (08), 626–633. doi: 10.1557/mrs2006.160
41. Carey, J.E. (2004) *Femtosecond-Laser Microstructuring of Silicon for Novel Optoelectronic Devices*, Harvard University.
42. Her, T.-H., Finlay, R.J., Wu, C., and Mazur, E. (2000) Femtosecond laser-induced formation of spikes on silicon.

Appl. Phys. A, **70** (4), 383–385. doi: 10.1007/s003390051052

43. Wu, C., Crouch, C.H., Zhao, L., Carey, J.E., Younkin, R., Levinson, J.A., Mazur, E., Farrell, R.M., Gothoskar, P., and Karger, A. (2001) Near-unity below-band-gap absorption by microstructured silicon. *Appl. Phys. Lett.*, **78** (13), 1850–1852. doi: 10.1063/1.1358846

44. Younkin, R., Carey, J.E., Mazur, E., Levinson, J.A., and Friend, C.M. (2003) Infrared absorption by conical silicon microstructures made in a variety of background gases using femtosecond-laser pulses. *J. Appl. Phys.*, **93** (5), 2626. doi: 10.1063/1.1545159

45. Sheehy, M.A., Tull, B.R., Friend, C.M., and Mazur, E. (2007) Chalcogen doping of silicon via intense femtosecond-laser irradiation. *Mater. Sci. Eng.*, **137** (1–3), 289–294. doi: 10.1016/j.mseb.2006.10.002

46. Smith, M.J., Winkler, M., Sher, M.-J., Lin, Y.-T., Mazur, E., and Gradečak, S. (2011) The effects of a thin film dopant precursor on the structure and properties of femtosecond-laser irradiated silicon. *Appl. Phys. A*, **105** (4), 795–800. doi: 10.1007/s00339-011-6651-2

47. Fuller, C.S. and Ditzenberger, J.A. (1956) Diffusion of donor and acceptor elements in silicon. *J. Appl. Phys.*, **27** (5), 544. doi: 10.1063/1.1722419

48. Canham, L.T. (1990) Silicon quantum wire array fabrication by electrochemical and chemical dissolution of wafers. *Appl. Phys. Lett.*, **57** (10), 1046. doi: 10.1063/1.103561

49. Lehmann, V. and Gösele, U. (1991) Porous silicon formation: a quantum wire effect. *Appl. Phys. Lett.*, **58** (8), 856. doi: 10.1063/1.104512

50. Williams, P., Lévy-Clément, C., Péou, J.-E., Brun, N., Colliex, C., Wehrspohn, R.B., Chazalviel, J.-N., and Albu-Yaron, A. (1997) Microstructure and photoluminescence of porous Si formed on N-type substrates in the dark. *Thin Solid Films*, **298** (1–2), 66–75. doi: 10.1016/S0040-6090(96)09162-6

51. Wehrspohn, R.B., Chazalviel, J.-N., Ozanam, F., and Solomon, I. (1997) Electrochemistry and photoluminescence of porous amorphous silicon. *Thin Solid Films*, **297** (1–2), 5–8. doi: 10.1016/S0040-6090(96)09362-5

52. Allongue, P., Kieling, V., and Gerischer, H. (1995) Etching mechanism and atomic structure of H-Si(111) surfaces prepared in NH4F. *Electrochim. Acta*, **40** (10), 1353–1360. doi: 10.1016/0013-4686(95)00071-L

53. Cullis, A.G., Canham, L.T., and Calcott, P.D.J. (1997) The structural and luminescence properties of porous silicon. *J. Appl. Phys.*, **82** (3), 909. doi: 10.1063/1.366536

54. Torres-Costa, V. and Martín-Palma, R.J. (2010) Application of nanostructured porous silicon in the field of optics. A review. *J. Mater. Sci.*, **45** (11), 2823–2838. doi: 10.1007/s10853-010-4251-8

55. Lehmann, V. (1992) Porous silicon preparation: alchemy or electrochemistry? *Adv. Mater.*, **4** (11), 762–764. doi: 10.1002/adma.19920041115

56. Lee, C.-L., Tsujino, K., Kanda, Y., Ikeda, S., and Matsumura, M. (2008) Pore formation in silicon by wet etching using micrometre-sized metal particles as catalysts. *J. Mater. Chem.*, **18** (9), 1015. doi: 10.1039/b715639a

57. Tsujino, K., Matsumura, M., and Nishimoto, Y. (2006) Texturization of multicrystalline silicon wafers for solar cells by chemical treatment using metallic catalyst. *Sol. Energy Mater. Sol. Cells*, **90** (1), 100–110. doi: 10.1016/j.solmat.2005.02.019

58. Chang, S.-W., Chuang, V.P., Boles, S.T., Ross, C.A., and Thompson, C.V. (2009) Densely packed arrays of ultra-high-aspect-ratio silicon nanowires fabricated using block-copolymer lithography and metal-assisted etching. *Adv. Funct. Mater.*, **19** (15), 2495–2500. doi: 10.1002/adfm.200900181

59. Harada, Y., Li, X., Bohn, P.W., and Nuzzo, R.G. (2001) Catalytic amplification of the soft lithographic patterning of Si. Nonelectrochemical orthogonal fabrication of photoluminescent porous Si pixel arrays. *J. Am. Chem. Soc.*, **123** (36), 8709–8717. doi: 10.1021/ja010367j

60. Peng, K., Huang, Z., and Zhu, J. (2004) Fabrication of large-area silicon nanowire P–n junction diode arrays. *Adv. Mater.*, **16** (1), 73–76. doi: 10.1002/adma.200306185
61. Zhang, M.-L., Peng, K.-Q., Fan, X., Jie, J.-S., Zhang, R.-Q., Lee, S.-T., and Wong, N.-B. (2008) Preparation of large-area uniform silicon nanowires arrays through metal-assisted chemical etching. *J. Phys. Chem. C*, **112** (12), 4444–4450. doi: 10.1021/jp077053o
62. Peng, K. and Zhu, J. (2003) Simultaneous gold deposition and formation of silicon nanowire arrays. *J. Electroanal. Chem.*, **558**, 35–39. doi: 10.1016/S0022-0728(03)00374-7
63. Nayak, B.K., Iyengar, V.V., and Gupta, M.C. (2011) Efficient light trapping in silicon solar cells by ultrafast-laser-induced self-assembled micro/nano structures. *Prog. Photovoltaics*, **19** (6), 631–639. doi: 10.1002/pip.1067
64. Sarnet, T., Halbwax, M., Torres, R., Delaporte, P., Sentis, M., Martinuzzi, S., Vervisch, V. *et al.* (2008) Femtosecond laser for black silicon and photovoltaic cells. *Proc. SPIE* (Commercial and Biomedical Applications of Ultrafast Lasers VIII), **6881**, 688119. doi: 10.1117/12.768516
65. Tsakalakos, L., Balch, J., Fronheiser, J., Shih, M.-Y., LeBoeuf, S.F., Pietrzykowski, M., Codella, P.J. *et al.* (2007) Strong broadband optical absorption in silicon nanowire films. *J. Nanophotonics*, **1** (1), 013552. doi: 10.1117/1.2768999
66. Nishioka, K., Sueto, T., and Saito, N. (2009) Formation of antireflection nanostructure for silicon solar cells using catalysis of single nano-sized silver particle. *Appl. Surf. Sci.*, **255** (23), 9504–9507. doi: 10.1016/j.apsusc.2009.07.079
67. Inomata, Y., Fukui, K., and Shirasawa, K. (1997) Surface texturing of large area multicrystalline silicon solar cells using reactive ion etching method. *Sol. Energy Mater. Sol. Cells*, **48** (1–4), 237–242. doi: 10.1016/S0927-0248(97)00106-2
68. Schnell, M., Lüdemann, R., and Schäfer, S. (2000) Plasma surface texturization for multicrystalline silicon solar cells. 28th IEEE Photovoltaic Specialists Conference, Anchorage, AK, IEEE, pp. 367–370. doi:10.1109/PVSC.2000.915841.
69. Zaidi, S.H., Ruby, D.S., and Gee, J.M. (2001) Characterization of random reactive ion etched-textured silicon solar cells. *IEEE Trans. Electron Devices*, **48** (6), 1200–1206.
70. Yoo, J., Yu, G., and Yi, J. (2009) Black surface structures for crystalline silicon solar cells. *Mater. Sci. Eng., B*, **159–160**, 333–337. doi: 10.1016/j.mseb.2008.10.019
71. Sai, H., Fujii, H., Arafune, K., Ohshita, Y., Yamaguchi, M., Kanamori, Y., and Yugami, H. (2006) Antireflective subwavelength structures on crystalline Si fabricated using directly formed anodic porous alumina masks. *Appl. Phys. Lett.*, **88** (20), 201113–201116.
72. Southwell, W.H. (1983) Gradient-index antireflection coatings. *Opt. Lett.*, **8** (11), 584–586. doi: 10.1364/OL.8.000584
73. Wilson, S.J. and Hutley, M.C. (1982) The optical properties of 'moth eye' antireflection surfaces. *Opt. Acta*, **29** (7), 993–1009. doi: 10.1080/713820946
74. Boden, S.A. and Bagnall, D.M. (2008) Tunable reflection minima of nanostructured antireflective surfaces. *Appl. Phys. Lett.*, **93** (13), 133108.
75. Yablonovitch, E. (1982) Statistical ray optics. *J. Opt. Soc. Am.*, **72** (7), 899–907.
76. Green, M.A. (1984) Limits on the open-circuit voltage and efficiency of silicon solar cells imposed by intrinsic auger processes. *IEEE Trans. Electron Devices*, **31** (5), 671–678. doi: 10.1109/T-ED.1984.21588
77. Oskooi, A.F., Roundy, D., Ibanescu, M., Bermel, P., Joannopoulos, J.D., and Johnson, S.G. (2010) Meep: a flexible free-software package for electromagnetic simulations by the FDTD method. *Comput. Phys. Commun.*, **181** (3), 687–702. doi: 10.1016/j.cpc.2009.11.008
78. Li, L. (1997) New formulation of the Fourier modal method for crossed surface-relief gratings. *J. Opt. Soc. Am. A*, **14** (10), 2758–2767.

79. Li, L.E. (1996) Formulation and comparison of two recursive matrix algorithms for modeling layered diffraction gratings. *J. Opt. Soc. Am. A*, **13** (5), 1024–1035.
80. Kim, K., Mandel, L., and Wolf, E. (1987) Relationship between Jones and Mueller matrices for random media. *J. Opt. Soc. Am. A*, **4** (3), 433–437. doi: 10.1364/JOSAA.4.000433
81. Kroll, M., Otto, M., Käsebier, T., Füchsel, K., Wehrspohn, R.B., Kley, E.-B., Tünnermann, A., and Pertsch, T. (2012) Black silicon for solar cell applications. *Proc. SPIE* (Photonics for Solar Energy Systems IV), **8438**, 843817. doi: 10.1117/12.922380
82. Tiedje, T., Yablonovitch, E., and Cody, G.D. (1984) Limiting efficiency of silicon solar cells. *IEEE Trans. Electron Devices*, **31**, 711–716.
83. Dingemans, G. and Kessels, W.M.M. (2012) Status and prospects of Al2O3-based surface passivation schemes for silicon solar cells. *J. Vac. Sci. Technol., A*, **30** (4), 040802. doi: 10.1116/1.4728205
84. Brendel, R. and Ernst, M. (2010) Macroporous Si as an absorber for thin-film solar cells. *Phys. Status Solidi RRL*, **4** (1–2), 40–42. doi: 10.1002/pssr.200903372
85. Yuan, H.-C., Yost, V.E., Page, M.R., Stradins, P., Meier, D.L., and Branz, H.M. (2009) Efficient black silicon solar cell with a density-graded nanoporous surface: optical properties, performance limitations, and design rules. *Appl. Phys. Lett.*, **95** (12), 123501. doi: 10.1063/1.3231438
86. Reed, M.L. and Plummer, J.D. (1988) Chemistry of Si-SiO2 interface trap annealing. *J. Appl. Phys.*, **63** (12), 5776–5793. doi: 10.1063/1.340317
87. Gatz, S., Plagwitz, H., Altermatt, P.P., Terheiden, B., and Brendel, R. (2008) Thermal stability of amorphous silicon/silicon nitride stacks for passivating crystalline silicon solar cells. *Appl. Phys. Lett.*, **93** (17), 173502. doi: 10.1063/1.3009571
88. De Wolf, S., Descoeudres, A., Holman, Z.C., and Ballif, C. (2012) High-efficiency silicon heterojunction solar cells: a review. *Green*, **2** (1), 7–24. doi: 10.1515/green-2011-0018
89. Montesdeoca-Santana, A., Ziegler, J., Lindekugel, S., Jiménez-Rodríguez, E., Keipert-Colberg, S., Müller, S., Krause, C., Borchert, D., and Guerrero-Lemus, R. (2011) A comparative study on different textured surfaces passivated with amorphous silicon. *Phys. Status Solidi C*, **8** (3), 747–750. doi: 10.1002/pssc.201000234
90. Ernst, M., Brendel, R., Ferré, R., and Harder, N.-P. (2012) Thin macroporous silicon heterojunction solar cells. *Phys. Status Solidi RRL*, **6** (5), 187–189. doi: 10.1002/pssr.201206113
91. Hezel, R. and Jaeger, K. (1989) Low-temperature surface passivation of silicon for solar cells. *J. Electrochem. Soc.*, **136** (2), 518–523. doi: 10.1149/1.2096673
92. Agostinelli, G., Delabie, A., Vitanov, P., Alexieva, Z., Dekkers, H.F.W., De Wolf, S., and Beaucarne, G. (2006) Very low surface recombination velocities on P-type silicon wafers passivated with a dielectric with fixed negative charge. *Sol. Energy Mater. Sol. Cells*, **90** (18–19), 3438–3443. doi: 10.1016/j.solmat.2006.04.014
93. Hoex, B., Heil, S.B.S., Langereis, E., van de Sanden, M.C.M., and Kessels, W.M.M. (2006) Ultralow surface recombination of c-Si substrates passivated by plasma-assisted atomic layer deposited Al2O3. *Appl. Phys. Lett.*, **89** (4), 042112. doi: 10.1063/1.2240736
94. Puurunen, R.L. (2005) Surface chemistry of atomic layer deposition: a case study for the trimethylaluminum/water process. *J. Appl. Phys.*, **97** (12), 121301. doi: 10.1063/1.1940727
95. Poodt, P., Lankhorst, A., Roozeboom, F., Spee, K., Maas, D., and Vermeer, A. (2010) High-speed spatial atomic-layer deposition of aluminum oxide layers for solar cell passivation. *Adv. Mater.*, **22** (32), 3564–3567. doi: 10.1002/adma.201000766
96. Poodt, P., Cameron, D.C., Dickey, E., George, S.M., Kuznetsov, V., Parsons, G.N., Roozeboom, F., Sundaram, G., and Vermeer, A. (2012) Spatial atomic

layer deposition: a route towards further industrialization of atomic layer deposition. *J. Vac. Sci. Technol., A*, **30** (1), 010802. doi: 10.1116/1.3670745
97. Naumann, V., Otto, M., Wehrspohn, R.B., and Hagendorf, C. (2012) Chemical and structural study of electrically passivating Al2O3/Si interfaces prepared by atomic layer deposition. *J. Vac. Sci. Technol., A*, **30** (4), 04D106. doi: 10.1116/1.4704601
98. Dingemans, G., Beyer, W., van de Sanden, M.C.M., and Kessels, W.M.M. (2010) Hydrogen induced passivation of Si interfaces by Al2O3 films and SiO2/Al2O3 stacks. *Appl. Phys. Lett.*, **97** (15), 152106. doi: 10.1063/1.3497014
99. Hoex, B., Schmidt, J., Bock, R., Altermatt, P.P., van de Sanden, M.C.M., and Kessels, W.M.M. (2007) Excellent passivation of highly doped P-type Si surfaces by the negative-charge-dielectric Al2O3. *Appl. Phys. Lett.*, **91** (11), 112107. doi: 10.1063/1.2784168
100. Hoex, B., Gielis, J.J.H., van de Sanden, M.C.M., and Kessels, W.M.M. (2008) On the c-Si surface passivation mechanism by the negative-charge-dielectric Al2O3. *J. Appl. Phys.*, **104** (11), 113703. doi: 10.1063/1.3021091
101. Hoex, B., van de Sanden, M.C.M., Schmidt, J., Brendel, R., and Kessels, W.M.M. (2012) Surface passivation of phosphorus-diffused N+-type emitters by plasma-assisted atomic-layer deposited Al2O3. *Phys. Status Solidi RRL*, **6** (1), 4–6. doi: 10.1002/pssr.201105445
102. Otto, M., Kroll, M., Käsebier, T., Salzer, R., Tünnermann, A., and Wehrspohn, R.B. (2012) Extremely low surface recombination velocities in black silicon passivated by atomic layer deposition. *Appl. Phys. Lett.*, **100** (19), 191603. doi: 10.1063/1.4714546
103. Gimpel, T., Guenther, K.-M., Kontermann, S., and Schade, W. (2011) Study on contact materials for sulfur hyperdoped black silicon. 37th IEEE Photovoltaic Specialists Conference, Seattle, WA, IEEE, pp. 002061–002065, doi: 10.1109/PVSC.2011.6186358.
104. Schaefer, S. and Lüdemann, R. (1999) Low damage reactive ion etching for photovoltaic applications. *J. Vac. Sci. Technol., A*, **17** (3), 749–754. doi: 10.1116/1.581644
105. Kern, W. and Puotinen, D. (1970) Cleaning solutions based on hydrogen peroxide for use in silicon semiconductor technology. *RCA Rev.*, **31**, 187–206.
106. Rein, S. (2005) *Lifetime Spectroscopy: A Method of Defect Characterization in Silicon for Photovoltaic Applications*, vol. 85, Springer, Berlin, Heidelberg. doi: 10.1007/3-540-27922-9
107. Meuris, M. and Mertens, P.W. (1995) The IMEC clean: a new concept for particle and metal removal on Si surfaces. *Solid State Technol.*, **38**, 109–112.
108. Toor, F., Branz, H.M., Page, M.R., Jones, K.M., and Yuan, H.-C. (2011) Multi-scale surface texture to improve blue response of nanoporous black silicon solar cells. *Appl. Phys. Lett.*, **99** (10), 103501. doi: 10.1063/1.3636105
109. Yonehara, T., Sakaguchi, K., and Sato, N. (1994) Epitaxial layer transfer by bond and etch back of porous Si. *Appl. Phys. Lett.*, **64** (16), 2108. doi: 10.1063/1.111698
110. Tayanaka, H. and Matsushita, T. (1996) Separation of thin epitaxial Si films on porous Si for solar cells (in Japanese). Proceedings of the 6th Sony Research Forum, p. 556.
111. Brendel, R., Feldrapp, K., Horbelt, R., and Auer, R. (2003) 15.4%-efficient and 25 μm-thin crystalline Si solar cell from layer transfer using porous silicon. *Phys. Status Solidi A*, **197** (2), 497–501. doi: 10.1002/pssa.200306552
112. Petermann, J.H., Zielke, D., Schmidt, J., Haase, F., Rojas, E.G., and Brendel, R. (2012) 19%-efficient and 43 μm-thick crystalline Si solar cell from layer transfer using porous silicon. *Prog. Photovoltaics Res. Appl.*, **20** (1), 1–5. doi: 10.1002/pip.1129
113. Bilyalov, R.R., Lüdemann, R., Wettling, W., Stalmans, L., Poortmans, J., Nijs, J., Schirone, L., Sotgiu, G., Strehlke, S., and Lévy-Clément, C. (2000) Multicrystalline silicon solar cells with porous silicon emitter. *Sol. Energy*

Mater. Sol. Cells, **60** (4), 391–420. doi: 10.1016/S0927-0248(99)00102-6

114. Lipiński, M., Bastide, S., Panek, P., and Lévy-Clément, C. (2003) Porous silicon antireflection coating by electrochemical and chemical etching for silicon solar cell manufacturing. *Phys. Status Solidi A*, **197** (2), 512–517. doi: 10.1002/pssa.200306555

115. Kim, B.S., Lee, D.H., Kim, S.H., An, G.-H., Lee, K.-J., Myung, N.V., and Choa, Y.-H. (2009) Silicon solar cell with nanoporous structure formed on a textured surface. *J. Am. Ceram. Soc.*, **92** (10), 2415–2417. doi: 10.1111/j.1551-2916.2009.03210.x

116. Jia, G., Steglich, M., Sill, I., and Falk, F. (2012) Core–shell heterojunction solar cells on silicon nanowire arrays. *Sol. Energy Mater. Sol. Cells*, **96**, 226–230. doi: 10.1016/j.solmat.2011.09.062

117. Garnett, E.C., Peters, C., Brongersma, M., Cui, Y., and McGehee, M. (2010) Silicon nanowire hybrid photovoltaics. 35th IEEE Photovoltaic Specialists Conference, IEEE, pp. 000934–000938, doi:10.1109/PVSC.2010.5614661.

118. Shen, X., Sun, B., Liu, D., and Lee, S.-T. (2011) Hybrid heterojunction solar cell based on organic-inorganic silicon nanowire array architecture. *J. Am. Chem. Soc.*, **133** (48), 19408–19415. doi: 10.1021/ja205703c

119. Füchsel, K., Schulz, U., Kaiser, N., Käsebier, T., Kley, E.-B., and Tünnermann, A. (2010) Nanostructured SIS solar cells. *Proc. SPIE* (Photonics for Solar Energy Systems III), **7725**, 772502–772502-8. doi: 10.1117/12.854694

120. Füchsel, K., Kroll, M., Käsebier, T., Otto, M., Pertsch, T., Kley, E.-B., Wehrspohn, R.B., Kaiser, N., and Tünnermann, A. (2012) Black silicon photovoltaics. *Proc. SPIE* (Photonics for Solar Energy Systems IV), **8438**, p. 84380M–84380M-8, doi: 10.1117/12.923748.

121. Yoo, J., Yu, G., and Yi, J. (2011) Large-area multicrystalline silicon solar cell fabrication using Reactive Ion Etching (RIE). *Sol. Energy Mater. Sol. Cells*, **95** (1), 2–6. doi: 10.1016/j.solmat.2010.03.029

122. Powell, D.M., Winkler, M.T., Choi, H.J., Simmons, C.B., Berney Needleman, D., and Buonassisi, T. (2012) Crystalline silicon photovoltaics: a cost analysis framework for determining technology pathways to reach baseload electricity costs. *Energy Environ. Sci.*, **5** (3), 5874. doi: 10.1039/c2ee03489a

123. Schirone, L., Sotgiu, G., and Califano, F.P. (1997) Chemically etched porous silicon as an anti-reflection coating for high efficiency solar cells. *Thin Solid Films*, **297** (1–2), 296–298. doi: 10.1016/S0040-6090(96)09436-9

6
Concentrator Optics for Photovoltaic Systems

Andreas Gombert, Juan C. Miñano, Pablo Benitez, and Thorsten Hornung

6.1
Fundamentals of Solar Concentration

6.1.1
Introduction

Most of us, if not all, have had personal experience with solar concentration by playing with magnifiers. The power handled by a simple magnifier under the sun is, at least, surprising: the simplicity, promptness, and cleanliness with which fire can be initiated using a magnifier in contrast with chemical or mechanical techniques. All these experiences provide us the basic ideas of solar concentration: we have to use an optical system aimed at the sun to concentrate the light. This optical system is called *concentrator*. A concentrator needs to have two well defined apertures: An entry aperture where the radiation from the sun is collected and an exit aperture through which the radiation leaves the concentrator toward the target. In a photovoltaic concentrator, the target is the place where the solar cell is placed and in most of the cases also coincides with the exit aperture.

Most of the concentrators can be accurately analyzed in the framework of Geometrical Optics (GO), where the light can be modeled with the concept of rays. The rays define the paths followed by the electromagnetic energy. In general, GO is not enough for a full analysis of the concentrator. For instance the analysis of antireflective (AR) coatings needs a Wave Optics framework. The most frequent case is that the GO and Wave Optics can be used separately and the results can be easily combined.

6.1.2
Concentration and Acceptance Angle

Minimizing energy cost (€/kWh) is the goal of Concentrated Photovoltaics (CPVs). Key to minimizing this cost is an efficient and low-cost optical design. These goals are best met with the fewest elements, the most relaxed tolerances, and the highest concentration (that allows for rapid amortization of the cost

Photon Management in Solar Cells, First Edition. Edited by Ralf B. Wehrspohn, Uwe Rau, and Andreas Gombert.
© 2015 Wiley-VCH Verlag GmbH & Co. KGaA. Published 2015 by Wiley-VCH Verlag GmbH & Co. KGaA.

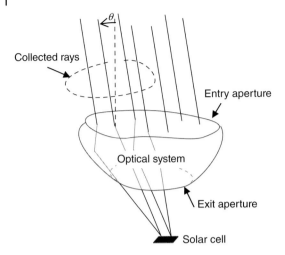

Figure 6.1 Concentrator geometry.

of high-efficiency solar cells). Nevertheless, it is not possible to achieve high concentration and high tolerance at the same time. There is a fundamental limit that links both variables (concentration and tolerance). It is not an easy task for a low-cost concentrator to be close to this limit and in practice most of the concentrators are far from this limit.

This fundamental limit relates the geometric concentration C_g and the acceptance angle α. C_g is the ratio of the concentrator entry aperture area and the target area, that is, the cell active area. C_g gives an upper bound of the irradiance on the cell relative to the irradiance received at the entry aperture. The acceptance angle α is defined as 1/2 of the angular span within which the concentrator collects more than 90% of the on-axis power [1]. For this definition we are assuming that the concentrator is illuminated by a plane wavefront, that is, a set of parallel rays which form a given angle θ with some preferential concentrator axis (see Figure 6.1). This axis is commonly normal to the entry aperture. Typically, the concentrator collects only a fraction of these incoming rays. This fraction is called *transmission T*. The transmission function is, in general, a function of two variables (those needed to characterize the direction of the incoming rays). In rotational symmetric systems, T can be written as a function of a single angular variable θ, which is the angle formed by the incoming parallel rays with the axis of rotational symmetry $T(\theta)$. Typically, $\theta = 0$ is also the direction of maximum transmission. In this case α is the angle fulfilling $T(\alpha)/T(0) = 0.9$. In the general case, α is the minimum tilting angle from the on-axis position such that the transmission does not fall below 90% (see Figure 6.2).

A more practical definition of the acceptance angle says that it is the sun's tilting angle at which the generated photocurrent is at 90% of the maximum (typically achieved at normal incidence) [2]. In terms of generated electric power, a more useful definition says that it is the angle at which the maximum electric power

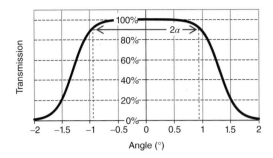

Figure 6.2 Definition of acceptance angle from the angular transmission curve of a concentrator.

generated by the cell (at the Maximum Power Point, MPP, of the $I-V$ curve) is at 90% of the maximum electric power generated by the cell at the best incidence angle. This is usually referred to as the *effective acceptance angle*. These last two definitions gather optical and electrical effects, that is, they take into account the solar cell behavior, the optical concentrator, and sun's angular extension. Henceforth, we will use the first definition of α.

The greatest interest of the acceptance angle concept is not because it may represent the tolerance of the tracking system but because most of the manufacturing tolerances of the system as well as other effects affecting the optical concentrator performance (such as the sun's angular size and soiling) can be expressed in terms of the angular acceptance. The acceptance angle α measures the total tolerance available to be apportioned among the different imperfections of the system: (i) shape errors and roughness of the optical surfaces, (ii) concentrator module assembly, (iii) array installation, (iv) tracker structure finite stiffness, (v) sun-tracking accuracy, (vi) solar angular diameter, (vii) lens warp, and (viii) soiling. Each of these imperfections can be expressed as a fraction of the tolerance angle, so that, all together, they comprise a "tolerance budget." Then, one optimization strategy is to look for the concentrator with the highest α with all other variables kept constant, in particular, the C_g. This optimization strategy may be incorrect if the concentrators to be compared are not similar; that is, when they are composed of different elements, as some of them may include step functions in the cost. Assume that we have a concentrator with an acceptance angle α high enough so there is still a tolerance rest after all different manufacturing and system tolerances have been allocated. Then we have to look for cheaper manufacturing procedures, which presumably consumes more of the tolerance budget, to exchange the tolerance rests for cost savings until the tolerance budget is exhausted.

Then, when comparing similar concentrators, the higher the acceptance angle, the better for the concentrator cost (keeping all other variables constant). Also, the higher the geometrical concentration, the better for the concentrator cost (again keeping all other variables constant) since a higher concentration allow us to generate the same power with a smaller cell. We are implicitly assuming that the cell cost per unit of cell area is much higher than the cost of the optical concentrator per unit of entry aperture area. Then, the most desirable case would be to get the highest concentration with the highest tolerance. Nevertheless, the acceptance angle and concentration cannot increase indefinitely. There is a fundamental limitation, derived from the conservation of étendue theorem [1], which can be written as

$$\sqrt{C_g}\sin(\alpha) \leq n \qquad (6.1)$$

where n is the refractive index of the medium surrounding a cell's illuminated face. This medium is sometimes air ($n \approx 1$), and sometimes a clear silicone ($n \approx 1.4$–1.5). Equation (6.1) assumes some simplifications that are not relevant to our analysis. In particular, it assumes that the transmission-curve of Figure 6.2 is a top-hat one with a sharp cut-off at the acceptance angle.

The left hand side of the inequality in Eq. (6.1) is called the concentration-acceptance product CAP, that is,

$$\text{CAP} = \sqrt{C_g}\sin(\alpha) \qquad (6.2)$$

Then we can say that the conservation of étendue theorem (Eq. (6.1)) establishes that CAP $\leq n$. This definition of CAP is close to the concept of numerical aperture (NA) in Imaging Optics, but is not exactly the same. NA is either the sine of the acceptance cone of an objective, or refers to the maximum angle of acceptance (or illumination) of a ray bundle transmitted by a fiber. NA and CAP definitions coincide when every point of the cell is illuminated by light cones with the same angle, which is an unlikely ideal case. The most likely case is when the cones illuminating different points of the cell are not identical and, in general, do not have a circular base (see Figure 6.3). The CAP does not provide the sine of the maximum angle of the cone of light transmitted to the cell, but the maximum angle of such an ideal illumination enclosing the same étendue.

What makes the CAP a useful parameter is that for a given concentrator architecture, the CAP is practically constant. By "concentrator architecture" we mean

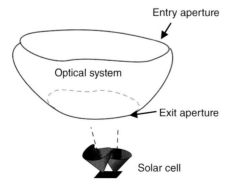

Figure 6.3 Cones illuminating different points of the cell.

the number and type of optical elements. For instance, a Fresnel lens focusing the light on the cell defines a concentrator architecture. A Fresnel lens plus an inverted truncated reflective pyramid, as secondary optical element (SOE), defines another architecture. The concentrators of the same architecture differ in the values of the parameters defining the different elements of the concentrator. For instance, the focal length, the f-number, and the cell size are parameters defining the elements of the concentrator in the case of the single Fresnel lens architecture. By changing these parameters we can have concentrators of the same architecture with different C_g and different α, but their best CAPs are approximately the same for a large range of C_g and α. Then the CAP can be used to characterize how good an architecture is in terms of the best achievable C_g and α.

6.1.3
Optical Efficiency

Because the acceptance angle α depends on the relative transmission of the concentrator and C_g is a purely geometric definition, these two parameters do not provide information about the amount of power that is transmitted by the concentrator. This is the role of the optical efficiency. A first definition of the optical efficiency, $\eta_{opt,1}$, is light power transmission efficiency through the concentrator [2] when its entry aperture is illuminated by a parallel beam of rays of constant radiance. All the light reaching the target must have entered through the concentrator's entry aperture. This is important for a consistent entry aperture definition (used when calculating C_g). The direction of the incoming parallel rays is the one providing maximum transmission and is usually referred to as *normal incidence*.

This optical efficiency definition is wavelength dependent. For CPV applications, it is averaged with the AM1.5 spectral distribution as the weighting function. This definition is typically used by optics manufacturers because it is cell independent.

For the second definition $\eta_{opt,2}$ we need that the nominal parallel incident rays produce 1 sun irradiance at the entry aperture. $\eta_{opt,2}$ is given by the ratio of the concentrator photocurrent (this is the photocurrent when the cell is in the concentrator) to the product of the geometrical concentration and the cell photocurrent when illuminated at 1 sun with the solar spectrum [2]. For the 1 sun photocurrent, the cell is assumed to receive the light at normal incidence inside the same refractive index medium as the one that surrounds the cell inside the concentrator. The solar spectrum classically used for this definition is the AM1.5D. When compared to $\eta_{opt,1}$, this second definition considers the spectral response of both the concentrator and the cell, and includes the possible angular dependence of the external quantum efficiency of the cell.

A third definition [3] $\eta_{opt,3}$ has been used by considering the photocurrent ratio, as in the second definition $\eta_{opt,2}$, but with two differences: (i) the rays incident on the concentrator have the angular extension of the sun and (ii) the cell photocurrent at 1 sun is obtained by encapsulating it as a flat module. This third definition is obviously suitable for measurements with real sun. Difference (i) is often not relevant, because the concentrator photocurrent is usually the same for any parallel rays impinging from inside the sun disk when the sun is centered at the nominal position. However, difference (ii) is not negligible at all, because the Fresnel and absorption losses on the glass cover of the flat module are around 4–5%. Therefore, this third definition can be seen as the ratio of the concentrator to the flat module type-$\eta_{opt,2}$ optical efficiencies, and then, it is a relative optical efficiency.

6.1.4
Effect of Spatial and Spectral Non-uniformities on the Cell Illumination

None of these optical definitions is such that the optical efficiency η_{opt} times the cell efficiency η_{cell} given by the cell manufacturers gives the light to electricity efficiency of the concentrator η_{conc}. Why? There are several reasons. The first one is because η_{cell} is measured under normal (or close to normal) incidence of light in air. Light reaching the cell in the concentrator has a wide angular range and cannot be considered close to be collimated. This is a fundamental consequence of concentration and cannot be avoided. The cell AR coating design should take this into account. This AR coating design is not an easy task. Note that the radiation has also a wide spectral range and that AR cost must be low. The use of silicone as encapsulant of the cells in some concentrators also changes the working conditions of the cells in the concentrators relative to the conditions found in the cell efficiency measurements. The use of encapsulants tends to improve the AR properties.

Another reason for differences between η_{conc} and $\eta_{cell} \times \eta_{opt}$ is the non-uniform irradiance on the cell within the concentrator compared to the one found in solar cell efficiency characterization. The solar cell grid is designed for uniform irradiance conditions. This is not because uniformity is the optimum irradiance distribution in terms of efficiency [4], but because it is simple and it seems to adapt better to arbitrary irradiance distributions.

Non-uniformities in the irradiance distribution can be a problem for tunnel-junction interconnected IIIV multijunction solar cells if the resulting local photocurrent exceeds the peak tunneling current density [5]. Otherwise it is not a severe problem, provided that the irradiance distributions corresponding to the spectral bands of the different junctions are matched, even if these distributions are not uniform [6, 7]. However, the non-uniformities should be kept low because some cell efficiency losses may be noticeable (due to the higher Joule losses) and also because the reliability of the cell or that of the encapsulant may be compromised [8, 9].

Quite a different and worse situation is found when the irradiances in the spectral bands of the different junctions are not matched for all points of the cell [10]. This could happen even with the total irradiance being constant across the cell surface and with total powers of the junction bands well matched. In this case internal currents spreading perpendicular to the main current flow must appear to balance the local photocurrent mismatch between junctions caused by the mismatch of irradiance in their bands. The sensitivity of modern IIIV cells to this spectral dependence of the irradiance distribution is expected to increase in the future, when four or even more junctions are used.

The concentrator design must provide sufficient spatial and spectral uniformity in a variety of situations. Some concentrators provide acceptable uniformity at nominal conditions, but it degrades rapidly when the conditions change. Examples of variable conditions are moderate mispointings to the sun (still inside the acceptance angle), alignment errors between optics and the cell, or temperature changes (which affects significantly the focal distance of SoG Fresnel lenses). This sensitivity will affect the yearly energy production under real operation conditions that may not be detected at the early prototype stages.

6.2 Optical Designs

The continuous efficiency increase of multijunction solar cells in the past decade has been paralleled by advances in optical designs and technologies. These advances have focused on improving not only system efficiency, but also reliability and cost.

Photovoltaic concentrators usually consist in a primary optical element (POE) which collects the light rays from the sun and a SOE which receives the light concentrated from the primary and sends it to the solar cell.

The POE can typically be either refractive (i.e., a lens) or reflective (i.e., a mirror). Contrary to conventional lenses used in imaging optics, POE lenses are usually made with rotational symmetric facets (called *Fresnel lenses*), which allows to make them very thin, and as a consequence, cheaper, lighter, and less absorptive. Generally speaking, mirrors provide a higher CAP than Fresnel lenses as POEs for two reasons. First, since there are no light rays existing the Fresnel lens form the optically inactive facets (which join adjacent refractive facets), and there is a decrease of the total available exit area of the deflected ray bundle, which causes an increase of the angular spread of the deflected ray bundle. Secondly, lens materials suffer from chromatic dispersion (so light rays are deflected in slightly different directions depending on their wavelength), and that causes an additional increase of the angular spread. Therefore, both effects produce an apparent increase of the transmitted étendue, which is responsible for the lower CAP.

Chronologically, three generations of CPV concentrators can be identified. The first generation, namely classical concentrators (Section 6.2.1), are essentially based on imaging optics POE designs without using an SOE. The second one (Section 6.2.2) appeared when nonimaging secondary optics were added to the imaging POEs to further concentrate or homogenize the cell illumination. And finally, the most recent generation, advanced concentrators (Section 6.2.3), in which the POE and SOE are jointly designed with the latest nonimaging tools

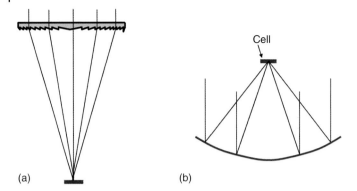

Figure 6.4 Fresnel lenses and parabolic reflectors are the classical concentrators in CPV.

making use of the latest optics shapes available at present: the freeform surfaces (i.e., surfaces without rotational or linear symmetry).

6.2.1
Classical Imaging Concentrators

The classical mirror POE for solar concentration has the shape of a parabola (see Figure 6.4), whose discovery is attributed to Menaechmus in the fourth century BC. For high concentration, the rotational symmetry is applied to obtain a parabolic dish, which has the optical property that all rays parallel to the paraboloid axis (i.e., coming from the center of the sun when it is perfectly tracked) are sharply imaged on a point, the parabola focus. For maximum CAP, the focus must be located at the center of the solar receiver. In some cases, the receiver is just a single solar cell (and the POE is limited to tenths of centimeters in size), while in others, a dense array of solar cells forming a module is used, allowing for multimeter size parabolic dishes as POEs (resembling the radio-frequency antenna receivers). The parabolic mirror suffers strongly from an optical aberration called *coma* (specially for compact designs), because of which the magnification is not constant and, thus, local acceptance angle of different sections of the parabola are variable; hence, the CAP is limited by region by the small acceptance angle at the paraboloid rim. However, it is not the CAP that is the main limitation of a parabolic dish as a CPV concentrator, but the strongly non-uniform illumination of the solar receiver. In solar thermal applications (in which the receiver for a dish is usually coupled to a Stirling engine), this non-uniformity does not affect the efficiency as much as it does for CPV receivers. In the case of a dense array module, the non-uniform illumination would cause some cells to run at higher concentration than designed and others at lower, dropping their efficiencies. The high mismatch in cell photocurrents advices against connecting them in series. As it is indicated in Section 6.2.2, homogenizing nonimaging secondary optics is usually added to the parabolic dish.

Fresnel lenses were first proposed by Georges de Buffon in 1748, and some decades later optimized by Augustine Fresnel for its use in lighthouses. The most popular in CPV is the flat Fresnel lens, in which the facets face the cell side (see Figure 6.4) [11]. As an imaging device, its discontinuous nature makes its magnification non-constant, producing non-uniform illumination, but besides it, the chromatic aberration of Fresnel lenses, when used without SOEs, cause very significant differences in the irradiance distributions for the different junction spectral bands, specially under variable operation conditions [9]. Nevertheless, due to its simplicity, Fresnel lenses without secondaries are still considered a competing solution at moderate concentrations. The company Soitec, using the technology developed by Concentrix, uses this approach.

In order to alleviate these problems, a domed Fresnel lens instead of a flat one can be used [12, 13]; the approach has been commercialized by the company Daido Steel [14]. Domed Fresnels have less geometrical and chromatic aberrations, but its manufacturing is more challenging. As in the case of parabolic mirrors, the addition of nonimaging SOE homogenizers to both flat and domed Fresnel lenses also helps (see Section 6.2.2), especially when the more advanced Köhler architectures are used (Section 6.2.5).

6.2.2
Nonimaging Secondary Optics

The pioneer nonimaging device introduced for high concentration photovoltaics was the Compound Parabolic Concentrator (CPC). It was proposed for further concentrating the radiation from classical POEs [15]. This approach considers the SOE and POE designs separately, and the SOE does not consider the actual rays exiting the primary, but considers the primary as a Lambertian source. The smaller the depth H to entry aperture diagonal D ratio, H/D, the lower accuracy of this approximation, and thus the CAP is reduced. As a consequence, to keep a CAP, this solution is necessarily non-compact (for example, a ratio $H/D \approx 1.5-2$ for a Fresnel lens with a CPC, the SOE is typical). Compactness is desirable to reduce transportation costs, especially when the selected cell size is in the 1 cm^2 range. More important is that compactness is preferred probably because CPC-type SOEs do not provide a good and equalized illumination homogeneity for the different junctions [16].

A slightly better result in terms of spectral homogeneity than the CPC (but with a smaller CAP) can be obtained with a hollow inverted truncated pyramid reflector (XTP), shown in Figure 6.5b. This can be manufactured from a flat mirror sheet via threefolds, and it is usually designed so that the centroid of the POE focal region from the primary is essentially on the cell, and the reflector mainly operates for tilted incidence angles. This solution is the one adopted, for instance, by the company Amonix.

Improving the homogeneity and CAP of the XTP can be achieved with the solid dielectric version, the RTP, which works by total internal reflection (TIR). In this case, the optimum pyramid angle is usually smaller than the one of the

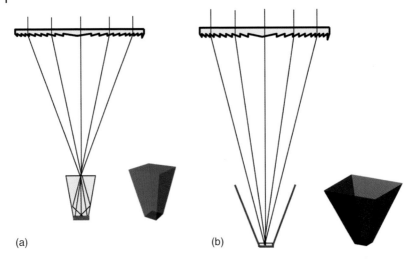

Figure 6.5 Inverted truncated pyramid secondary concentrators: solid version (RTP) (a), and hollow one (XTP) (b).

XTP, and the centroid of the POE focus is usually allocated closer to the entry of the pyramid. Since the entry of this SOE is larger than the exit, it produces not only homogenization but also concentration. The homogenization performed by multiple reflection on the sides of the pyramid generate multiple apparent images of the focus (as a kaleidoscope does). It must be remarked that the optimization of an RTP must be performed simulating the cell performance under its specific non-uniform spatial and spectral illumination, because there is evidence that the cell efficiency in this device can be significantly lower than one could expect from a first sight, as will be shown in Section 6.2.4. This type of SOE is being used by multiple companies around the world, as Daido Steel (Japan), Solfocus (US), or Suntrix (China).

6.2.3
Advanced Concentrator Designs

Nonimaging optics design has experienced a dramatic advance in the past decade, and is particularly reflected in CPV by the invention of multiple new concentrator architectures. Owing to the challenging constrains of the CPV application (high efficiency, high CAP, low cost, reliability, etc.) many of those will probably not survive in the selection process that the market will carry out, but it has been a breeding ground for innovation.

The progress in nonimaging optics has evolved toward freeform optical surfaces, which compared to classical rotational or cylindrical ones, are more complex to design but not necessarily more expensive to mass produce. Modern high-precision tooling, as single point diamond turning, has made the machining of molds with freeform surfaces possible and affordable.

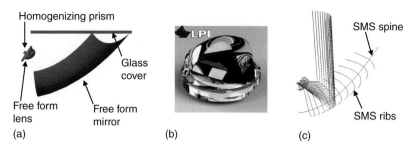

Figure 6.6 XR free-form concentrator.

6.2.4
Freeform SMS Concentrators

Breaking the rotational symmetry allows to achieve performance levels unattainable with classical approaches. A good example is to find a solution to avoid the shading produced by the necessary heat sink in the classical parabolic dish configuration (Figure 6.4). In order to maintain the rotational symmetry, one approach is to reduce the heat sink size. This could be achieved either by active cooling or by vertical heat pipes but those solutions have their specific challenges. Another approach uses a two-mirror aplanatic Cassegrain type configuration [17] which is commercialized by the US company Solfocus. In this option, the SOE mirror also produces some shading and the additional reflection lowers the optical efficiency. However, a CPV concentrator with a highly asymmetric mirror such as the POE can completely avoid the shading by hiding the cell receiver and heat sink behind the adjacent mirror (Figure 6.6).

In order to achieve a high CAP, this extremely asymmetric problem is best solved using freeform optical surfaces instead of rotational or cylindrical ones. The SMS (simultaneous multiple surface) 3D design method [18], proprietary technology of the company LPI, has been used in recent years to design the high performance asymmetric mirror–lens combination shown in Figure 6.6a (called *XR*, in the SMS nomenclature, where X stands for reflection and R for refraction). The SMS 3D designs the two freeform surfaces by calculating curves contained in them, called *spines* and *ribs* (on the right of Figure 6.6c). From a classical optical designer view, it is particularly striking that the resulting SOE lens surface is highly freeform, that is, it deviates strongly from a rotational surface (see Figure 6.6b). The homogenization is performed by a very short homogenizing TIR prism that protrudes from the lens back (its length is similar to that of the cell side). This XR achieves an acceptance angle of $\alpha = \pm 1.85°$ at $C_g = 1000\times$, which implies a CAP ~ 1, the highest ever reported to the authors' knowledge. Module solar-to-electrical efficiencies over 33% without temperature correction have been measured (with AR coated SOE and cover glass) [19].

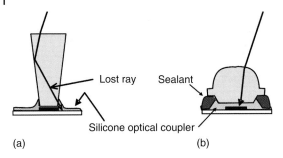

Figure 6.7 Since the total internal reflection is frustrated by the contact with silicone or sealants, the cell protection may be compromised with the optical efficiency kaleidoscope homogenizers (a). This risk is eliminated in Köhler concentrators (b).

6.2.5
Multifold Köhler Concentrators

The SMS XR concentrator and RTP concentrators use kaleidoscope-type homogenization using a prism or an inverted pyramid by TIR. However, the optical coupling between the cell and the SOE, usually performed with a transparent silicone, becomes critical: silicone overflow meniscus would frustrate the TIR causing significant losses, while absence of such overflow could compromise the protection of the cell against moisture (see Figure 6.7b).

An alternative way to kaleidoscopic homogenization that solves that compromise is a concept called *Köhler illumination*. It was developed for microscopes by August Köhler at Carl Zeiss at the end of the nineteenth century. The basic Köhler concentrator for CPV would consist in two imaging optical elements (POE and SOE) with positive focal length (that is, producing a real image of an object at infinity, as a magnifying glass does). The SOE is placed at the focal plane of the POE, and the SOE images the primary on the cell. This configuration makes that the POE images the sun on the SOE aperture, and thus the SOE contour defines the acceptance angle of the concentrator. As the POE is uniformly illuminated by the sun, the irradiance distribution is also uniform and the illuminated area will have the contour of the POE, and it will remain unchanged when the sun moves within the acceptance angle (equivalently when the sun image moves within the SOE aperture). If the POE is tailored in a square shape, the cells will be uniformly illuminated in a squared area. The squared aperture is usually the preferred contour to tessellate the plane when making the modules, while the squared illuminated area on the cell is also usually preferred because it fits the cell's shape.

The first Köhler concentrator in CPV was proposed by James at Sandia Labs around 1990 [20], and it was commercialized later by the company Alpha Solarco. This approach was based on the basic concept introduced above. The imaging POE was a flat Fresnel lens and the imaging SOE was a single-surface lens (called *SILO*, from *SIngLe Optical* surface) that was optically coupled to the cell.

The performance of the SILO is very limited, especially for high concentration (>500×), since its CAP of the SILO is not high and its spatial and spectral uniformity is only acceptable for perfect tracking conditions, and degrades rapidly when moving off-axis. These limitations are caused by the incapability of the SILO SOE to image properly the whole POE, whose angular size as seen from the SOE (the SOE "field of view" in imaging terminology) and the required magnification (POE to cell ratio) are too large for a single refractive surface to manage, specially for the wide spectrum required.

The solution to the SILO limitations has been found recently also by the company LPI with a new technology called *multifold Köhler concentrators* [21, 22]. It consists in dividing the POE and SOE in sectors which provide independent Köhler channels, so each SOE sector will need to manage a correspondingly smaller field of view and also provide a smaller magnification. Additionally, the multichannel approach provides a further improvement and robustness due to the superposition principle: the degradation for any reason of one of the POE images is less noticeable than in the SILO due to its smaller contribution to the total irradiance produced.

The most developed multifold Köhler concentrators is known as *FK concentrator* [21], which is fourfold symmetric, uses a flat Fresnel lens POE and whose SOE is a single-surface lens as the SILO (see Figure 6.8). A standard version of this optics, called *Ventana^{TM}*, made available recently with PMMA (polymethylmethacrylate) as POE material, operates at 1024× geometrical concentration with an acceptance angle of ±1.1°. Module solar-to-electrical efficiencies closed over 32% (without AR coatings) have been measured [23].

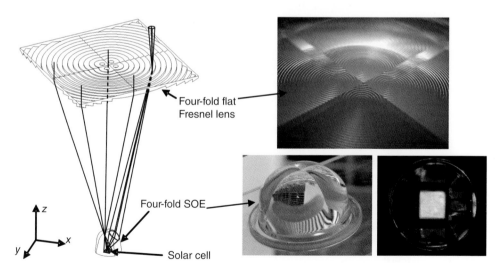

Figure 6.8 The FK is an advanced fourfold Köhler concentrator. In the picture at the bottom-right, actual CCD image of the cell plane under outdoor sun tracking. It a white square illumination, indicating the excellent spatial and spectral uniformity provided by the FK.

Other families of multifold Köhler are possible [22]. For instance, the Köhler version of the XR concentrator (Section 6.2.4) has been developed for both symmetric and asymmetric cases. Another example is the XXR, which is a fourfold symmetric two-mirror plus lens solution, or the D-FK, which is the FK-type design with a fourfold domed Fresnel lens. Finally, it is remarkable that the FK-RXI concentrator, whose POE is flat Fresnel lens and whose SOE is a fourfold freeform SMS 3D concentrator which counts with two surfaces to image the POE sectors, leads to a CAP = 0.85, the highest reported for flat Fresnel POEs to the authors' knowledge.

Different numbers of Köhler channels can be useful depending on the application: for very high concentration, a ninefold FK could provide a further superior spatial and spectral uniformity on the cell, which might be needed when four or more junction cells become available.

6.2.6
Comparison

In order to compare different concentrators, the definite merit function that should count is the final cost of the produced electricity. However, as the CPV industry is still in a nascent stage, there is still a lack of reliable data on manufacturing costs and tolerances, and annual electricity production of complete power plants, making that definite merit function not available. Therefore, we have to limit the comparison to partial aspects, i.e. the CAP and the expected solar-to-electrical efficiencies of potential modules as indicated by simulated IV curves.

We can compare two concentrators with different CAP either assuming that they have the same α (and consequently different C_g) or assuming that they have the same C_g (and consequently different α). In both cases the concentrator with higher CAP has a potential lower cost. This cost comparison is easier if the acceptance angle and entry aperture area of the concentrators under comparison is the same because in this situation the only sources of cost differences are the cell size and the optical components cost. The actual impact of CAP on costs has been analyzed in Ref. [21] by comparing several flat Fresnel lens-based systems containing no more than two optical elements (a Fresnel lens as POE and, optionally an SOE). Because all the architectures being compared have a Fresnel lens as POE with the same size (and consequently with the same cost), the only source of cost differences are the cell size and SOE costs. All other costs (housing, tracking, structure, assembling, etc.) are very similar. For instance, housing cost depends fundamentally on the POE area (constant in the comparison) and assembling tolerance (part of the tolerance budget, which is also constant). Although the Fresnel lens area is also constant in the comparison the design of their grooves is not, but this difference does not affect the lens cost because the manufacturing techniques are basically insensitive to the optical design, at least for mass production.

Here, we will restrict the CAP comparison to the flat Fresnel-based architectures whose cross-sections are shown in Figure 6.9, in addition to the D-FK. All these concentrators have the same Fresnel lens entry aperture area (625 cm^2)

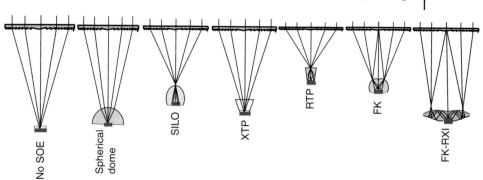

Figure 6.9 Cross section of the selected concentrator architectures using a flat 625 cm² Fresnel lens as POE.

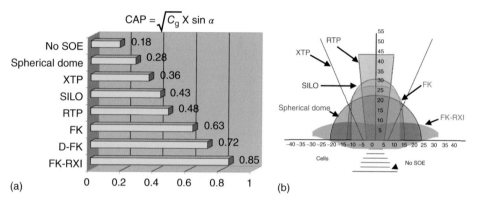

Figure 6.10 (a) CAPs achieved by the architectures of Figure 6.9 plus the D-FK. (b) Cross-section of the SOEs of the concentrators shown in Figure 6.9 (units in millimeters). Those dielectric filled are color filled in the figure. The different cells diameters are also shown at the bottom as colored segments (the collar is that of its SOE). Their actual positions are centered at the origin of the drawing (they have been spread downwards for clarity).

and the same acceptance angle $\alpha = \pm 1°$. The main differences among them are in the SOEs which cause quite different geometrical concentrations C_g. The cross-sections of the SOEs with flat Fresnel lenses are shown in Figure 6.10b. The first one in Figure 6.9 is a Fresnel lens without any SOE (but the cell is protected with a flat cover, as for the XTP). The different CAPs achieved by these architectures are shown in Figure 6.10a.

For the calculation of the acceptance angle α in the CAP values shown in Figure 6.10a, we have included the angular size of the sun and computed the 10% drop of the triple junction cell photocurrent (@AM1.5 spectrum). However, the calculation has neglected the possible cell efficiency loss due to the spatial and spectral non-uniformities. Taking those into account is equivalent to computing the acceptance angle α using the power at the MPP of the $I-V$ curve versus

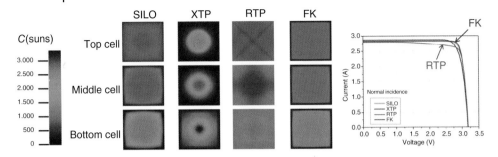

Figure 6.11 (a) Photocurrent densities produced when the sun is perfectly tracked by selected 850× concentrators. (b) Resulting simulated *I–V* curves.

incidence angle. This definition is not usually taken because it depends on a detailed modeling of the cell and it will be cell dependent.

Detailed cell models are becoming more available, and the time in which optics and cell modeling merge seems close. For instance, in Ref. [24] we integrated ray-tracing simulations, the irradiance, and spectrum maps produced on the solar cell surface by four 850× concentrators with flat Fresnel POE (FK, SILO, XTP, and RTP), with a 3D fully distributed circuit model that simulates the electrical behavior of a state-of-the-art Ge-based triple-junction solar. The results were very interesting and some of them unexpected.

At normal incidence of the sun, that is, when it is perfectly tracked (Figure 6.11), the FK is the one that produces the most spatially uniform illumination and the most spatially uniform illumination for all three junction bands. The SILO shows a slight increase (up to about 2000×) for the bottom cell, but that does not affect the *I–V* curve in the simulation, which is similar to that of the FK. The most surprising is the result for the RTP, whose spatial and spectral non-uniformity seem very acceptable at visual inspection, but the cell simulation reveals an important reduction of the *I–V* curve fill factor due to a strong drop of the current at which the MPP, characteristic effect of the spectral unbalance. Finally, another surprise is given by the XTP, since it shows a remarkable spatial and spectral non-uniformity (above 3000 peaks), but the fill factor reduction is caused only by a voltage drop at the MPP, which indicates that only the higher Joule losses spatial non-uniformities (not the spectral ones) affect the performance.

When the sun is mistracked by 0.6° (Figure 6.12), the difference between the concentrators become more evident. Regarding light loss at such incidence angle relative to normal incidence, the FK loses less than 1%, while RTP, SILO, and XTP lose 3, 5, and 11% respectively (due to their smaller CAPs). Regarding the drop of short-circuit current relative to normal incidence, the cell simulation shows that FK, SILO, and XTP lose the same fraction as they lose of light, but, unexpectedly, the RTP shows a 13% short-circuit current drop to be compared to the 3% loss of light. This is caused by the spectral non-homogeneity of the irradiance [10]. Regarding the reduction of the fill factor, now both the SILO and the RTP show an important drop of the current of the MPP, which reveals that the spectral

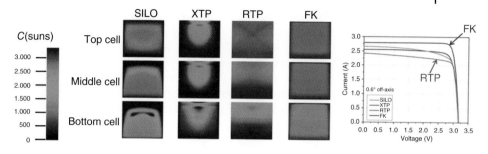

Figure 6.12 (a) Photocurrent densities produced when the sun is not perfectly tracked by selected 850× concentrators. (b) Resulting simulated $I-V$ curves.

non-uniformity is severely limiting their performance. The XTP fill factor remains similar to that at normal incidence, so the spectral non-uniformities are again not causing additional efficiency drops. The FK is the only concentrator whose $I-V$ curve fill factor remains very high, as it was at normal incidence, thanks to its capability to keep the excellent irradiance spatial and spectral uniformity also at 0.6°.

6.3
Silicone on Glass Fresnel Lenses

The majority of high concentrating photovoltaic systems on the market use Fresnel lenses as primary concentrator. Currently, two different types of Fresnel lens materials are deployed. PMMA was already used for some of the first concentrator systems developed by the Sandia project in the 1970s [25]. Another option, silicone on glass (SOG), was also developed in the 1970s [26, 27] but rarely used until it entered the market in 2005 with the founding of Concentrix Solar GmbH as a spin-off from Fraunhofer ISE. During the past years, many CPV companies decided to use SOG Fresnel lenses as primary concentrators. Accordingly, SOG Fresnel lenses have gained a significant market share.

A SOG Fresnel lens is a two component system in which a low iron float glass serves as a protective cover to the environment outside of the CPV module. This glass pane simultaneously provides a mechanical stable support structure for the Fresnel lenses that are molded into a silicone layer on the surface facing the solar cell. The SOG technology gives some important advantages over the PMMA Fresnel lenses but is also affected by some disadvantages. The most obvious disadvantage is the high density of glass that makes SOG Fresnel lenses roughly twice as heavy as PMMA Fresnel lenses. The most frequently mentioned advantage of SOG Fresnel lenses is their robustness to harsh environmental conditions. The glass surface is much more scratch resistant and glass and silicone are believed to be less prone to degradation from ultraviolet light than PMMA. A review covering these and other degradation aspects of PMMA and SOG Fresnel lenses can be found in Ref. [28].

The material combination SOG possesses a high optical transmission that expands far into the infrared region where substantial absorption occurs in PMMA Fresnel lenses. Owing to a high excess current in the bottom sub cell of currently available multijunction solar cells this difference in transparency has almost no influence on the generation of electrical energy. The situation may change in the near future when new cell concepts with only little excess current in the bottom sub cell or with more than three junctions become commercially available.

6.3.1
Physical Influence of Lens Temperature

On some high plains suitable for CPV systems, temperatures well below freezing point are common in winter; at other suitable sites in hot deserts, air temperatures often rise above 40 °C. The temperature of a Fresnel lens concentrator in a CPV system additionally depends on wind speed and solar irradiation. It seems reasonable to ask how this wide range of temperatures influences the optical properties of Fresnel lens concentrators. For SOG Fresnel lenses first theoretical and experimental works on temperature influences have been published by Schult et al. [29], Hornung et al. [30], and by Rumyantsev et al. [31].

Thermal expansion of the Fresnel lens plate changes the relative distance between the centers of different Fresnel lenses if a single Fresnel lens plate contains more than one Fresnel lens. This might cause issues with the lens to cell alignment if not considered properly in module design. Assuming isotropic thermal expansion, the thermal expansion also scales each individual Fresnel lens which leads to a scaling of the focal length. The linear coefficient of thermal expansion of float glass (9×10^{-6} K^{-1} [32]) is significantly lower than that of PMMA (6.5×10^{-5} K^{-1} [33]). Therefore, one might be misled to conclude that thermal expansion is less important for SOG Fresnel lenses than for PMMA Fresnel lenses, but this is only true for the large scale mechanical expansion of whole Fresnel lenses or Fresnel lens plates.

Although the thermal behavior of a SOG lens plate as a whole is dominated by the properties of the glass pane, the lens structures themselves are made of silicone with a coefficient of linear thermal expansion of the order of 3×10^{-4} K^{-1} [34], which is much higher than that of glass or PMMA. The thermal expansion of the silicone in lateral direction is restricted to the thermal expansion of the glass pane. This restricts the thermal expansion at the base of each individual prism of the Fresnel structure, whereas the remaining sides of the prism may freely expand. In consequence, this results in a deformation of the Fresnel structure (Figure 6.13). If used as a concentrator lens, this deformation will smear out the focal spot and may lead to optical losses by spilling light off the solar cell. This behavior is schematically pictured in Figure 6.14.

It should be noted that these arguments will still hold if the silicone layer is thicker than the Fresnel lens structure that is molded into it, that is, if a closed silicone layer remains between the glass surface and the valleys of the Fresnel lens

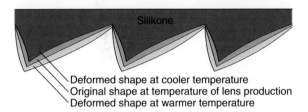

Figure 6.13 Schematic visualization of the deformation of a silicone on glass Fresnel lens structure that is caused by the mismatch of the coefficient of thermal expansion of the lens materials. For purpose of clarity, the small thermal expansion of the glass plate is not included in the picture. If the lens temperature equals the temperature during the molding of the Fresnel lens, the lens is in a stress free state. Lower or higher temperatures cause concave or convex deformations, respectively. The real deformations typically are so small that they would not be visible in this picture.

structure. The glass pane restricts the lateral expansion of the silicone layer. For reasons of symmetry the closed silicone layer expands the same way at every point. Thus, for a typical silicone layer thickness no influence on the deformation of the Fresnel lens is expected. This theoretical argument has also been verified by measurements [35].

Figure 6.15 also includes a less obvious but also important influence of temperature. Thermal expansion changes the density of an optical material and thereby also alters its optical density which is specified by the refractive index. In short, the refractive index depends on temperature. A theoretical model for this behavior was proposed independently by Lorentz [36, 37]. If the temperature of a Fresnel lens rises, its refractive index will decrease and the focal length of the Fresnel lens will increase. This increase is higher for silicone than for PMMA due to its higher coefficient of thermal expansion.

Additionally, due to typical Poisson's ratios close to 0.5 for solar silicones any reduced thermal expansion in lateral direction is mostly compensated by an enlarged expansion in the direction normal to the glass surface. Thus, the density of the silicone layer is not noticeably influenced and still only changes according to the coefficient of thermal expansion. The same accounts for the optical density, that is, the refractive index. Refractometer measurements of a silicone layer on a glass pane were performed at Fraunhofer ISE, which validated the assumption that for a typical solar silicone and a layer thickness of a few hundred micrometers, which is typical for SOG Fresnel lenses, the lateral restriction of the thermal expansion does not influence the refractive index of the silicone. Contrary findings have been reported in literature for very thin films [38].

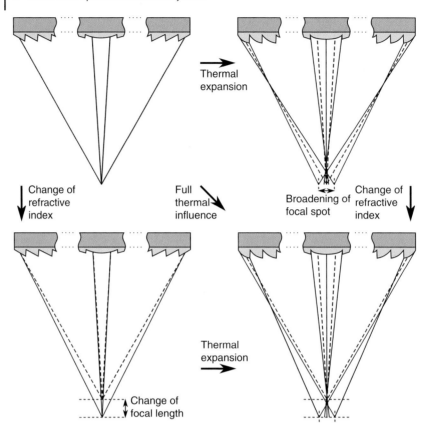

Figure 6.14 Schematic overview of the influence of lens temperature on the optical behavior of a silicone on glass Fresnel lens. The variation of the refractive index changes the focal length with temperature (bottom left figure). Simultaneously, the mismatch of the coefficients of thermal expansion of silicone and glass deforms the Fresnel lens structure which broadens the focal spot (top right figure). The combined influence of both effects is shown in the bottom right figure.

6.3.2
Influence of Lens Temperature on Efficiency

Ray tracing software is usually used for optical simulation of concentrator Fresnel lenses. Different refractive indices for different lens temperatures can easily be included in these simulations. The change of refractive index with temperature can either be measured or (less accurate) be calculated from theory by the Lorentz–Lorenz equation [36, 37] or by empirical formulas like the Eykman equation [39] and the Gladstone–Dale equation [40].

The deformation of the Fresnel structure of a SOG Fresnel lens, which is caused by thermal expansion is usually determined with applying the finite element method (FEM). The resulting complex deformed lens structure has to

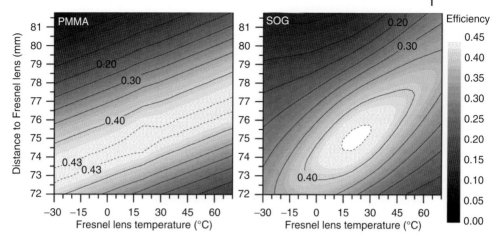

Figure 6.15 Calculated efficiency of the generation of electrical energy by a concentrator system consisting of a Fresnel lens ((a) PMMA and (b) silicone on glass) and a triple junction solar cell. The map shows the dependence of efficiency on temperature and on the distance between Fresnel lens and solar cell. The Fresnel lenses were assumed to be manufactured at 20 °C and were modeled in physical detail. Their maximum optical efficiencies are 82.4 and 83.6% for the PMMA and SOG Fresnel lens, respectively. In contrast, the solar cell was idealized to the Shockley–Queisser-Limit. With a lossless optical concentrator its efficiency would reach 52.2%.

be included in ray tracing simulations to investigate the influence on the optics. Although FEM and ray tracing simulations are usually very time-consuming tasks for complex structures like concentrator Fresnel lenses, knowledge of the properties of Fresnel structure (e.g., rotational symmetry) can be used to speed up simulations.

This allows generating a full map of the sunlight to electrical energy conversion efficiency as a function of lens to cell distance and lens temperature. An example of a distance-temperature-map is presented in Figure 6.15. The detailed optical simulations were performed for a 509× concentrating system without SOE. The simulated system consists of a single Fresnel lens and a triple junction solar cell with band gaps of 1.87, 1.41, and 0.66 eV at 298 K. The top junction was assumed to be 5.5% transparent (independent of wavelength). In contrast to the detailed physical simulation of the concentrator Fresnel lens, the efficiency of the solar cell was calculated with the program etaOpt [41] which determines the theoretical maximum solar cell efficiency given by the Shockley–Queisser limit [42]. The ASTM G173 direct normal plus circumsolar irradiation spectrum [43] and mean angular irradiation profiles [44] with the additional assumption of 0.1° tracking tolerance were used to calculate the overall system efficiency for each point on the distance-temperature-map. More details on the simulated system can be found in Refs. [45, 46].

This map contains valuable information on optimizing a CPV system. For each temperature there is a lens to cell distance that results in optimal efficiency. The connection of these distances on the distance-temperature-map gives a straight line that shows the change of focal length with temperature. This shall be called the line of optimal performance of the lens. Its slope is dominated by the change of refractive index with temperature. In the case of PMMA Fresnel lenses, a small fraction of it can also be attributed to the thermal expansion of the lens geometry. This expansion is negligible for the small thermal expansion of the glass pane of a SOG Fresnel lens.

The efficiency of a system using a SOG Fresnel lens changes along the line of optimal performance because of the deformation of the Fresnel lens structure. The PMMA Fresnel lens scales isotropically in these simulations and, therefore, shows no deformation of the lens structure. In consequence, no significant change of efficiency is present along the line of optimal performance of a PMMA Fresnel lens.

The clear distinction between the change of focal distance and deformation of the Fresnel lens structure that becomes visible in this plot can be used for lens efficiency measurements. If the detector in the measurement setup is always positioned at the distance of best optical efficiency during measurement, only changes caused by Fresnel lens structure deformation become visible. Knowledge of the dependence of the lens to cell distance on lens temperature simultaneously allows conclusions about the temperature dependence of the refractive index. This can be used to validate FEM simulations and the change of refractive index separately [30, 47].

In a concentrator photovoltaic module, the Fresnel lens remains at a fixed distance from the solar cell. This is equivalent to a horizontal line in the distance-temperature map. Obviously, the distance between Fresnel lens and solar cell or SOE should be selected according to some direct normal irradiance (DNI) weighted average lens temperature that is expected during the operation of the system. Variation of lens temperature will still cause losses but the sensitivity of the system to lens temperature variation will be minimized and the overall system efficiency will be maximized. In contrast, simulations show significantly lower annual efficiencies at locations where the average lens temperature differs from design temperature [45, 46]. For SOG lens systems, additional attention has to be paid to the fact that the lens shape gets deformed at any temperature that differs from the temperature during molding of the Fresnel lens [31, 35, 48].

A close examination of Figure 6.15 reveals that the highest energy conversion efficiencies of the system using a SOG Fresnel lens are not centered on the lens temperature of 20 °C which was chosen as the temperature during lens production. In fact, the highest efficiencies can be achieved at temperatures slightly above the temperature at which the Fresnel lens structure is in its original state. Figure 6.13 already provides the indication for this astonishing feature. Owing to thermal expansion, the lens area occupied by the optically inactive draft facet of each Fresnel prism decreases with increasing temperature and the area covered by the active slope of each Fresnel prism increases accordingly. This increase in active area over-compensates spillage losses due to the deformation of the lens structure at small temperature differences to the production temperature.

In conclusion, Fresnel lenses may exhibit high concentration efficiencies if temperature effects are considered thoroughly. The specific challenges that well designed CPV systems needs to account for are different for SOG and PMMA Fresnel lenses.

6.4 Considerations on Concentrators in HCPV Systems

This chapter is on considerations which should be made from a module manufacturer's perspective when designing the concentrator of a HCPVs module (highly concentrating photovoltaics). In the following we will concentrate on passively cooled HCPV modules; design with a central receiver are excluded because they have very different constraints typically. Passively cooled modules are typically designed as a more or less planar device where a multitude of concentrator elements and concentrator cells are arranged in an array. An image of such a HCPV module is shown in Figure 6.16.

6.4.1 General Requirements on CPV Concentrator Optics

The main application of CPV modules is in utility scale PV power plants in regions with a high direct solar irradiation. From a customer's perspective, the most interesting viewpoints are the specific cost (e.g., $US/W) and the performance ratio PR

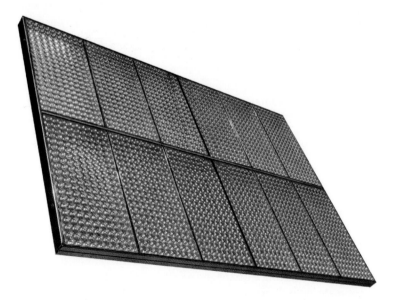

Figure 6.16 Example of a HCPV module (Soitec CX-M500).

of the installed CPV system because these define to a large part the investment and the return on investment for a given solar resource. Reliability is mandatory, which allows for 25 years operation at degradation rates which are usual in the PV market. These market expectations on CPV systems are typically translated by HCPV module manufacturers to the following requirements on the optical concentrator:

- high optical efficiency
- low cost
- high durability
- suitability for volume manufacturing.

The high optical efficiency requirement is a direct consequence of the HCPV approach itself. The reason to concentrate is to opt for highest solar-to-grid conversion efficiency which implies the use of multijunction solar cells (MJC) and is supported by the solar concentration itself leading to a higher voltage of the solar cell. The optical efficiency of the concentrator and the efficiency of the solar cell define almost completely the efficiency of the module.

The cost of the concentrator is determined by the market prices for PV modules as well as by the efficiency of the module. As of 2012, the PV module cost was in the range of 0.60–0.90 $US/Wp and the HCPV module efficiency in the range of 30% at CSTCs (concentrator standard testing conditions). It is expected that the PV module cost will further drop to 0.30–0.50 $US/Wp. The targeted cost for CPV modules are even less. This means that the CPV module cost is in the range of few tens of $US per square meter at the current efficiency. As a result, the cost for the optical concentrator can be in the range of a few $US per square meter which is a challenge for the designer and manufacturer of the optical concentrator.

The durability of the optical concentrator must allow for 25 years life time in the field as already mentioned. During this period the optical efficiency may not degrade more than 20% and – depending on the module design – probably much less in order to allow for degradation of other module components because the minimum accepted module power after 25 years is 80% of the initial value.

The suitability for volume manufacturing goes beyond cost constraints. Quality control, yield, shipping, handling, assembly, and recycling are crucial in volume manufacturing. A reasonable size for a module production site is in the range of several hundred MW capacity per year. For example, 300 MW annual production capacity means 1 million m^2 of optical concentrator per annum to be manufactured and assembled at a module efficiency of 30%.

6.4.2
Design Considerations

There are many design concepts for optical concentrators of CPV modules. A very fundamental decision is between a central receiver design where one CPV system consists of one large optical concentrator and one PV receiver (Figure 6.17) versus an assembly of many optical concentrators and PC cells in one module and the assembly of a certain number of modules on a tracker to realize one CPV system

Figure 6.17 Photo of a CPV system with a central receiver (Zenith Solar Z20; photo: courtesy D. Faiman).

(Figure 6.18). In the central receiver design with an active cooling of the receiver the separation of optical concentrator and of the PV receiver is obvious and there is little interference between the two components. As most of the central receiver designs use a large parabolic mirror (so-called "dish"), the only consideration is how to reduce the shading from the central receiver and it's mounting structure.

For a module assembly of many concentrators and PV cells, it is one of the most important considerations how to integrate the optical concentrator into a manufacturable module design because the cells have to be located at positions which are defined by the concentrator. So, the typically passive thermal management of the cell and the electrical interconnection between the cells cannot be optimized independently from the concentrator design. In the case of a Fresnel lens array as shown in Figure 6.16 as concentrator in a one-stage optical design, the result is a planar design where the plane of the concentrators is completely separated from the plane of the PV cells. This gives the highest freedom to optimize the concentrator and the receiver independently. Adding an SOE directly mounted to the cells may already increase the difficulty of the electrical interconnection of the cells because the accessibility of the front contact can be significantly reduced. In case of a Fresnel lens, the back contact electrical interconnection allows for optimization of the thermal management of the cells without major constraints.

In concentrator designs using mirrors, the thermal management and the electrical interconnection of the cells may become more difficult. For example, on-axis

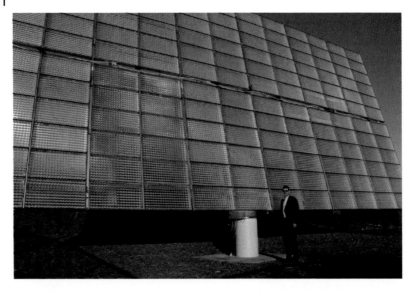

Figure 6.18 Photo of a CPV system having CPV modules with passive cooling (Soitec Solar CX-S530).

parabolic mirrors as in the central receiver design are not possible in case of passive cooling because the cooling device cannot be made small enough to avoid significant losses by shading. A much better solution with no shading losses is therefore the use of off-axis paraboloids as mirrors. One can also use Cassegrain type optical concentrators to allow for a cell arrangement in one plane but this implies inevitable shading around the optical axis due to the hyperbolic second mirror.

6.4.3
Experiences with Concentrator Designs

As pointed out earlier, there are many different concentrator designs possible and a concentrator design also depends on the optimization strategy. Therefore, only a very general review on experiences with concentrator designs can be made.

It could be observed that the majority of CPV module manufacturers used a Fresnel lens based POE. Five years ago, this Fresnel lens was most likely made out of PMMA. More recently, the share of SOG lenses increased significantly. Regarding the size of the Fresnel lens POE, there is still a huge divergence with lens areas ranging from 1600 to 100 000 mm^2. Accordingly, the CPV module thickness varies significantly as a function of the focal length of the respective Fresnel lens POE. The larger lenses are designed for more standardized receiver packages with solar cells having an area of approximately 100 mm^2. So, in case of a geometrical concentration ratio of approximately 1000, this will result automatically in the mentioned size of the POE. Reflector POEs were also designed for the 100 mm^2 solar cell size primarily.

One of the basic lessons learnt is to take into consideration all those design criteria which are not purely optical ones for the concentrator design. As already mentioned the thermal management, the manufacturability, the whole chain of process tolerances, the variation of the operating conditions, for example, the temperature variation, and last but not least the sensitivity to soiling all play a role. All commercially available CPV modules have a housing which protects the concentrator optics from soiling and have a transparent cover which can be easily cleaned. In general, it is anticipated that a CPV module has to be cleaned with cleaning periods which are site dependent.

6.5
Conclusions

An overview of the fundamentals, the optimization strategies, and specific optical designs of concentrators for CPV systems was given. Several designs of SOE were compared with respect to their CAP and the achievable photocurrent densities. When using SOG Fresnel lens POE special attention has to be given to the temperature effects of this POE type. It was shown how these effects can be minimized.

Concentrator optics for CPV modules are lowest-cost high-volume optical elements. This puts very strong constraints on the parameter space of theoretically feasible optical designs. Despite this, the field of concentrator optics for CPV is still emerging and it is only recent that very important feedback from the operation of CPV modules in high volumes can be gained. For the future, it is expected that out of the many possible design variants the most promising will be selected and incorporated in the next generation CPV module designs.

References

1. Winston, R., Miñano, J.C., and Benítez, P. (2005) *Nonimaging Optics*, Elsevier Academic Press, Burlington, MA.
2. Benitez, P. and Miñano, J.C. (2004) Concentrator Optics for the next generation photovoltaics, in *Next Generation Photovoltaics: High Efficiency through Full Spectrum Utilization* (eds A. Marti and A. Luque), Taylor & Francis, London.
3. Luque, A., Sala, G., Arboiro, G.C., Bruton, T., Canningham, D., and Mason, M. (1997) Some results of the EUCLIDES photovoltaic concentrator prototype. *Prog. Photovoltaics Res. Appl.*, **5**, 195–212. doi: 10.1002/(SICI)1099-159X(199705/06)5:3<195::AID-PIP166>3.0.CO;2-J
4. Benitez, P. and Mohedano, R. (1999) Optimum irradiance distribution of concentrated sunlight for photovoltaic energy conversion. *Appl. Phys. Lett.*, **74** (2543), 2543–2546. doi: 10.1063/1.123892
5. Olson, J.M. (2010) Simulation of nonuniform irradiance in multi-junction IIIV solar cells. 35th IEEE Photovoltaic Specialists Conference (PVSC), 2010, pp. 000201–000204. doi: 10.1109/PVSC.2010.5614523
6. Braun, A., Hirsch, B., Katz, E.A., Gordon, J.M., Guter, W., and Bett, A.W. (2009) Localized irradiation effects on tunnel diode transitions in multi-junction concentrator solar cells. *Sol.*

Energy Mater. Sol. Cells, **93**, 1692–1695. doi: 10.1016/j.solmat.2009.04.022

7. Korech, O., Hirsch, B., Katz, E.A., and Gordon, J.M. (2007) High-flux characterization of ultrasmall multijunction concentrator solar cells. Appl. Phys. Lett., **91**, 064101. doi: 10.1063/1.2766666

8. Miller, D.C., Muller, M.T., Kempe, M.D., Araki, K., Kennedy, C.E., and Kurtz, S.R. (2012) Durability of polymeric encapsulation materials. Prog. Photovolt: Res. Appl., **21**, 631–651. doi: 10.1002/pip.1241

9. Victoria, M., Herrero, R., Domínguez, C., Antón, I., Askins, S., and Sala, G. (2011) Characterization of the spatial distribution of irradiance and spectrum in concentrating photovoltaic systems and their effect on multi-junction solar cells. Prog. Photovolt: Res. Appl., **21**, 308–318. doi: 10.1002/pip.1183

10. García, I., Espinet-González, P., Rey-Stolle, I., and Algora, C. (2011) Analysis of chromatic aberration effects in triple-junction solar cells using advanced distributed models. IEEE J. Photovoltaics, **1**, 219–224. doi: 10.1109/JPHOTOV.2011.2171671

11. Luque, A. (1989) *Solar Cells and Optics for Photovoltaic Concentration*, Adam Hilger, Bristol.

12. Kritchman, E.M., Friesem, A.A., and Yekutieli, G. (1979) Highly concentrating Fresnel lenses. Appl. Opt., **18**, 2688–2695.

13. Piszczor, M.F., Swartz, C.K., O'Neill, M.J., McDanal, A.J., and Fraas, L.M. (1990) The mini-dome Fresnel lens photovoltaic concentrator array: current status of components and prototype panel testing. Proceeding of 21st IEEE Photovoltaic Specialists Conference, doi: 10.1109/PVSC.1990.111817

14. Araki, K., Yano, T., and Kuroda, Y. (2010) 30 kW concentrator photovoltaic system using dome-shaped fresnel lenses. Opt. Express, **18** (S1), A53–A63. doi: 10.1364/OE.18.000A53

15. Welford, W.T. and Winston, R. (1980) Design of nonimaging concentrators as second stages in tandem with image forming first-stage concentrators. Appl. Opt., **19** (3), 347–351.

16. Victoria, M., Domínguez, C., Antón, I., and Sala, G. (2009) Comparative analysis of different secondary optical elements for aspheric primary lenses. Opt. Express, **17**, 6487–6492.

17. Gordon, J.M., Feuermann, D., and Young, P. (2008) Unfolded aplanats for high-concentration photovoltaics. Opt. Lett., **33**, 1114–1116.

18. Benitez, P., Miñano, J.C., Blen, J., Mohedano, R., Chaves, J., Dross, O., Hernández, M., and Falicoff, W. (2004) Simultaneous multiple surface optical design method in three dimensions. Opt. Eng., **43** (7), 1489–1502.

19. Plesniak, A., Jones, R., Schwartz, J., Martins, G., Hall, J., Narayanan, A., Whelan, D., Benítez, P., Miñano, J.C., Cvetkovíc, A., Hernandez, M., Dross, O., and Alvarez, R. (2009) Photovoltaic Specialists Conference (PVSC), 2009 34th IEEE.

20. James, L.W. (1989) Use of Imaging Refractive Secondaries in Photovoltaic Concentrators. SAND89-7029, Alburquerque, New Mexico.

21. Benítez, P., Miñano, J.C., Zamora, P., Mohedano, R., Cvetkovic, A., Buljan, M., and Hernández, M. (2010) High performance Fresnel-based photovoltaic concentrator. Opt. Express, **18**, A25–A40, http://www.opticsinfobase.org/oe/abstract.cfm?uri=oe-18-101-A25 (accessed 4 November 2014).

22. Benitez, P., Miñano, J.C., Zamora, P., Buljan, M., Cvetkovíc, A., Hernandez, M., R. Mohedano, J. Chaves, and Dross, O. (2010) Free-form Köhler nonimaging optics for photovoltaic concentration. Optical Design and Testing IV, SPIE/COS Photonics Asia, Beijin, 2010, Vol. 7849.

23. Benítez, P., Miñano, J.C., and Alvarez, R. (2012) Ventana^{TM} optical train: a high-performance off-the-shelf HCPV optics solution. CPV China 2012 and 4th International CPV Workshop, Jiaxing, China.

24. Espinet, P. Algora, C., García, I., Rey-Stolle, I., Benítez, P., Chaves, J., Cvetkovic, A., Hernández, M., Miñano, J.C., Mohedano, R., and Zamora, P. (2012) Triple-junction solar cell

performance under Fresnel-Based concentrators taking into account chromatic aberration and off-axis operation. 8th International Conference on Concentrating Photovoltaic Systems, ISBN: 978-1-4244-7865-1/11
25. Swanson, R.M. (2003) in *Handbook of Photovoltaic Science and Engineering* (eds A. Luque and S. Hegedus), John Wiley & Sons Ltd, Chichester, pp. 449–503.
26. Lorenzo, E. and Sala, G. (1979) in *Proceedings of the International Solar Energy Society* (eds K.W. Böer and B.H. Glenn), Pergamon, Atlanta, GA, pp. 536–539.
27. Sala, G. and Lorenzo, E. (1979) in *Proceedings of the 2nd European Photovoltaic Solar Energy Conference* (eds R.V. Overstraten and W. Palz), D. Reide, Berlin, pp. 1004–1010.
28. Miller, D.C. and Kurtz, S.R. (2011) Durability of Fresnel lenses: a review specific to the concentrating photovoltaic application. *Sol. Energy Mater. Sol. Cells*, **95** (8), 2037–2068.
29. Schult, T., Neubauer, M., Beßler, Y., Nitz, P., and Gombert, A. (2009) Temperature dependence of fresnel lenses for concentrating photovoltaics. 2nd International Workshop on Concentrating Photovoltaic Power Plants: Optical Design and Grid Connection, March 9–10, Darmstadt, Germany.
30. Hornung, T., Bachmaier, A., Nitz, P., and Gombert, A. (2010) in *6th International Conference on Concentrating Photovoltaic Systems* (eds A.W. Bett, F. Dimroth, R.D. McConnell, and G. Sala), AIP American Institute of Physics, Freiburg, pp. 85–88.
31. Rumyantsev, V.D., Davidyuk, N.Y., Ionova, E.A., Pokrovskiy, P.V., Sadchikov, N.A., and Andreev, V.M. (2010) in *6th International Conference on Concentrating Photovoltaic Systems* (eds A.W. Bett, F. Dimroth, R.D. McConnell, and G. Sala), AIP American Institute of Physics, Freiburg, pp. 89–92.
32. Häuser, K., Kramer, B., Schmid, R.W., and Walk, R. (2002) *Gestalten Mit Glas*, 6th edn, Interpane Glas Industrie AG, pp. 24–26.
33. Weber, M.J. (2002) *Handbook of Optical Materials*, CRC Press Inc, Boca Raton, FL.
34. Dow Corning (2007) Sylgard® 184 Silicone Elastomer: Product Information.
35. Askins, S., Victoria, M., Herrero, R., Dominguez, C., Anton, I., and Sala, G. (2011) in *7th International Conference on Concentrating Photovoltaic Systems* (eds F. Dimroth, S. Kurtz, G. Sala, and A.W. Bett), AIP American Institute of Physics, Las Vegas, NV, pp. 57–60.
36. Lorentz, A.H. (1880) Über die Beziehung zwischen Fortpflanzungsgeschwindigkeit des Lichts und der Körperdichte. *Ann. Phys. Chem.*, **9** (4), 641–665.
37. Lorenz, L. (1880) Über die Refractionsconstante. *Ann. Phys. Chem.*, **11** (9), 70–103.
38. Norris, A.W., Jon DeGroot, J., Nishida, F., Pernisz, U., Kushibiki, N., and Ogawa, T. (2002) Silicone Polymers for Optical Films and Devices.
39. Eykman, J.F. (1895) Recherches réfractométriques (suite). *Recl. Trav. Chim. Pays-Bas*, **14** (7), 185–202.
40. Gladstone, J.H. and Dale, T.P. (1862) Researches on the refraction, dispersion, and sensitiveness of liquids. Proceedings of the Royal Society of London, vol. 12, pp. 448–453.
41. Létay, G. and Bett, A.W. (2001) EtaOpt – a programm for calculating limiting efficiency and optimum bandgap structure for multi-bandgap solar cells and TPV cells. 17th European Photovoltaic Solar Energy Conference, pp. 178–181.
42. Shockley, W. and Queisser, H.J. (1961) Detailed balance limit of efficiency of p-n junction solar cells. *J. Appl. Phys.*, **32** (3), 510–519.
43. American Society for Testing and Materials ASTM G173-03. (2003) Standard Tables for Reference Solar Spectral Irradiances: Direct Normal and Hemispherical on 37° Tilted Surface, ASTM International, West Conshohocken, PA.
44. Neumann, A., Witzke, A., Jones, S.A., and Schmitt, G. (2002) Representative terrestrial solar brightness profiles. *J. Sol. Energy Eng.*, **124** (2), 198–204.
45. Hornung, T., Steiner, M., and Nitz, P. (2012) Estimation of the influence of Fresnel lens temperature on energy generation of a concentrator photovoltaic

system. *Sol. Energy Mater. Sol. Cells*, **99**, 333–338.

46. Hornung, T., Steiner, M., and Nitz, P. (2011) in *7th International Conference on Concentrating Photovoltaic Systems* (eds F. Dimroth, S. Kurtz, G. Sala, and A.W. Bett), AIP American Institute of Physics, Las Vegas, NV, pp. 97–100.

47. Hornung, T., Bachmaier, A., Nitz, P., and Gombert, A. (2010) Temperature and wavelength dependent measurement and simulation of Fresnel lenses for concentrating photovoltaics, in *SPIE Photonics Europe: Photonics for Solar Energy Systems III* (eds R.B. Wehrspohn and A. Gombert), SPIE, Brussels.

48. Hornung, T., Neubauer, M., Gombert, A., and Nitz, P. (2011) in *7th International Conference on Concentrating Photovoltaic Systems* (eds F. Dimroth, S. Kurtz, G. Sala, and A.W. Bett), AIP American Institute of Physics, Las Vegas, NV, pp. 66–69.

7
Light-Trapping in Solar Cells by Directionally Selective Filters

Carolin Ulbrich, Marius Peters, Stephan Fahr, Johannes Üpping, Thomas Kirchartz, Carsten Rockstuhl, Jan Christoph Goldschmidt, Andreas Gerber, Falk Lederer, Ralf Wehrspohn, Benedikt Bläsi, and Uwe Rau

7.1
Introduction

The radiative recombination limit described by Shockley and Queisser (SQ) [1] (see Chapter 2.1) provides the maximum efficiency for the conversion of sunlight into electricity by a solar cell. This maximum efficiency depends only on the radiation balance between the sun, the solar cell, and the ambient. In this concept, radiation emitted from the solar cell into the ambient is a power loss and reduces the achievable efficiency. Therefore, the maximum efficiency is achieved if the radiative interactions are restricted to those between the sun and the cell, thereby omitting losses to the ambient [2]. This is the case for a concentrator solar cell under maximum concentration and equally for a confinement of the angle of acceptance (and emittance) of a non-concentrating cell to the solid angle calculated from the sun's diameter and distance [3]. The concept of angular restriction implies not only the suppression of radiative emission but also (for non-perfect absorbers) the suppression of re-emission of non-absorbed light, which is then redirected into the solar cell, thereby enhancing the capability of the cell to trap light in the wavelength range of weak absorption [4, 5].

Commonly, a light path enhancement in solar absorbers is achieved by incorporating scattering interfaces. Wafer solar cells are etched to introduce a less reflective, scattering texture. The inverted pyramids etched into the surface of monocrystalline silicon solar cells, for example, incorporated into the PERL-cell [6], increase together with a reflecting rear side the effective optical thickness by a factor of 40 [7] for weakly absorbed light. For thin-film solar cells, the absorber layers are thin by nature. In amorphous silicon based cells, the absorber layer is, therefore, not etched itself as in crystalline silicon (c-Si) cells but it is conformally grown onto a rough substrate that usually consists of glass covered with a transparent conductive oxide. In 1983, Tiedje *et al.* [8] reported an enhancement of the effective light path at long wavelengths by a factor of 25 in amorphous silicon deposited on a rough reflecting substrate. Current approaches to further

Photon Management in Solar Cells, First Edition. Edited by Ralf B. Wehrspohn, Uwe Rau, and Andreas Gombert.
© 2015 Wiley-VCH Verlag GmbH & Co. KGaA. Published 2015 by Wiley-VCH Verlag GmbH & Co. KGaA.

increase the light-trapping in thin solar cells span from geometric light-trapping schemes (for absorber thicknesses exceeding the wavelength) to wave optics targeting a specific wavelength range.

A geometric approach in waver cells is, for example, the implementation of crossed V-shaped grooves or pyramids on both sides of the absorber that allows to exceed the Lambertian limit for specific angles of incidence [9, 10].

Wave optics is applied using grating couplers [11–15], surface plasmon enhanced light-trapping [16–19], scattering into guided modes by nanoparticles [20–22], and photonic crystals (in 1D [23–26], 2D [22, 23], or 3D [27–29]). The response of gratings is strongly dependent on the incidence angle of the illumination, the orientation of the electric field and on the wavelength. This is disadvantageous compared to random textures for non-tracked systems [30]. Using silver nanoparticles at the rear side, Eminian *et al.* [31] report an enhancement of the light path by a factor of 10 in thin-film amorphous silicon solar cells in nip-configuration. Ferry *et al.* [32] show an absorption enhancement for a conformally grown a-Si:H solar cell for well-designed plasmonic nanostructures. Using an ultra-thin photonic crystal, Mallick *et al.* [33] measured an enhancement by a factor of 8 in the optical absorption in the relevant spectral range for an ultra-thin photonic crystal c-Si solar cell over a solid slab of equivalent volume. Apart from light-trapping, a suppression of the radiative recombination current relative to the isotropic emission assumed by the generalized Planck formula was observed in a quantum well solar cell [34] as an effect of compressive strain.

The effect of *directional* and *energy selective* optical filters attached to solar cells is discussed in this chapter. Such surfaces may result in both the enhancement or deterioration of the photovoltaic performance of a solar cell. On the one hand, restricting the cell acceptance to the small incidence angle of direct- and circumsolar irradiation enhances the maximum path length of the light in a solar cell with Lambertian surfaces even above the Yablonovitch limit [35] (ultra-light-trapping). On the other hand, under natural irradiation conditions, restrictions to small acceptance angles imply losses of diffuse sunlight, even for tracked solar cells.

Furthermore, the requirements for appropriate optical filters are severe and not every theoretical approach can be applied in their manufacture. Nonetheless, the effect of an angle restriction was successfully shown in experiments that prove a current increase by the improved light-trapping [36, 37].

The theory of energy and solid angle restriction and the theoretical limits of light-trapping and cell efficiency are summarized in the Section 7.2. As exemplary cases, the calculated efficiency limits of c-Si solar cells with directionally and spectrally selective filters are presented. In the same section and for the same type of solar cells, the gain in Annual Yield is estimated taking into account the loss in diffuse irradiation. The considered Filter Systems presented in the Section 7.3 are a 1D layer stack (Rugate filter) and a 3D opal structure. The Section 7.4 reports an Experimental Realization of a spectrally and directionally selective Bragg-like filter covering the front glass of hydrogenated amorphous silicon solar cells and of a similar filter covering the front of a germanium solar cell. The chapter eventually concludes with a Summary and Outlook.

7.2 Theory

7.2.1 Radiative Efficiency Limit

As mentioned in the introduction, an angular restriction of the transmission of photons emitted by a solar cell suggests an increase in the theoretical efficiency limit. Note that this specific effect of a suppression of the radiative recombination is only usable in the case of solar absorbers that work in the radiative limit. For non-radiative limited solar cells, the predominant effect of a spectral and angular selective filter is an absorption enhancement due to an increase in light-trapping and, therefore, a suppression of the loss of non-absorbed solar photons.

To calculate the efficiency limit in the radiative limit under angular restriction it is instructive to assume that under normal incidence all impinging photons are transmitted into the solar cell. The solar absorber is assumed to be perfect and hence the increase in light-trapping is not regarded here. Thus, the only effect of angular restriction considered here is the suppression of the emission of radiative recombination under oblique angles. It will be shown in the following text that the efficiency limit under angular restriction corresponds to the efficiency limit under optical concentration [4, 38–41].

The efficiency η of a solar cell is given by

$$\eta = \frac{J_{SC} V_{OC} FF}{P_{opt}}. \tag{7.1}$$

Here, P_{opt} is the optical power irradiated onto the solar cell, J_{SC} is the short circuit current density converted by the solar cell, V_{OC} is its open circuit voltage, and FF is the fill factor.

Imagine an idealized solar cell with unity absorption that is limited by radiative losses. For perpendicularly incident light, the cell absorbs all impinging photons above a certain band gap. Because of the perpendicular incidence, J_{SC} is not affected by an *angular restriction*. For such an idealized solar cell, the saturation current density J_0 is determined by Planck's black body radiation $\tilde{\varphi}_{bb}$,

$$\tilde{\varphi}_{bb} = \frac{2\pi E^2}{h^3 c_0^2} \frac{1}{\exp\left(\frac{E}{kT}\right) - 1} \tag{7.2}$$

$$J_0 = e \int \tilde{\varphi}_{bb} dE. \tag{7.3}$$

Here, E stands for the photon energy, c_0 is the speed of light, h Planck's constant, k Boltzmann's constant, T the temperature, and e the elementary charge. An angular selectivity or restriction of the solar cell emission reduces the radiative recombination current by the factor s to

$$J_{0,s} = \frac{J_0}{s}, \qquad (7.4)$$

which implies an open circuit voltage $V_{OC,s}$ of

$$V_{OC,s} = \frac{kT}{e} \ln\left[\frac{J_{SC}}{J_{0,s}} + 1\right] \approx \frac{kT}{e} \ln\left[\frac{J_{SC}}{J_{0,s}}\right] = \frac{kT}{e}\left(\ln\left[\frac{J_{SC}}{J_0}\right] + \ln s\right)$$

$$= V_{OC} + \frac{kT}{e} \ln s. \qquad (7.5)$$

The open circuit voltage $V_{OC,s}$ rises with the angular restriction. The fill factor FF is slightly increased by an angular restriction as a consequence of the increased open circuit voltage. For simplicity, the increase in FF is not regarded here.

The solar cell efficiency under angular restriction is thus given by

$$\eta_s = \frac{J_{SC} V_{OC,s} FF}{P_{opt}} \approx \frac{J_{SC}(V_{OC} + kT/e \ln s)FF}{P_{opt}}. \qquad (7.6)$$

Now consider the case of *optical concentration*, where the optical power P_{opt} irradiated onto the solar cell is increased by the concentration factor c. The short circuit current density $J_{SC,c} = c J_{SC}$ increases by the same factor c. As a consequence, also the open circuit voltage $V_{OC,c}$ increases with increasing concentration. Not regarding the slight increase in the fill factor FF, the efficiency changes under optical concentration to

$$\eta_c = \frac{J_{SC,c} V_{OC,c} FF}{P_{opt,c}} \approx \frac{c J_{SC}(V_{OC} + kT/e \ln c)FF}{c P_{opt}} = \frac{J_{SC}(V_{OC} + kT/e \ln c)FF}{P_{opt}}. \qquad (7.7)$$

Under optical concentration the maximum broadening of the angle range of incident illumination would widen from pointing only at the sun, θ_{sun}, to the entire halfspace, $\theta_{halfspace}$. The maximum concentration is then given by the quotient

$$c_{max} = \frac{\Omega^{proj}(\theta_{halfspace})}{\Omega^{proj}(\theta_{sun})} \qquad (7.8)$$

of the solid angles Ω projected onto the plane of the solar cell. The projection is accounted for introducing a $\cos(\theta)$-term while the angle integration brings about a $\sin(\theta)$.

$$\Omega^{proj} = \int_{\phi=0}^{2\pi}\int_{\theta=0}^{\theta_i} \sin(\theta)\cos(\theta) d\theta d\tilde\varphi = 2\pi\left[-\frac{1}{2}\cos^2(\theta)\right]_0^{\theta_i} = \pi[1 - \cos^2(\theta_i)]$$

$$= \pi \sin^2(\theta_i) \qquad (7.9)$$

Introducing the respective values $\theta_{halfspace} = \pi/2$ and $\theta_{sun} = 4.63$ mrad yields

$$c_{max} = \frac{\Omega^{proj}(\theta_{halfspace})}{\Omega^{proj}(\theta_{sun})} = \frac{\pi\sin^2(\theta_{halfspace})}{\pi\sin^2(\theta_{sun})} = \frac{\sin^2(\pi/2)}{\sin^2(4.63\,\text{mrad})} = 46\,582. \qquad (7.10)$$

Both, the confinement (angular restriction) and the broadening (optical concentration) of the angle range of incident light are given and limited by the geometry of the setup. In both cases the angle ranges of interest are the entire halfspace above the cell surface and the narrow cone pointing at the sun. Concentration and angular confinement are complementary processes in respect to the angles of incidence and emission. With concentration the angle of incidence is increased and with angular confinement the angle of emission is decreased. The maximum *angular selectivity* of the emission, s_{max}, is equal to the maximum *optical concentration*, c_{max},

$$s_{max} = c_{max}. \tag{7.11}$$

The maximum efficiencies under optical concentration and directional selectivity are thus the same. Note that the neglected increase in the fill factor is also the same for both techniques.

Note also that the two techniques can be combined to come closer to the limit of Eqs. (7.6) and (7.7) but they cannot exceed this limit [42, 43]. The confinement is restricted by the requirement that the angle of emission cannot be smaller than the angle of incidence.

7.2.2
Ultra-Light-Trapping

7.2.2.1 Universal Light-Trapping Limit for Completely Randomized Light

Next, we derive the universal limit for light-trapping named after Yablonovitch [35] for the case of a directional selective device. The photovoltaically active material emits black body radiation. The emission obeys Planck's radiation law

$$\phi(E)dEd\Omega = \frac{2E^2}{h^3(c_0/n)^2} \frac{1}{e^{E/(kT)} - 1} dEd\Omega. \tag{7.12}$$

The variable n labels the refractive index of the considered material. For simplicity, we introduce the variable

$$\phi_{bb}(E) = \frac{2E^2}{h^3 c_0^2} \frac{1}{e^{E/(kT)} - 1} \tag{7.13}$$

for the black body spectrum. In thermal equilibrium and for ideal light-trapping, the volume of a photovoltaic absorber is penetrated by a perfectly randomized flux of photons per energy interval within the solar absorber of thickness d as shown in Figure 7.1, that is,

$$\tilde{\phi}(E)dE = 4\pi n^2 \phi_{bb}(E)\alpha(E)d dE. \tag{7.14}$$

The integration over all angles yields 4π, the factor n^2 is not included in the abbreviation for the black body radiation (Eq. (7.13)) and is thus maintained. The flux scales with the absorption coefficient $\alpha(E)$ and the absorber thickness d.

The total amount of radiation (at energy E) produced in this absorber per surface area amounts to

$$\phi_{tot}(E) = 4\pi n^2 \phi_{bb}(E)\alpha(E)d. \tag{7.15}$$

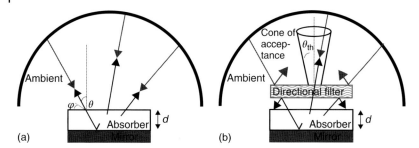

Figure 7.1 (a) In thermal equilibrium the flux emitted by the absorber is equal to the flux incident from the ambient. (b) A directional filter restricts the out-coupling to an acceptance cone defined by the opening angle θ_{th}. Figure adapted from [37].

For an optimum coupling of the solar cell to the ambient and in the limit $\alpha(E)d \ll 1$ all this radiation ϕ_{tot} would be emitted through the front surface of the solar cell. Owing to Kirchhoff's law and in thermal equilibrium, the number of emitted photons equals the radiation received from the ambient. The emitted radiation flux ϕ_{em} at energy E into an element of the spherical angle ($n = 1$) reads thus

$$d^2\phi_{em}(E, \theta, \varphi) = a(E, \theta, \varphi)\phi_{bb}(E) \cos(\theta) \sin(\theta) d\theta d\varphi, \qquad (7.16)$$

where $a(E,\theta,\varphi)$ is the absorptance and the spherical angle is defined by the pair (θ,φ) indicated in Figure 7.1a. The absorptance $a(E,\theta,\varphi)$, is contrary to the absorption coefficient $\alpha(E)$, which was considered in the absorber material itself, directionally dependent and accounting for an angular selectivity of the absorber surface. The factor $\sin(\theta)$ is introduced to account for the spherical coordinates and $\cos(\theta)$ accounts for the area reduction under oblique angles. Equalizing the emitted radiation to the radiation ϕ_{tot} produced in the solar cell yields

$$4\pi n^2 \phi_{bb}(E)\alpha(E)d = \int_0^{2\pi}\int_0^{\pi/2} a(E, \theta, \varphi)\phi_{bb}(E) \sin(\theta) \cos(\theta) d\theta d\varphi. \qquad (7.17)$$

For an absorptance that is independent from the angles of incidence we have

$$4\pi n^2 \phi_{bb}(E)\alpha(E)d = \pi a(E)\phi_{bb}(E) \qquad (7.18)$$

and, hence, an absorptance

$$4n^2\alpha(E)d = a(E). \qquad (7.19)$$

This result is equivalent to the thermodynamic limit for light-trapping derived by Yablonovitch [35] assuming a Lambertian light-trapping scheme for the complete randomization of the incident light. Independent of the exact morphology of the surface texture the upper limit is always valid for an optimum rear side reflection [44]. Miñano and Luque [45–47] have shown that this limit applies to any geometrical light-trapping scheme.

Equation (7.17) is the Yablonovitch limit, the ultimate limit for light-trapping. Note, that the derivation accounts for imperfect out-coupling and reabsorption.

A Lambertian light-trapping scheme for the complete randomization of the incident light inside the absorber is an example for an angle independent absorptance $a(E)$. Light paths exceeding the limit given in Eq. (7.19) do not contradict the statement. Equation (7.17) applies to the integrated absorption over all path lengths w corresponding to respective photon energies. We assume an arbitrary path length distribution $f(w)$. The average path length is

$$\overline{w} = \int_0^\infty f(w)w\,dw. \tag{7.20}$$

Using Beer's law, the absorptance can be defined as

$$a = 1 - \int_0^\infty f(w)e^{-\alpha(E)w}\,dw \tag{7.21}$$

or

$$a \approx 1 - \int_0^\infty f(w)(1 - \alpha(E)w)\,dw = 1 - \int_0^\infty f(w)\,dw + \alpha(E)\int_0^\infty f(w)w\,dw$$
$$= \alpha(E)\overline{w} \tag{7.22}$$

for weakly absorbed light [42], $\lim(E \to 0)\alpha(E)w = 0$, and the term $\exp(-\alpha(E)w)$ can be simplified to $1 - \alpha(E)w$. Thus, Eq. (7.19) is valid for the average path length

$$4n^2\alpha(E)d = a(E) = \alpha(E)\overline{w}$$
$$4n^2 d = \overline{w} \tag{7.23}$$

Angular confinement, as sketched in Figure 7.1b, makes it possible to overcome the absorptance and light path enhancement limits for angle independent absorptance if non-isotropic illumination is regarded [48]. Miñano [45] calculated the light path enhancement of an idealized step-function-like angle selective filter with transmittance $t(E,\theta < \theta_{th}) = 0$ and $t(E, \theta > \theta_{th}) = 1$ with an acceptance angle θ_{th}. On top of a scattering surface the light path enhancement amounts to $4n^2/\sin^2(\theta_{th})$. Equation (7.17) applies to this case as well, for such a step function of the absorptance $a(E,\theta)$ it yields the same result

$$4\pi n^2 \alpha(E)d = 2\pi \int_0^{\pi/2} a(E,\theta)\sin(\theta)\cos(\theta)d\theta d\varphi$$
$$= 2\pi \frac{\sin^2(\theta_{th})}{2} a(E, \theta > \theta_{th})$$
$$= \frac{4n^2\alpha(E)d}{\sin^2(\theta_{th})} = a(E, \theta > \theta_{th}). \tag{7.24}$$

An angle selective filter can be used to restrict the transmittance of photons to the small threshold angle of $\theta_{th} = 2.5°$. This angle restriction excludes most diffuse light while letting the direct solar and circumsolar light pass [49]. It leads to a light path enhancement of up to a factor of 26 000 for silicon. Note that it is necessary to trade off the blocked emitted light against the blocked incident light impinging onto the solar cell under oblique angles.

In general, an enhancement in light-trapping and, thus, cell absorptance allows for the use of thinner solar cells, which results in lower material consumption. This brings about another important implication: thinner cells have a better carrier collection, such that lower quality and cheaper photovoltaic absorber materials may be used. The minority charge carrier generation rate per unit volume and thus the steady state carrier concentration increases as well. As a consequence, less entropy is produced per photon, leading to a larger open circuit voltage [40].

If the absorber thickness is smaller or in the order of the photon wavelength (or a grating of the same dimensions is used), the limit derived above is no longer applicable. The solar cell can then not be described by geometric optics; instead, wave optics has to be taken into account. The upper limit for the absorption enhancement factors is determined by the local density of optical states in the absorber, which can be modified by proper design of the structure [50, 51].

In practice, the modification of the density of optical states can be done by inserting a thin low-refractive index layer between two high-refractive index layers [52, 53]. The density of optical states in the thin layer is then increased above the value n_H of the cladding layers if the thin layer has a sufficiently low thickness and refractive index, enabling a massive absorption enhancement in the thin cladding with refractive index n_L if a scattering mechanism enables isotropic filling of the optical density of states. According to Yu *et al.* and Green [52, 53], the maximum absorption enhancement F is then

$$F = 4n_L^2 \left(\frac{2n_H}{3n_L} + \frac{n_H^5}{5n_L^5} \right). \tag{7.25}$$

Realizing this concept in photovoltaics is, however, difficult because it requires an extremely thin (~10 nm) active layer of a usually organic material (to get the low refractive index $n < 2$) and a high index but ideally transparent cladding layer. Transparency in the visible range is, however, usually incompatible with a sufficiently high refractive index $n > 2$.

7.2.3
Annual Yield for Directionally Selective Solar Absorbers

This chapter discusses the influence of a directional and spectral selective absorber on the annual yield of a solar cell under outdoor conditions. The effect of a directional and spectral selective absorber has to withstand an examination that considers the outdoor situation in order to be applicable. Such an examination has to take into account the apparent solar course at a specified location on Earth, the spectral and directional distribution of the incident solar illumination and the directional and spectral selectivity of the transmission into the solar absorber. The quantity of interest is the annual yield, that is, the sum of the energy converted by the solar cell during 1 year. It is especially interesting to assess the advantages as well as the disadvantages of directional and spectral selective absorbers.

To keep the calculations as general as possible, an idealized c-Si solar cell located at the equator [54] is assumed. The position and irradiation of the sun are modeled

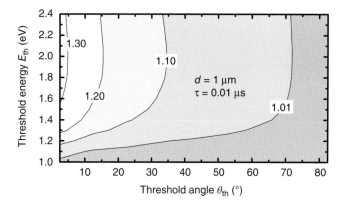

Figure 7.2 Enhancement factor of the annual energy yield of a solar cell with directional selective absorber, normalized to the case without directional selectivity. The simulation was performed for the equator for perfect tracking and crystalline silicon solar cells of the thickness 1 μm and minority carrier lifetime of 0.01 μs. The values depend on threshold energy and threshold angle of the directional selectivity. The maximum enhancement in annual energy density of 32.5% is reached for a narrow threshold angle of $\theta_{th} = 2.5°$ that assures a good light-trapping, and a threshold energy of $E_{th} = 2.0$ eV. Too small threshold energies make the filter ineffective, high threshold energies block more diffuse irradiation. A filtering surface of opening angle 90° corresponds to the case of no angular selectivity, that is, unity transmittance for the entire upper halfspace. Figure adapted with permission from [37, 54].

with SOLPOS [55] and SPTRL2 [56, 57]. Several threshold angles and energies, tracking modes as well as solar absorber thicknesses are considered.

Figure 7.2 shows a graph of the simulation result for a 1 μm thin silicon absorber with a lifetime of the minority carriers of 0.01 μs chosen to represent an intermediate material quality. Here, the solar cell is assumed to be perfectly tracked. Scattering within the cell is assumed to be Lambertian and the absorption follows Lambert–Beer's law, for details on the simulation see [54]. A threshold angle of 90° is equivalent to no restriction in the surface transmittance, that is, the isotropic reference case. Under a maximum restriction to the threshold angle $\theta_{th} = 2.5°$ (corresponding to the solar disk plus the angle range of circumsolar irradiation) and to the threshold energy $E_{th} = 2.0$ eV a maximum enhancement in the annual energy density of 32.5% is reached. For high threshold energies, the restriction is not effective while for low threshold energies a big amount of diffuse irradiation is lost.

As a side remark, it is interesting to note that in case of no tracking, the increase shrinks to 1.4% in case of the 1 μm thick absorber and that, in general, thicker absorbers profit less from the increase in light-trapping. Light-trapping schemes that bear an unintentional angular selectivity may still enable a high annual yield provided the angular restrictions are relatively wide or they are limited to a certain range of photon energies [36]. For the untracked systems the tolerance for spectral and angular selectivity increases for thinner cells.

An increase in the annual yield could also be implemented by tailoring the angular and spectral dependence of light-incoupling to the local solar illumination conditions. Such a sophisticated transmission geometry would optimize gains and minimize losses in a non-tracked system.

7.3
Filter Systems

7.3.1
1D Layer Stack Rugate Filters

One way for producing a directionally and spectrally selective reflectance profile is to use a coating system with a continuously varying refractive index [58]. The solid line in Figure 7.3 shows schematically the refractive index n as a function of the optical thickness $T_{opt} = \int n(z)dz$. The graph can be divided into three regions labeled I–III. Across the entire coating system, the refractive index is modulated as an exponential function of an argument which varies sinusoidally with the optical thickness [59]. This periodic modulation gives rise to the desired reflection band around the design wavelength λ_0. The specific choice of an exponential function with a sinusoidal argument allows to minimize the reflectance for higher order reflection bands at shorter wavelengths, which would otherwise deteriorate the light incoupling in the spectral range with a high photon flux. Since the entire coating system has a finite thickness, so-called side lobes, that is, small oscillations of the reflectance over the inverse wavelength, would be present throughout the entire spectral range. However, these side lobes can be minimized by a continuous increase or decrease of the amplitude of the modulation in regions I and III, respectively [60]. The amplitude of the modulation governs the width and strength

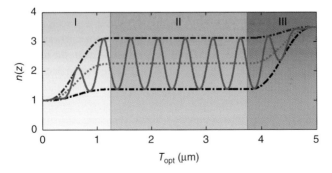

Figure 7.3 Schematic graph of the refractive index as a function of the optical thickness T_{opt}. A filter with such a profile yields closely the desired angle and wavelength selective reflectivity. Figure reproduced with permission from [58].

of the resulting reflection band, whereby the total optical thickness determines the strength as well as the antireflection properties at shorter wavelengths.

Analytical formulas describe the maximum (upper dash-dotted line), minimum (lower dash-dotted), and mean (dotted) refractive indices, as well as the refractive index $n(z)$ itself; the derivation is given in Ref. [58]. Since these values are governed by only a small number of parameters and the calculation of the angle and wavelength resolved reflectivity requires only a reasonable amount of computational resources, a complete optimization of these parameters is feasible. Figure 7.3 shows a highly idealized coating system made from dispersionless materials. It was assumed, that almost any value of the refractive index between 1.0 and 3.5 could be achieved. In order to simulate a possibly manufacturable angle and wavelength selective filter, it is instructive to consider the co-deposition [61] of the two materials SiO_2 and TiO_2, which allows for a continuous variation of the refractive index between ~1.5 (lower limit governed by SiO_2) and ~2.5 (upper limit governed by TiO_2). Both materials are transparent in the desired spectral range and allow for arbitrary stoichiometric composition.

For an optimization, the dispersion of both materials has to be taken into account. The analytical refractive index profile is then valid only for a chosen predefined design wavelength, whereas for wavelengths further apart the material dispersion will lead to deviations. Therefore, the optimization routine has to be modified to also take this design wavelength into account. Finally, we performed several simplex optimizations [62] in order to maximize the annual yield. Optimized refractive index profiles as a function of the thickness z are shown in Figure 7.4a for a wavelength λ of 300, 700, and 1100 nm. The resulting transmittance t as a function of wavelength λ and angle of incidence θ is shown in Figure 7.4b. The entire coating system is ~4.0 μm thick, which allows for several modulations of the refractive index and, hence, a clear reflection band at longer wavelengths. For oblique incidence this reflection band is blue shifted, which leads to a high reflection for large angles and a wavelength of around 1100 nm, whereas normally incident light of the same wavelength can still couple into the solar cell. However, this large thickness in combination with the small but non-vanishing absorption coefficient of TiO_2 yields significant parasitic absorption at short wavelengths. Hence, the light incoupling is reduced, which limits the overall cell efficiency. Nonetheless, the broadband reduction of reflection losses in combination with the suppression of emission at long wavelengths leads to an annual efficiency of $\eta_{annual} = 25.8\%$.

7.3.2
3D Photonic Crystal Opal Structures

A 3D photonic crystal is another possible directional selective filter. Since 1987, significant progress has been made toward the theoretical understanding and the fabrication of 3D photonic crystals [63, 64]. Especially, their feature of a complete photonic band gap has generated considerable attention. This effect allows

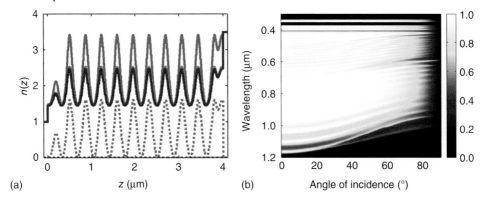

Figure 7.4 (a) Optimized refractive index profiles of an angle and wavelength selective filter made of SiO_2 and TiO_2. Due to the material dispersion the refractive index profiles differ for varying wavelengths (blue- 300 nm, orange-700 nm, dark red-1100 nm). The real part of $n(z)$ is shown as solid lines, whereas the imaginary part is dotted. (b) Corresponding transmittance as a function of wavelength and angle of incidence which approaches zero in the IR due to the reflection band and in the UV due to the absorption in TiO_2.

for realizing a number of novel concepts such as the suppression of spontaneous emission.

As before, there are two key features the 3D photonic crystal has to comply with to be useful for ultra-light-trapping. The photonic crystal should exhibit a high transmission for a large spectral range defined by the external quantum efficiency (EQE) in the crystal direction pointing to the sun so that the photons can reach the solar cell where the opening cone defines the angle of acceptance of the cell. In all other directions the filter should be opaque to avoid the escape of light that leaves the solar absorber without being absorbed. Under optimum conditions these opaque areas would cover a large share of the 2π upper halfspace, except for the angle of acceptance. Within the cone, the filter has a perfect transmission of $t = 1$ and outside the cone it provides a perfect reflection $t = 0$, in the wavelength range of low absorption of the solar cell. This is for silicon approximately in the range from $\lambda = 870$ up to 1100 nm, which is determined by the absorption properties and thus also by the thickness of the silicon solar cell.

For the following considerations, a thickness of 10 µm for the silicon solar cell was assumed. For wavelengths below $\lambda = 870$ nm the transmission in all directions should be as high as possible, in order to collect as much diffuse radiation as possible. Thus, the reduced acceptance cone is desired only in the spectral region of the electronic band gap of the absorber material where the absorption is weak [54]. This region is defined by the solar cell properties and can be determined by EQE measurements.

As mentioned above, the ideal directional selective filter has an acceptance (escape) cone of less than 5° in order to collect the circumsolar light. In Figure 7.5, the isofrequency surfaces in the region of spectral interest near the electronic band gap of the absorber is plotted for a 1D photonic crystal (Rugate filter) and

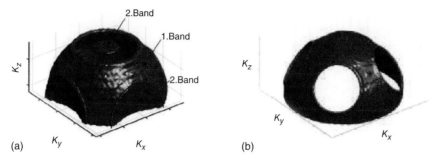

Figure 7.5 (a) Isofrequency surfaces of a 1D photonic crystal (Rugate filter) and (b) a 3D photonic crystal (inverted opal in Γ–X direction). Figure reproduced with permission from [66].

a 3D photonic crystal (inverted opal in the Γ–X direction). For the 1D photonic crystal, the acceptance cone is determined by the second band just above the first photonic band gap that closes with increasing angle due to the blueshift of the photonic stop gap. For even higher angles, the blueshift of the first band leads to photonic states and determines the outer radius for the ring-shaped acceptance area (Figure 7.5a). For 3D photonic crystals, different structures are possible that fulfill these criteria. Here, we focus on inverted opals due to their well-known properties and their fabrication simplicity. For the ultra-light-trapping concept, the inverted opal structure can be used in different high symmetry directions. Crystals grown in the Γ–L direction with the stop gap pointing to the sun use the shift of the stop gap for ultra-light-trapping, similarly to Rugate filters [65]. For this configuration, the stop gap is designed to have a center frequency f_c below the electronic band gap of the absorber (especially the spectral area of low absorption). This ensures that the stop gap reflects less light from oblique angles that can be converted by the solar cell. For non-absorbed light escaping the solar cell system with an angle larger than $\theta_{th} = 2.5°$ the stop gap shifts into the electronic band gap region of the absorber and reflects it back so that the light has another chance to be absorbed. However, the use of Γ–L inverted opals results in very strong parasitic reflections from higher-order stop gaps for higher photon energies. Unlike this, crystals grown in the Γ–X direction use the stop gap itself and not only the spectral shift for ultra-light-trapping.

The stop gap is designed directly in the spectral region of low absorption. Thus, the stop gap reflects radiation from a large part of the upper halfspace, as shown in the isofrequency surface (Figure 7.5b) where this fraction is visualized for the largest extension of the stop gap at the lower band edge of the third band. In the direction pointing to the sun, the stop gap in the Γ–X direction is closed.

For an ideal structure using only the isofrequency surface above as the filter function, the predicted annual gain can amount to about 7.5% [54]. However, for a more realistic filter function, the full spectral transmission of an inverted opal was used for the simulation. To optimize the lattice constant of the photonic crystal the diameter of the spheres forming the opal has been changed. In this way,

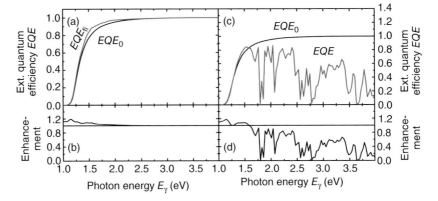

Figure 7.6 (a) Simulated quantum efficiency of a solar cell for which an inverted opal was grown in Γ–X direction on top of it. For photon energies above the photonic stop gap in Γ–X direction 100% transmission is assumed. In (b) the enhancement is given. (c) Same parameters as (a) but with a realistic transmission for photon energies above the photonic band gap based on the dispersion relation of the inverted opal (Adopted from Ref. [66], d.) gives the calculated enhancement. Figure adapted with permission from [66].

the spectral position of the stop gap shifted against the electrical band gap of the absorber. The electrical simulation [54] yields the spectrally resolved quantum efficiency for the thin film silicon solar cell (Figure 7.6a) when a perfect transmission of the entire solar spectrum through the inverted opal is assumed. The ultra-light-trapping effect is clearly visible close to the electronic band gap of the absorber. Figure 7.6b gives the enhancement. When the transmission for the solar irradiation above the photonic band gap of the realistic filter is included, the EQE for higher photon energies is strongly decreased (Figure 7.6c). Nevertheless, the direct ultra-light-trapping effect of the stop gap of the inverted opal is still visible (Figure 7.6d) as an enhancement at low energies. Thus, the directional selective filter based on an inverted opal grown in the Γ–X direction is able to increase the quantum efficiency within the spectral range of the electronic band gap of the absorber. Although this verifies the fundamental concept of ultra-light-trapping using inverted opals, the numerical simulations also show a strong decrease of the quantum efficiency for higher photon energies. This reduced quantum efficiency is due to the high fraction of parasitic reflection back to the sun caused by the flat photonic bands of the inverted opal. Therefore, the inverted opal filter for ultra-light-trapping will currently decrease the efficiency of the solar cell under solar irradiation, even if it would be possible to fabricate a perfect 3D photonic crystal on large scale. However, there might be other 3D photonic crystal structures such as graded index ones that might compensate for this low transmission above the first band gap.

7.4 Experimental Realization

This chapter describes two experimental realizations of a directional and spectral selective filter, one for a hydrogenated amorphous silicon solar cell and one for a germanium solar cell. A spectrally and directionally selective Bragg-like filter on top of a solar absorber is used to improve the light-trapping within a solar cell.

7.4.1 Bragg Filter Covering a Hydrogenated Amorphous Silicon Solar Cell

A light path enhancement is mandatory especially for hydrogenated amorphous silicon (a-Si:H) solar cells. Here, the absorber layer thickness is limited by the low mobility-lifetime-product to a few hundred nanometers [65].

Figure 7.7 sketches the cross section of a hydrogenated amorphous silicon (a-Si:H) solar cell with a directionally selective optical filter on top. State-of-the-art a-Si:H solar cells consist of a glass substrate, the etched transparent conductive oxide layer (TCO) that scatters the transmitted radiation into the photovoltaic absorber material and helps to reduce reflection losses, the absorber consisting of a p-type (p), an intrinsic (i), and an n-type (n) layer, and a second TCO layer and a silver (Ag) layer that form the back contact.

In the experimental realization in Ref. [36], the non-periodic multi-layer filter consists of 73 alternating layers of SiO_2 and Ta_2O_5 with a total thickness of the

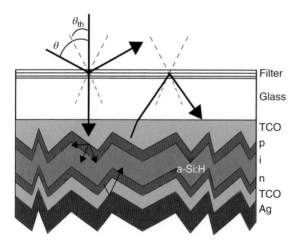

Figure 7.7 The investigated solar cells consist of a multi layer stack filter on top of the front glass, a transparent conductive oxide (TCO) layer, a pin-absorber of hydrogenated amorphous silicon (a-Si:H), and a back contact of TCO and silver (Ag). The Bragg-like filter is applied on top of the front glass after a first characterization of the cell without filter. The filter reflects light with angles exceeding $\theta_{th}(\lambda)$ and transmits all radiation impinging under incidence angles $\theta(\lambda) < \theta_{th}(\lambda)$. (Thicknesses are not to scale.) Figure reproduced with permission from [36, 37, 72].

filter of 7.5 μm. The filter was fabricated by the company mso Jena Mikroschichtoptik GmbH in a plasma-ion assisted deposition process.

Owing to the Bragg-effect, the transmission and reflection of the filter are spectrally and directionally selective. Traversing the different layers of the solar cell, the incoming radiation is partly absorbed. The non-absorbed portion of the light (large wavelengths) is re-emitted and impinges onto the filter under angles deviating – in most cases – from the perpendicular incident angle. According to the Bragg characteristic, in a certain wavelength range, the filter is transparent for perpendicularly incident and opaque for obliquely incident photons. Thus, most non-absorbed photons are reflected back into the solar cell. The light paths for photons with large wavelengths thus increase.

The reflectance and EQE are effective tools to quantify the light-trapping effect. In order to measure the impact of the filter, the solar cells are characterized prior to and after the application of the filter. In Ref. [36], the filter was tested on solar cells deposited on etched ZnO:Al on Corning glass [67].

Figure 7.8a shows the total reflectance of the a-Si:H solar cell measured for perpendicularly incident light. The absorber layer thickness of the sample amounted to 322 nm. The reflectance of the solar cell without filter (reflectance $r_0(\lambda)$, black line) rises for larger wavelengths. Here, the light path becomes smaller than the absorption length of photons in the solar cell material and the photons are re-emitted. After the deposition of the filter, the reflectance ($r_{fi}(\lambda)$, orange line) is substantially decreased in the wavelength range 650 nm $< \lambda <$ 770 nm due to the directional selectivity of the filter suppressing the re-emission of photons that

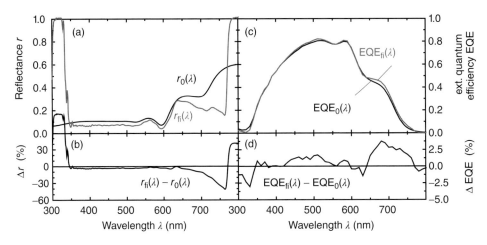

Figure 7.8 (a) The reflectance $r_{fi}(\lambda)$ with filter is decreased in comparison to the reflectance $r_0(\lambda)$ of a solar cell without filter in the wavelength range 650 nm $< \lambda <$ 770 nm. The difference $\Delta r = r_{fi}(\lambda) - r_0(\lambda)$ in (b) points out this effect even stronger. (c) In the same wavelength range the external quantum efficiency EQE is increased. Graph (d) shows the difference ΔEQE of the measurements with and without filter. The effect is due to a decrease in the re-emission by an increase in light-trapping. Figure reproduced with permission from [36, 37, 72].

leave the solar cell after a reflection at the rear without being absorbed. For wavelengths $\lambda \geq \lambda_0 = 767$ nm, the reflectance $r_{fi}(\lambda)$ rises steeply to unity once the transmission of the layer stack drops. In the wavelength range 350 nm $< \lambda <$ 650 nm, the reflectance decreases slightly due to the anti-reflective properties of the filter. The filter induces a high reflectance for $\lambda < 350$ nm, a spectral range that is negligible in the absorptance of glass-covered a-Si:H solar cells. Figure 7.8b shows the difference in the reflectances $\Delta r(\lambda) = r_{fi}(\lambda) - r_0(\lambda)$ highlighting a reduction of the reflectance by 40% at the wavelength $\lambda = 764$ nm.

Figure 7.8c depicts the EQE of the same sample with $EQE_{fi}(\lambda)$ and without $EQE_0(\lambda)$ filter as well as in Figure 7.8d the difference thereof, $\Delta EQE(\lambda) = EQE_{fi}(\lambda) - EQE_0(\lambda)$. The difference $\Delta EQE(\lambda)$ essentially reflects the features already observed in Figure 7.8a. The wavelength interval, where the quantum efficiency decreased significantly, corresponds to the wavelength range where the filter is opaque, that is, $\lambda < 350$ nm. The anti-reflective effect of the filter results in an increase of EQE_{fi} with respect to EQE_0 in the wavelength range 420 nm $< \lambda <$ 550 nm. An enhancement of EQE_{fi} is also observed in the range 650 nm $< \lambda <$ 780 nm due to the directional selectivity of the filter. Integrating the standardized AM1.5g solar spectrum over the EQE with and without filter results in short circuit current densities $J_{SC} = 13.40$ and 13.66 mA cm^{-2}, respectively. Thus, a total gain of 0.26 mA cm^{-2} is detected, thereof 0.20 mA cm^{-2} in the wavelength range 650 nm $< \lambda <$ 770 nm due to the use of the directional filter.

Note that unlike many other attempts [68, 69] to increase the light-trapping in solar cells, the open circuit voltage and the fill factor are not affected by the deposition of the filter on top of the front glass.

In order to quantify an improvement of light-trapping, Lambert–Beer's law can be used. The reflectance corresponds to the ratio of the reflected flux Φ_r to the incoming photon flux Φ_0

$$r(\lambda) = \frac{\Phi_r}{\Phi_0} = e^{-\alpha(\lambda)w_0}. \tag{7.26}$$

Here, it is assumed that light impinges under normal incidence, that the front reflectance is negligible, and that Lambert–Beer's law is used to define the experimental optical path length w_0. The relation $w_0 = k_{0/fi}w$ between optical path length w_0 and geometrical thickness w defines the path length enhancement factor $k_{0/fi}$ for the solar cell with/without filter. The optical path length w_0 includes parasitic losses as those at the back reflector. Equation (7.26) can be rewritten as

$$k_{0/fi} = -\frac{\ln(r_{0/fi}(\lambda))}{\alpha(\lambda)w}. \tag{7.27}$$

Since in the experiment the reflectances of the same device with and without filter are measured, the absorption $\alpha(\lambda)$ as well as the geometrical thickness w remain unchanged. Thus, the ratio

$$\kappa_r := \frac{k_{fi}}{k_0} = -\frac{\ln(r_{fi}(\lambda))}{\ln(r_0(\lambda))} \tag{7.28}$$

can be determined directly from the reflectance data. With similar arguments follows the definition for the quantum efficiency

$$\kappa_{EQE} := \frac{\ln(1 - EQE_{fi}(\lambda))}{\ln(1 - EQE_0(\lambda))}. \tag{7.29}$$

This evaluation method is applicable to any angle and energy selective structure on top of or below a photovoltaic absorber. The quantities κ_r and κ_{EQE} can be considered as improvement factors for the additional light-trapping provided by the filter.

Figure 7.9 exemplarily shows these factors obtained with the help of Eqs. (7.28) and (7.29) from the reflectance of the same sample as in Figure 7.8a and the EQE data in Figure 7.8b, respectively. For 350 nm $< \lambda <$ 650 nm, κ_r is above unity due to the reduction of direct reflection by the filter, whereas κ_{EQE} is very close to unity indicating no change in the EQE. For $\lambda >$ 650 nm both quantities increase because of the additional light-trapping due to the directional selectivity of the filter. In this range, κ_r represents a factor quantifying the additional light path prolongation in the device. The peak value $\kappa_{r,max} = 3.5$ in Figure 7.9 implies that the light path in the device is enhanced close to the threshold wavelength, at $\lambda_0 = 765$ nm, due to the filter. The improvement factor κ_{EQE} obtained from the EQE is significantly below κ_r with a maximum value of $\kappa_{EQE,max} = 1.5$. The quantity κ_{EQE} represents that portion of the light path prolongation that is useful for generating additional short circuit current density. The difference between the improvement factor κ_r and κ_{EQE} is due to parasitic absorption in the TCO and at the back contact. This imposes limitations to any effort to maximize light-trapping in solar cells [70]. At

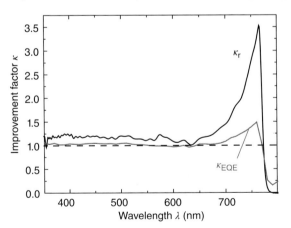

Figure 7.9 The optical path length enhancement κ_r decreases to zero for $\lambda >$ 780 nm, where the filter is opaque. For wavelengths 350 nm $< \lambda <$ 650 nm, the increased path length is attributed to better anti-reflective properties of the covering glass. The improvement of the light-trapping due to the blue shift of the reflection band in the aperiodic layer stack is visible between 650 and 780 nm. The effective path length enhancement κ_{EQE} shows the same wavelength dependence but a weaker increase. Figure reproduced with permission from [36, 37, 72].

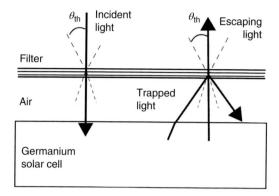

Figure 7.10 Sketch of the measurement setup used to characterize the light-trapping in a germanium solar cell. A directionally selective filter with cut-off angle θ_{th} is placed loosely on top of the germanium solar cell. Non-absorbed and scattered light or light emitted by the solar cell either escapes from the system through the escape cone or is trapped, if it hits the filter with an angle greater than the cut-off angle.

the peak values, the light path is prolonged by $\kappa_{r,max} - 1 \approx 250\%$ whereas the useful enhancement in the device is only $\kappa_{EQE,max} - 1 \approx 50\%$, that is, only 20% of the additional light confinement is used for generating additional short circuit current density and 80% is parasitically absorbed.

7.4.2
Bragg Filter Covering a Germanium Solar Cell

The effect of a spectral and angular selective filter has also been detected for a germanium solar cell [71]. For a germanium solar cell, the filter range of high transmittance must be very wide. Germanium absorbs in a spectral range between 600 and 1850 nm.

In the experimental realization, a germanium solar cell with a flat front- and rear surface was used. No scattering mechanisms were intentionally added to the cell. Light is, however, scattered at the non-polished surface to some extent. The used germanium solar cell had a thickness of 220 μm and was equipped with an antireflection coating and a rear side reflector.

To induce directional selectivity, a Bragg filter was deposited on a glass substrate – not directly onto the front glass as for the thin-film solar cell before – that was then loosely placed on top of the solar cell. This procedure was more appropriate for the wafer-based technology. Note that this implies that additional optical losses are caused by the rear side reflection of the glass substrate. The Bragg filter itself is similar to the one shown in Figure 7.4b. Yet, the transition from transmission to reflection is designed to be at 1850 nm to match the band gap of the germanium absorber. A sketch of the measurement setup is shown in Figure 7.10.

To assess the light-trapping effect of the filter, the difference between the reflectance r, $\Delta r = r_{fi} - r_0$, and EQE measurements, $\Delta EQE = EQE_{fi} - EQE_0$, with

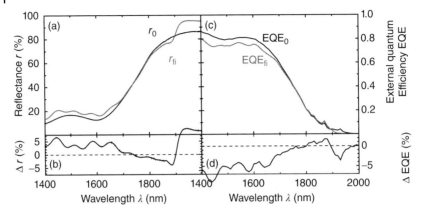

Figure 7.11 (a) Reflectance r and (c) external quantum efficiency EQE of a germanium solar cell with and without directional and spectrally selective filter. The data in (b) give the differences $\Delta r = r_{fi} - r_0$ and in (d) $\Delta EQE = EQE_{fi} - EQE_0$. The enhanced absorption induced by the filter results in a reduced reflection and an increased EQE for the system with filter. Figure adapted with permission from [71].

and without filter were calculated. The acquired data are shown in Figure 7.11a–d, respectively.

Since the solar cell does not transmit light, it can be assumed that all light that is not reflected is absorbed in the device. Reflectance and EQE are therefore opposing quantities. For wavelengths in the range ~1750 nm $< \lambda <$ 1900 nm the reflection losses diminish and the photons that enter the solar absorber are trapped. The cut-off wavelength shifts from larger angles to shorter wavelengths as shown in Figure 7.4b. The photons that were not absorbed and are scattered at the solar cell surface are partially reflected by the directionally selective filter and therefore re-enter the solar absorber. The reflectance becomes lower and the quantum efficiency higher with a filter than without a filter.

As was shown before for the a-Si:H thin-film silicon solar cell, the path length improvement can be calculated using Eqs. (7.28) and (7.29). The resulting graph is shown in Figure 7.12.

Other than for the case of amorphous silicon (Figure 7.9), the path length improvement factors κ_r and κ_{EQE} are comparable. As was mentioned before, in amorphous silicon solar cells light is parasitically absorbed in layers that do not contribute to the generation of photo current. In a germanium solar cell, only little light is absorbed parasitically.

7.5
Summary and Outlook

A directional selectivity of the surface structure of solar cells implies the same high efficiency limits as optical concentration. In the case of low absorption, directional

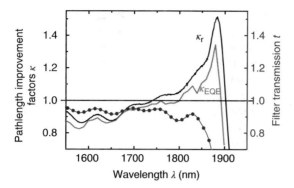

Figure 7.12 Calculated path length improvement factors κ_r and κ_{EQE}. The full circles and the respective connecting line represent the filter transmission. Photons of wavelengths close to the band gap are transmitted by the filter when normally incident but are trapped after the passage through the device where they are reflected at the rear and scattered at its front surface. Since there is little parasitic absorption in a germanium solar cell, κ_r and κ_{EQE} are comparable. Figure adapted with permission from [71].

selective filters theoretically result in an increase in light-trapping that overcomes the Yablonovitch limit for solar cells (an even higher increase in light-trapping is possible in the wave optical range). An increased light-trapping compared to a standard textured solar cell has been shown in experiments. Normally, for incident light and the standardized spectrum the augmentation in light-trapping in a hydrogenated amorphous silicon solar cell with angle selective surface results in a measurable increase in the short circuit current. The total benefit of this potential is limited by parasitic absorption in the device. An absorption and quantum efficiency enhancement has also been shown for a less-parasitic wafer-based germanium solar cell. Depending on the local spectrum and filter configuration simulations predict a maximum increase of 32.5% in the annual gain for a tracked 1 μm thick c-Si solar cell. Strong angular limitations are, however, only applicable to tracked systems. As concentrator systems are tracked, the combination of both concepts, that is, optical concentration and angular restriction, is very promising. Another opportunity for an efficiency increase is to tailor the angular and spectral dependence of light-incoupling to the local solar illumination conditions.

References

1. Shockley, W. and Queisser, H.J. (1961) Detailed balance limit of efficiency of p-n junction solar cells. *J. Appl. Phys.*, **32**, 510–519.
2. Markvart, T. (2008) The thermodynamics of optical étendue. *J. Opt. A: Pure Appl. Opt.*, **10** (1), 015008:1–015008:7.
3. Green, M.A. (1987) *High Efficiency Solar Cells*, Trans Tech Publications, Aedermannsdorf.
4. Campbell, P. and Green, M.A. (1986) The limiting efficiency of silicon solar cells under concentrated sunlight. *IEEE Trans. Electron Devices*, **33** (2), 234–239.

5. Araújo, G.L. and Martí, A. (1994) Absolute limiting efficiencies for photovoltaic energy conversion. *Sol. Energy Mater. Sol. Cells*, **33** (2), 213–240.
6. Zhao, J., Wang, A., Altermatt, P., and Green, M. (1995) 24 % efficient silicon solar cells with double-layer antireflection coatings and reduced resistance loss. *Appl. Phys. Lett.*, **66** (26), 3636–3638.
7. Zhao, J., Wang, A., Green, M., and Ferrazza, F. (1998) 9.8 % efficient 'honeycomb' textured multicrystalline and 24.4 % monocrystalline silicon solar cells. *Appl. Phys. Lett.*, **73** (14), 1991–1993.
8. Tiedje, T., Abeles, B., Cebulka, J.M., and Pelz, J. (1983) Photoconductivity enhancement by light trapping in rough amorphous silicon. *Appl. Phys. Lett.*, **42** (8), 712–714.
9. Campbell, P. (1993) Enhancement of light absorption from randomizing and geometric textures. *J. Opt. Soc. Am. B*, **10** (12), 2410–2415.
10. Campbell, P. and Green, M.A. (1987) Light trapping properties of pyramidally textured surfaces. *J. Appl. Phys.*, **62** (1), 243–249.
11. Morf, R.H. and Kiess, H. (1989) Submicron gratings for light trapping in silicon solar cells: a theoretical study. Proceeding of the Ninth International Conference on Photovoltaic Solar Energy, Freiburg, Germany, 25–29 September, pp. 313–315.
12. Gale, M.T., Curtis, B.J., Kiess, H.G., and Morf, R.H. (1990) Design and fabrication of submicron grating structures for light trapping in silicon solar cells. *Proc. SPIE* (Optical Materials Technology for Energy Efficiency and Solar Energy Conversion IX, March 12, 1990, The Hague, Netherlands), **1272**, 60–66.
13. Zaidi, S. H., Gee, J. M. and Ruby, D. S. (2000) Diffraction grating structures in solar cells. Proceeding of the Twenty-Eighth IEEE Photovoltaic Specialists Conference, Anchorage, AK, September 15–22, 2000, pp. 395–398.
14. Llopis, F. and Tobías, I. (2005) The role of rear surface in thin silicon solar cells. *Sol. Energy Mater. Sol. Cells*, **87** (1–4), 481–492.
15. Yu, Z., Raman, A., and Fan, S. (2010) Fundamental limit of light trapping in grating structures. *Opt. Express*, **18** (S3), A366–A380.
16. Schaadt, D.M., Feng, B., and Yu, E.T. (2005) Enhanced semiconductor optical absorption via surface plasmon excitation in metal nanoparticles. *Appl. Phys. Lett.*, **86** (6), 063106-1–063106-3.
17. Pillai, S., Catchpole, K.R., Trupke, T., Zhang, G., Zhao, J., and Green, M.A. (2006) Enhanced emission from Si-based light-emitting diodes using surface plasmons. *Appl. Phys. Lett.*, **88** (16), 161102-1–161102-3.
18. Derkacs, D., Lim, S.H., Matheu, P., Mar, W., and Yu, E.T. (2006) Improved performance of amorphous silicon solar cells via scattering from surface plasmon polaritons in nearby metallic nanoparticles. *Appl. Phys. Lett.*, **89** (9), 093103.
19. Beck, F.J., Polman, A., and Catchpole, K.R. (2009) Tunable light trapping for solar cells using localized surface plasmons. *J. Appl. Phys.*, **105** (11), 114310.
20. Ferry, V.E., Sweatlock, L.A., Pacifici, D., and Atwater, H.A. (2008) Plasmonic nanostructure design for efficient light coupling into solar cells. *Nano Lett.*, **8** (12), 4391–4397.
21. Saeta, P.N., Ferry, V.E., Pacifici, D., Munday, J.N., and Atwater, H.A. (2009) How much can guided modes enhance absorption in thin solar cells? *Opt. Express*, **17** (23), 20975–20990.
22. Atwater, H.A. and Polman, A. (2010) Plasmonics for improved photovoltaic devices. *Nat. Mater.*, **9** (3), 205–213.
23. Bermel, P., Luo, C., Zeng, L., Kimerling, L.C., and Joannopoulos, J.D. (2007) Improving thin-film crystalline silicon solar cell efficiencies with photonic crystals. *Opt. Express*, **15** (25), 16986–17000.
24. Zhou, D. and Biswas, R. (2008) Photonic crystal enhanced light-trapping in thin film solar cells. *J. Appl. Phys.*, **103** (9), 093102.
25. Mutitu, J.G., Shi, S., Chen, C., Creazzo, T., Barnett, A., Honsberg, C., and Prather, D.W. (2008) Thin film solar cell design based on photonic crystal and diffractive grating structures. *Opt. Express*, **16** (19), 15238–15248.

26. Park, Y., Drouard, E., El Daif, O., Letartre, X., Viktorovitch, P., Fave, A., Kaminski, A., Lemiti, M., and Seassal, C. (2009) Absorption enhancement using photonic crystals for silicon thin film solar cells. *Opt. Express*, **17** (16), 14312–14321.
27. Nishimura, S., Abrams, N., Lewis, B.A., Halaoui, L.I., Mallouk, T.E., Benkstein, K.D., van de Lagemaat, J., and Frank, A.J. (2003) Standing wave enhancement of red absorbance and photocurrent in dye-sensitized titanium dioxide photo-electrodes coupled to photonic crystals. *J. Am. Chem. Soc.*, **125** (20), 6306–6310.
28. Huisman, C.L., Schoonman, J., and Goossens, A. (2005) The application of inverse titania opals in nanostructured solar cells. *Sol. Energy Mater. Sol. Cells*, **85** (1), 115–124.
29. Mihi, A. and Míguez, H. (2005) Origin of light-harvesting enhancement in colloidal-photonic-crystal-based dye-sensitized solar cells. *J. Phys. Chem. B*, **109** (33), 15968–15976.
30. Tobías, I., Luque, A., and Martí, A. (2008) Light intensity enhancement by diffracting structures in solar cells. *J. Appl. Phys.*, **104** (3), 034502.
31. Eminian, C., Haug, F.-J., Cubero, O., Niquille, X., and Ballif, C. (2010) Photocurrent enhancement in thin film amorphous silicon solar cells with silver nanoparticles. *Prog. Photovolt.: Res. Appl.*, **19** (3), 260–265.
32. Ferry, V.E., Verschuuren, M.A., Li, H.B.T., Verhagen, E., Walters, R.J., Schropp, R.E.I., Atwater, H.A., and Polman, A. (2010) Light trapping in ultrathin plasmonic solar cells. *Opt. Express*, **18** (S2), A237–A245.
33. Mallick, S.B., Agrawal, M., and Peumans, P. (2010) Optimal light trapping in ultra-thin photonic crystal crystalline silicon solar cells. *Opt. Express*, **18** (6), 5691–5706.
34. Adams, J.G.J., Browne, B.C., Ballard, I.M., Connolly, J.P., Chan, N.L.A., Ioannides, A., Elder, W., Stavrinou, P.N., Barnham, K.W.J., and Ekins-Daukes, N.J. (2010) Recent results for single-junction and tandem quantum well solar cells. *Prog. Photovolt.: Res. Appl.*, **19** (7), 865–877.
35. Yablonovitch, E. (1982) Statistical ray optics. *J. Opt. Soc. Am.*, **72** (7), 899–907.
36. Ulbrich, C., Peters, M., Bläsi, B., Kirchartz, T., Gerber, A., and Rau, U. (2010) Enhanced light trapping in thin-film solar cells by a directionally selective filter. *Opt. Express*, **18** (S2), A133–A138.
37. Ulbrich, C. (2011) Spectral and directional dependence of light-trapping in solar cells. PhD thesis. RWTH Aachen.
38. Martí, A., Balenzategui, J.L., and Reyna, R.F. (1997) Photon recycling and Shockley's diode equation. *J. Appl. Phys.*, **82** (8), 4067–4075.
39. Badescu, V. (2005) Spectrally and angularly selective photothermal and photovoltaic converters under one-sun illumination. *J. Phys. D: Appl. Phys.*, **38** (13), 2166.
40. Markvart, T. (2007) Thermodynamics of losses in photovoltaic conversion. *Appl. Phys. Lett.*, **91** (6), 064102.
41. Peters, M., Goldschmidt, J.C., and Bläsi, B. (2009) How angular confinement increases the efficiency of photovoltaic systems. Proceeding of 24th European Photovoltaic Solar Energy Conference, Hamburg, Germany, September 21-25, 2009.
42. Martí, A. and Araújo, G.L. (1996) Limiting efficiencies for photovoltaic energy conversion in multigap systems. *Sol. Energy Mater. Sol. Cells*, **43** (2), 203–222.
43. Peters, M., Goldschmidt, J.C., and Bläsi, B. (2010) Angular confinement and concentration in photovoltaic converters. *Sol. Energy Mater. Sol. Cells*, **94** (8), 1393–1398.
44. Brendel, R. (1995) Coupling of light into mechanically textured silicon solar cells: a ray tracing study. *Prog. Photovolt.: Res. Appl.*, **3** (1), 25–38.
45. Miñano, J.C. (1990) Optical confinement in photovoltaics, in *Physical Limitations to the Photovoltaic Solar Energy Conversion* (eds A. Luque and G.L. Araújo), Adam Hilger, Bristol.
46. Miñano, J.C. and Luque, A. (1988) in *Proceeding of the of 8th European Photovoltaic Solar Energy Conference* (eds I. Solomon, B. Equer, and P. Helm),

Kluwer Academic, Florence, Dordrecht, p. 1387.

47. Luque, A. (1989) *Solar Cells and Optics for Photovoltaic Concentration*, Adam Hilger, Bristol, p. 508.

48. Luque, A. (1990) in *Physical Limitations to Photovoltaic Energy Conversion* (eds A. Luque and G.L. Araújo), Adam Hilger, Bristol, p. 1, 28.

49. Buie, D., Dey, C.J., and Bosi, S. (2003) The effective size of the solar cone for solar concentrating systems. *Solar Energy*, **74** (5), 417–427.

50. Stuart, H.R. and Hall, D.G. (1997) Thermodynamic limit to light trapping in thin planar structures. *J. Opt. Soc. Am. A*, **14** (11), 3001–3008.

51. Callahan, D.M., Munday, J.N., and Atwater, H.A. (2012) Solar cell light trapping beyond the ray optic limit. *Nano Lett.*, **12** (1), 214–218.

52. Green, M.A. (2011) Enhanced evanescent mode light trapping in organic solar cells and other low index optoelectronic devices. *Prog. Photovolt.: Res. Appl.*, **19** (4), 473–477.

53. Yu, Z., Raman, A., and Fan, S. (2010) Fundamental limit of nanophotonic light trapping in solar cells. *Proc. Natl. Aacd. Sci. U.S.A.*, **107** (41), 17491–17496.

54. Ulbrich, C., Fahr, S., Üpping, J., Peters, M., Kirchartz, T., Rockstuhl, C., Wehrspohn, R., Gombert, A., Lederer, F., and Rau, U. (2008) Directional selectivity and ultra-light-trapping in solar cells. *Phys. Status Solidi. A*, **205** (12), 2831–2843.

55. Reda, I. and Andreas, A. (2004) Solar position algorithm for solar radiation application. *Solar Energy*, **76** (5), 577–589.

56. Bird, R.E. and Riordan, C.J. (1984) Simple Solar Spectral Model for Direct and Diffuse Irradiance on Horizontal and Tilted Planes at the Earth's Surface for Cloudless Atmospheres. Technical Report No. SERI/TR-215-2436, Solar Energy Research Institute, Golden, CO.

57. Bird, R.E. and Riordan, C.J. (1986) Simple solar spectral model for direct and diffuse irradiance on horizontal and tilted planes at the Earth's surface for cloudless atmospheres. *J. Clim. Appl. Meteorol.*, **25** (1), 87–97.

58. Fahr, S., Ulbrich, C., Kirchartz, T., Rau, U., Rockstuhl, C., and Lederer, F. (2008) Rugate filter for light-trapping in solar cells. *Opt. Express*, **16** (13), 9332–9343.

59. Southwell, W.H. (1997) Extended-bandwidth reflector designs by using wavelets. *Appl. Opt.*, **36** (1), 314–318.

60. Southwell, W.H. (1989) Using apodization functions to reduce sidelobes in rugate filters. *Appl. Opt.*, **28** (23), 5091–5094.

61. Gunning, W.J., Hall, R.L., Woodberry, F.J., Southwell, W.H., and Gluck, N.S. (1989) Codeposition of continuous composition rugate filters. *Appl. Opt.*, **28** (14), 2945–2948.

62. Nelder, J.A. and Mead, R. (1965) A simplex method for function minimization. *Comput. J.*, **7** (4), 308–313.

63. Yablonovitch, E. (1987) Inhibited spontaneous emission in solid-state physics and electronics. *Phys. Rev. Lett.*, **58** (20), 2059–2062.

64. John, S. (1987) Strong localization of photons in certain disordered dielectric superlattices. *Phys. Rev. Lett.*, **58** (23), 2486–2489.

65. Deng, X. and Schiff, E.A. (2003) Amorphous silicon-based solar cells, in *Handbook of Photovoltaic Science and Engineering* (eds A. Luque and S. Hegedus), John Wiley & Sons, Ltd, Chichester.

66. Üpping, J., Ulbrich, C., Helgert, C., Peters, M., Steidl, L., Zentel, R., Pertsch, T., Rau, U., and Wehrspohn, R.B. (2010) Inverted-opal photonic crystals for ultra light-trapping in solar cells. *Proc. SPIE* (Photonics for Solar Energy Systems III, April 12-16, 2010, Brussels, Belgium), **7725**, 772519.

67. Löffl, A., Wieder, S., Rech, B., Kluth, O., Beneking, C., and Wagner, H. (1997) Al-doped ZnO films for thin-film solar cells with very low sheet resistance and controlled texture. Proceeding of the 14th European PV Solar Energy Conference, Barcelona, Spain, June 30–July 4, 1997.

68. Feitknecht, L., Steinhauser, J., Schluchter, R., Fay, S., Domine, D., Vallat-Sauvin, E., Meillaud, F., Ballif, C., and Shah, A. (2005) Investigations on fill-factor drop of microcrystalline silicon p-i-n

solar cells deposited onto highly surface-textured ZnO substrates. Proceeding of 15th International Photovoltaic Science and Engineering Conference, Shanghai, China, October 10-15.
69. Bender, H., Szlufcik, J., Nussbaumer, H., Palmers, G., Evrard, O., Nijs, J., Mertens, R., Bucher, E., and Willeke, G. (1993) Polycrystalline silicon solar cells with a mechanically formed texturization. *Appl. Phys. Lett.*, **62** (23), 2941–2943.
70. Deckman, H.W., Wronski, C.R., Witzke, H., and Yablonovitch, E. (1983) Optically enhanced amorphous silicon solar cells. *Appl. Phys. Lett.*, **42** (11), 968–970.
71. Peters, M., Ulbrich, C., Goldschmidt, J.C., Fernandez, J., Siefer, G., and Bläsi, B. (2011) Directionally selective light trapping in a germanium solar cell. *Opt. Exp.*, **19**, A136–A145.
72. Ulbrich, C., Peters, M., Tayyib, M., Bläsi, B., Kirchartz, T., Gerber, A. and Rau, U. (2010) Enhanced Light Trapping in Thin Amorphous Silicon Solar Cells by Directionally Selective Optical Filters, Proceedings of SPIE 7725, p. 77250.

8
Linear Optics of Plasmonic Concepts to Enhance Solar Cell Performance

Gero von Plessen, Deepu Kumar, Florian Hallermann, Dmitry N. Chigrin, and Alexander N. Sprafke

8.1
Introduction

Thin-film solar cells, with typical film thicknesses in the range of $1-2\,\mu m$, offer the benefit of reduced costs of materials and fabrication as compared to conventional silicon solar cells. However, they have the disadvantage of small absorbance at near-bandgap wavelengths (for example, in the red and near-infrared for silicon), resulting in small photocurrent densities. Common strategies to increase the absorbance are a reduction of reflection losses at the cell surface and a trapping of light in the absorbing layer, both through structuring of the cell surfaces. In recent years, interest has grown in replacing the randomly textured layers conventionally used for this purpose by nanostructures made of metals such as silver and aluminum. They offer the advantages of very compact dimensions, large scattering/diffraction efficiencies, and potentially useful optical near-field effects.

The promising optical properties of nanostructures made of silver, aluminum, and a few other metals, for example, gold and copper, result from the fact that they support *surface plasmons*, which are collective excitations of conduction electrons at the metal surface and able to interact very strongly with the light field [1]. In metal nanoparticles, they are also called *particle plasmons* (PPs) or *localized surface plasmons*; on structured metal films, they are also known as *surface-plasmon polaritons (SPPs)*. Their strong interactions with the light field show themselves, for example, as huge optical cross-sections for light absorption and scattering. The large values of these cross-sections, which are caused by the great number of conduction electrons oscillating in concert even in relatively small optically excited metal nanoparticles [2], make them interesting for the incoupling of light into photovoltaic layers. In addition, the nanoparticles show a rich spectral behavior. Their spectra can be controlled by various factors: the choice of the metal, their shape and size, their dielectric environment, and the electromagnetic coupling between them. This control provides the potential for tuning the plasmon excitations into spectral regions where they are the most useful for applications in solar cells. An additional focus of interest are the optical near fields at and near the metal surface;

Photon Management in Solar Cells, First Edition. Edited by Ralf B. Wehrspohn, Uwe Rau, and Andreas Gombert.
© 2015 Wiley-VCH Verlag GmbH & Co. KGaA. Published 2015 by Wiley-VCH Verlag GmbH & Co. KGaA.

they are very strong, typically up to one order of magnitude higher than the incident field. These high fields have the potential of enhancing, through near-field energy transfer, light absorption in absorber materials in the vicinity of the metal nanostructures.

The interesting optical properties of metal nanostructures are a direct consequence of the fact that surface plasmons are collective electron excitations that can interact resonantly with optical fields. On the downside, their electronic character also implies that they are subject to dissipative processes in the metal such as electron scattering, resulting in optical absorption. The optical absorption of the metal has the potential to adversely affect the solar cell efficiency unless great care is taken as to where and in what form the metal nanostructures are integrated into the cell.

This contribution provides an overview of some concepts for linear optical applications of plasmonic metal nanostructures in thin-film solar cells. It describes the optical properties of the plasmonic nanostructures, with a special focus on metal nanoparticles, and discusses various parameters that can be used to control their properties to enhance solar-cell performance. Illustrative examples from the recent research literature and some results from our own work are provided where appropriate.

8.2
Metal Nanoparticles

The optical properties of metal nanoparticles are central to an understanding of the plasmonic nanostructures integrated into thin-film solar cells, be it random nanoparticle arrangements, periodic nanoparticle arrays, or protrusions on metal contacts. This section gives a description of various optical aspects of plasmonic nanoparticles.

8.2.1
Optical Excitations in Metal Nanoparticles

In metal nanoparticles, essentially two types of electron excitations can be excited by light: interband transitions and PPs.

Interband transitions are due to transitions of individual electrons between different energy bands of the metal. For example, in silver, electrons can be excited from occupied states of the 4d band into empty states of the 5sp band above the Fermi level; these transitions show themselves as an absorption edge above 3.8 eV [1]. In gold, transitions occur from the 5d into the 6sp band and set in at 1.8 eV, with an additional contribution above 2.38 eV. In aluminum, interband transitions lie in a narrow range around 1.5 eV; they occur between two parallel bands near the Σ axis of the $\Gamma-K-W-X$ plane [3]. Interband transitions often play an undesired role in applications, because they primarily result in absorption and give only small contributions to light scattering and near-field effects. While they can also luminesce, their luminescence efficiencies are, due to the very fast carrier–carrier scattering in metals, so low that for most practical purposes the luminescence from metal nanoparticles is negligible [4].

The excitations in metal nanoparticles that are interesting for applications are the PPs. The PP modes consist of collective oscillations of conduction electrons relative to the positively charged ion background. The oscillations can be triggered by high-frequency external electric fields, for example, the optical field of an incident plane light wave or the local optical field of an adjacent dipole emitter. These fields exert a periodic force on the conduction electrons. As shown in Figure 8.1a, the resulting periodic displacement of the electron liquid with respect to the positive ions causes polarization charges of opposite signs at opposite surfaces of the nanoparticle. The PP modes show themselves as resonance maxima in optical absorption and light-scattering spectra measured on metal-nanoparticle ensembles or computed in terms of the absorption and scattering cross-sections, σ_a and σ_s, respectively [1]. Their resonant character is due to the fact that the positive ions exert, via the Coulomb attraction, a restoring force on the oscillating conduction electrons. In general, various PP modes can appear in a metal nanoparticle, which differ with respect to the spatial distribution of polarization charges and to their frequencies. For instance, in spherical nanoparticles the PP modes can be classified in terms of multipole orders according to the symmetry of the polarization charges and their emission patterns. For example, a dipolar plasmon mode has, at a given moment, one positive and one negative polarization-charge antinode (Figure 8.1a) and the emission characteristics of a dipole. A quadrupolar mode has two positive and two negative polarization-charge antinodes (Figure 8.1b) and the emission characteristics of a quadrupole. Analogous properties apply to higher multipolar orders. Of all multipole modes, the dipolar mode has the lowest frequency; with increasing multipolar order, their resonance frequencies increase and lie closer together. As described below, higher multipolar orders are excited by plane light waves only in large nanoparticles. In the optical spectra of small spherical nanoparticles, the resonance of the dipolar mode is dominant and is therefore often called *the* PP resonance.

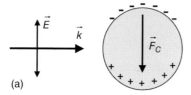 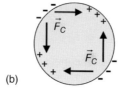

Figure 8.1 Schematic representations of polarization charges during the collective oscillation of conduction electrons associated with different particle-plasmon modes in a spherical metal nanoparticle. (a) Dipolar mode. (b) Quadrupolar electric field $\vec{E}$ and wave vector $\vec{k}$. The conduction electrons are periodically displaced with respect to the positive ion cores. Polarization charges appear where the negative und positive charge clouds do not overlap. The ions exert an attractive Coulomb force $\vec{F}_C$ on the oscillating electrons; this is the restoring force of the oscillation.

The linewidth of the PP resonances is determined by lifetime broadening and – in particle ensembles – by inhomogeneous broadening. Part of the lifetime broadening is caused by scattering of electrons off phonons and impurities; in nanoparticles smaller than ~5 nm, electronic interactions with the nanoparticle surface may contribute additionally [1]. Another contribution to the lifetime broadening stems from the decay of PPs into electron–hole-pair excitations, a process called *interband damping* [5]. This type of decay occurs when the PP resonance is energetically degenerate with interband transitions, and is due to the fact that the Coulomb field between ions and oscillating electrons excites interband transitions. Another contribution to the lifetime broadening is *radiation damping*, which is caused by the collective oscillation of conduction electrons radiating energy into the far field [5, 6]. The combined effect of all these lifetime broadenings results in homogeneous linewidths of (depending on situation) ~0.1–1.0 eV in noble-metal and aluminum nanoparticles. An important contribution to the total width of the PP line of particle ensembles are inhomogeneous broadening effects, which are primarily caused by variations of the PP energy due to fabrication-related differences between the particles [5].

One of the most intriguing properties of metal nanoparticles are the near-field effects associated with the collective electron oscillations. The polarization charges at the nanoparticle surface create an optical near field, which rapidly decays from the particle surface, typically within a few tens or hundreds of nanometers. It is maximal at or near the PP resonance frequency and can, for instance at specific points at the surface of spherical gold nanoparticles, reach enhancement factors of more than 10 with respect to the incident light field [7]. Two effects contribute to these high enhancement factors: first, the resonator nature of the collective electron oscillations allows the electromagnetic field energy to accumulate over various cycles of the exciting field [8]. Second, the electromagnetic field is concentrated spatially due to the strong curvature of the nanoparticles surface, which results in high polarization charge densities; this is known as the *lightning-rod effect* [9]. The field enhancement due to the near field can be used to enhance the excitation of absorbing species (for example, molecules or semiconductor materials) exposed to the near field. The metal nanoparticle can thus be used as a kind of "nanoantenna" for the coupling of the incident field into the absorbing species.

8.2.2
Control of Optical Properties

The optical properties of interband transitions are determined by the bandstructure of the metal and the optical selection rules. In small metal clusters they can be influenced by quantum size effects; however, this is usually not possible in metal nanoparticles with sizes above approximately 5 nm. In contrast to interband transitions, it is a remarkable feature of the PP resonance that its optical properties can be tailored, within wide limits, using several electrodynamic effects. This control is of great interest for applications and involves the spectral positions, widths, and peak values of the resonances as well as their scattering quantum efficiencies, radiation patterns, and near-field distributions. In the following sections, we will briefly discuss how to control each of these properties.

8.2.2.1 Resonance Energies of Particle Plasmons

The resonance energies of the PPs can be controlled using the following effects:

- *Selection of metal:* The PP resonance frequency depends, above all, on the dielectric properties of the metal used. These properties are determined by the density of conduction electrons and the partial screening of the Coulombic restoring force of the electron oscillation by interband transitions [10]. Thus, some control over the PP energy can be exerted via a selection of the metal. For example, a small nanosphere of aluminum embedded in glass has a PP energy of 4.8 eV (in the near UV), while a similar nanosphere of silver or gold has an energy of 2.9 (in the blue spectral region) or 2.3 eV (in the green), respectively (Figure 8.2a).

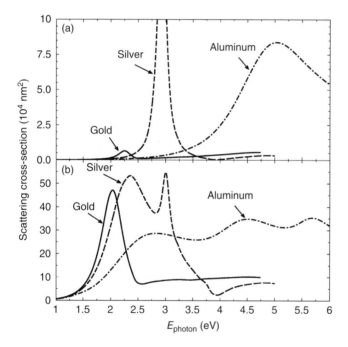

Figure 8.2 Calculated scattering cross-sections of gold, silver, and aluminum nanospheres embedded in glass (refractive index 1.5), for two different nanosphere radii: (a) $R=20$ nm. (b) $R=50$ nm. The spectra were calculated using Mie theory; dielectric data were taken from Ref. [11]. The peak at the lowest photon energy E_{photon} in each spectrum is the dipolar particle-plasmon resonance. Peaks at higher energies in (b) are associated with higher multipole resonances. The scattering background on the high-energy side of the dipolar resonance in the spectra of the gold nanospheres are caused by interband transitions.

- *Dielectric environment:* In addition, the dielectric properties of the nanoparticles environment exert an effect on the frequency of the PP resonance. For instance, the Coulomb field of the polarization charges at the particle surface induces, in turn, polarization charges in the surrounding medium (e.g., glass), which partially screen the Coulombic restoring force and thus result in a red shift of the PP resonance by up to several hundred millielectron volts with respect to a free (i.e., non-embedded) small nanosphere [12]. If only a part of the particle surface touches the medium, the screening is correspondingly weaker and the red shift smaller.
- *Particle size:* In large nanoparticles, effects caused by the electromagnetic retardation are important [1]. Owing to the finite velocity of light, the phase of the electromagnetic field is, in contrast to the case of a small free nanosphere, not constant over the extension of the particle. This applies both to the exciting light and the restoring Coulombic field. The non-uniform phase of the exciting field has the effect that higher multipolar modes beyond the dipolar mode are excited [13]. Simultaneously, the dipolar resonance experiences a shift to lower frequencies because the interference of the Coulomb fields from various regions of the nanoparticle is not entirely constructive; thus, the restoring force of the oscillation is weakened. This red shift as compared to a small nanosphere can amount to several hundred millielectron volts in large nanoparticles (cf. Figures 8.2a,b). This effect is known as *dynamic depolarization* [14] and is of a polaritonic nature. For this reason, the PP of large nanoparticles is also sometimes called a *particle-plasmon polariton*.
- *Particle shape:* A particularly strong influence on the spectral position of the PP resonance is exerted by the particle shape. For nonspherical particle shapes, the single dipolar plasmon resonance of the spherical case splits into several resonances that are associated with different PP modes [1]. For instance, the longitudinal PP mode in prolate spheroidal particles, which have one long and two short axes, is characterized by a collective electron oscillation along the long particle axis. In the transversal mode, the conduction electrons oscillate along one of the short axes. These two modes can be separately excited with light that is polarized along the respective axis. The longitudinal PP resonance is usually redshifted with respect to the PP resonance of a spherical nanoparticle of equal volume. This difference is caused by the greater distance between the polarization charges at opposite particle ends and by the resulting weaker Coulombic restoring force of the electron oscillation. This redshift can reach values of several hundred millielectron volts and larger in nanoparticles with high aspect ratios of the axis lengths. In contrast, the resonance of the transversal mode is usually blueshifted with respect to that of the spherical particle; this results from the smaller distance between the polarization charges. *Nanorods* are similar to prolate spheroidal particles in that they also support longitudinal and transverse plasmon modes [5]. Similarly to prolate spheroidal particles, their longitudinal PP resonance can be shifted far into the red and near-infrared [15]. In the limit of very large aspect ratios of the particle axes, the nanorods become *nanowires,* which have a transversal plasmon resonance similar to that of short nanorods

but no discrete longitudinal one [16]. Pronounced shape dependencies of the resonances have also been found for other particle shapes, for example, flat star-shaped particles, nanorings, and spherical core–shell particles [17–21].
- *Particle–particle coupling:* Another possibility for controlling PP resonances results from electromagnetic interactions between adjacent nanoparticles. When two or more nanoparticles lie close to each other (typically at a surface-to-surface distance below one particle diameter), the PPs of the individual nanoparticles interact via their optical near fields and form coupled oscillation modes. This effect is known as *plasmon hybridization* [22]. The coupled PP modes show characteristic distributions of oscillating polarization charges in the individual nanoparticles. They can experience considerable spectral shifts with respect to the modes of the uncoupled nanoparticles. For example, in a nanoparticle pair, the longitudinal symmetric mode, in which the conduction electrons of the two nanoparticles oscillate along the connecting axis and in phase with each other, may be redshifted by several hundred millielectron volts at small particle spacings [23]. The reason is that the opposite polarization charges at the nanoparticle surfaces to both sides of the gap attract each other across the gap, and the attractive force between them weakens the Coulombic restoring force inside each of the nanoparticles. Another example is the transversal symmetric mode, in which the conduction electrons in the two normal particles oscillate perpendicularly to the connecting axis and in phase with each other. Its resonance is blueshifted, because the surface charges of equal sign are closer to each other than those of opposite sign; thus, the Coulombic restoring force is, in total, enhanced by a repelling part. Spectrally shifted coupled PP modes appear also in more complex nanoparticle arrangements [24]. For large spacings (typically above surface-to-surface spacings of one particle diameter), radiative interactions are usually stronger than near-field coupling. In this situation, the retarded radiation field of one particle can, depending on distance and relative phase, enhance or reduce the Coulombic restoring force in another particle and thus blue-shift or red-shift the resonance, respectively [25]. Usually these shifts are small and amount only to several millielectron volts or tens of millielectron volts.

8.2.2.2 Linewidths of Particle-Plasmon Resonances

Another possibility to control the optical properties of metallic nanoparticles is based on adjusting the linewidth of the PP resonance. Here, one can use the following mechanisms:

- *Interband damping:* A lower limit for the linewidth is set by the scattering rate of the conduction electrons in the metal. For instance, a minimal linewidth of 71 meV results from the Drude scattering time of 9.3 fs in gold [5]. However, this minimum value is only reached when additional broadening effects such as interband damping play no role. This situation can be realized by redshifting the PP resonance from the range of its energetic degeneracy with the interband transitions.

- *Radiation damping:* A very efficient method for enhancing the linewidth is by enhancing the radiation damping of the PP resonance. This can be achieved by increasing the particle volume [5]. The reason for this size dependence is a superradiant effect in which the radiative decay rate of the collective electron oscillation is proportional to the total number of oscillating electrons. For instance, a 100 nm sized spherical gold nanoparticle in glass has a plasmonic linewidth of approximately 0.5 eV, while the radiative contribution to the linewidth of a small nanosphere of approximately 10 nm diameter is negligibly small. If higher multipole orders besides the dipolar mode appear in large nanoparticles, they generally have a lower radiation damping, because, due to the symmetry of their polarization charge distributions, they couple less strongly to the light field than the dipolar mode. The radiative linewidth of the PP resonance is also influenced by interactions with the particle environment. For instance, the coupling to other nanoparticles through the near field can exert such an influence. The radiative linewidth of coupled PP modes can be superradiantly increased or subradiantly reduced, depending on their symmetries. Similarly, in geometries where the particle spacing is too large for an interaction via the near field, the interference between the emissions of the nanoparticles can influence the linewidth superradiantly or subradiantly, depending on the particular spacing [25]. This is an effect that can, for example, be observed in particle gratings [26].
- *Inhomogeneous broadening:* Another important parameter is the inhomogeneous contribution to the linewidth of particle ensembles. Its minimum value is caused by inevitable inaccuracies of the fabrication process, which lead to shape and size differences between the particles and thus to variations in the PP line position from one particle to the next [5]. An intentional enhancement of the inhomogeneous contribution is possible by enhancing these differences in the fabrication process.

8.2.2.3 Peak Heights of Particle-Plasmon Resonances

The peak value of the PP in absorption and scattering spectra can be maximized by making the particles larger, because the oscillator strength of the PP increases with the number of oscillating conduction electrons. In particular, the oscillator strength can assume very large values for the PP mode oscillating along the long axis of a prolate or oblate nanoparticle [5]. Another obvious way of increasing the plasmonic absorption and light scattering of particle ensembles is to increase the number density of particles in two or three dimensions.

8.2.2.4 Scattering Quantum Efficiencies

The scattering quantum efficiency of the PP, that is, the probability of a PP in an optically excited nanoparticle to be transformed into an emitted photon is determined by the ratio of radiative to total decay rate of the PP. It can be computed from the absorption and scattering cross-sections of the nanoparticle [27]. For small metal particles in the size range of a few nanometers, the scattering quantum efficiency is close to zero, which means that such a nanoparticle absorbs almost all of the energy taken up from the exciting light field and transforms it into heat. Owing to the superradiant nature of the emission, the radiative decay rate and thus the scattering quantum efficiency increase with particle volume. For instance, a 100 nm-sized gold nanosphere embedded in glass exhibits a radiative quantum efficiency of more than 80%. Thus, a large volume of the nanoparticle is advantageous when a high scattering quantum efficiency is to be achieved. The scattering quantum efficiency can also be increased through constructive interference between the emitted fields from nanoparticles that lie close to each other [25]. In addition, the radiative quantum efficiency may be increased by minimizing the nonradiative decay rate, which contributes to the total decay rate. For instance, this can be achieved by tuning the PP resonance away from the interband transitions so that the probability for PPs to be transformed into nonradiative electron excitations is relatively low.

8.2.2.5 Light-Scattering Patterns

The light-scattering pattern of a resonantly excited small nanoparticle is essentially dipolar, that is, as much light is scattered into the propagation direction of the exciting wave as is in the opposite direction. If preferential scattering in the forward direction is desired, this can be achieved by making the nanoparticle larger [1]. In nanoparticles large enough to make the retardation of the light field felt, the exciting light wave generates a superposition of PP modes of various orders whose emissions interfere constructively in forward direction and destructively in backward direction. Another possibility to influence the light-scattering pattern of metal nanoparticles relies on the observation that emitters positioned at the interface between a low and a high refractive index medium emit the larger fraction of their emission power into the high refractive index medium [28]. For instance, a metal nanoparticle deposited on a transparent substrate will preferentially scatter incident light into the substrate [29]. This effect can be understood in a simple classical picture: An emitter emits the most power into the medium in which the field lines detach most easily from the emitter due to a relatively low value of the velocity of light. The light-scattering pattern can also be influenced by interferences with the emission of other nanoparticles. An example is a regular array of nanoparticles with a period of more than half a wavelength; the interference between the waves scattered off the individual nanoparticles results in the generation of different diffraction orders, as usual for gratings.

8.2.2.6 Near-Field Effects

The spectral position of the maximum local field enhancement can be tuned via the PP energy. Furthermore, the spectral width of the field enhancement is influenced by the lifetime broadening of the PP line: A large lifetime broadening results not only in a spectrally broad field enhancement but also in a reduced peak value of the enhancement [5]. Enhancement factors higher those that for spherical nanoparticles can be achieved by using shape anisotropies of the nanoparticles. For instance, enhancement factors of up to ∼30 are possible at the tips of prolate nanoparticles when the longitudinal PP mode is excited [30].

This difference with respect to spherical nanoparticles is a result primarily of the lightning-rod effect (Section 2.1), which is very pronounced for large curvatures of the nanoparticles surface, for example, at the tips of prolate nanoparticles. The local field enhancement can even reach almost 3 orders of magnitude in narrow (on the order of a few nanometers) gaps between nanoparticles, if the exciting field is polarized along the connecting line between the nanoparticles [30]. The reason for this additional enhancement is that the Coulomb attraction between the opposite polarization charges on both sides of the gap concentrates these charges at the surface points closest to each other and thus maximizes the field between these points.

8.2.2.7 Combinations of Effects

As described in the previous sections, there are various possibilities to tailor the spectral properties of metal nanoparticles. However, the optical properties can usually not be adjusted independently of each other. For instance, if it is desired to increase the peak value of the PP line by increasing the particle volume, this is not possible without a retardation-induced red shift and a superradiance-induced lifetime broadening. If, for the same purpose, the particle is given an anisotropic shape, this results in a spectral splitting of the PP line and a polarization dependence of light scattering and absorption. If the same goal is to be achieved by increasing the number density of particles in a particle ensemble, this may result in a mutual coupling of particles through the near field or in interference phenomena in the far field, both of which may lead to undesired line shifts and broadenings. However, in many cases an improvement of specific optical properties, while minimizing undesired side effects, can be realized by systematically varying parameters in model calculations and selecting the most advantageous parameter combinations.

8.3
Surface-Plasmon Polaritons

In contrast to metal nanoparticles, extended metal surfaces support another type of plasmonic excitations, SPPs. These mixed plasmonic–photonic excitations consist of longitudinal surface waves of conduction electrons at the metal surface, accompanied by copropagating near fields [31]. On flat, smooth metal surfaces, the SPP has a different dispersion from that of light in free space, with a greater momentum at each frequency. This momentum mismatch means that it cannot couple to light propagating in free space. Instead, incident light will simply be reflected off the surface and existing SPPs can propagate along the surface without radiating. However, the missing momentum for coupling of light to the SPPs can be provided by a periodic modulation of the metal surface. In such situations, SPPs can absorb and emit light in directions controlled by momentum conservation. The periodic modulation of the metal surface giving rise to the momentum matching is usually a modulation of the surface profile, either by periodically varying the film thickness or the film curvature. The modulated surface profile spatially modulates the Coulombic restoring force of the local electron oscillations that form the SPP, and thus results in Bragg scattering at the zone edges of the electromagnetic Brillouin zone, giving rise to an SPP band structure [32]. An intriguing property of SPPs is that they may, similarly to PPs in metal nanoparticles, cause local-field enhancements at the metal surface, due to the large polarizability of the electron system at the resonance frequency [33]. An additional important property of structured metal films that is advantageous for devices is that they are electrically conductive and can be used for carrier injection and extraction. A problem for applications involving SPP band gap structures are the dissipative energy losses in the metal. For instance, they limit the propagation length of SPPs on flat silver surfaces to a few tens of microns in the visible wavelength range [33]. Dissipation in the metal is also involved in quenching the fluorescence of emitters in the vicinity of the metal very efficiently; here, energy is transferred from the emitters to SPPs and electrical currents in the metal via near-field coupling, and is subsequently lost due to the dissipation. This effect can severely limit the fluorescence yield of fluorescence markers and light-emitting diodes [34].

8.4
Front-Side Plasmonic Nanostructures

The low absorption of the photovoltaic layer of thin-film solar cells and the high reflectivity of the surface, which prevents part of the solar light from entering the solar cell, are central problems for thin-film photovoltaics. Many studies of recent years have focused on the potential offered by plasmonic metal nanostructures for a solution to these two problems [8, 27, 35–37, 29, 38–47].

To reduce the reflectivity of the surface and couple in more light, one can attempt to use the forward scattering of metal nanoparticles deposited on or close to the absorbing layer (Figure 8.3a) [40]. As described in Section 2.2.5, metal nanoparticles located on a refractive medium scatter light incident from air preferably into the medium. If the relative fraction of the light power forward-scattered by the particles is greater than the transmissivity of the bare surface, the effect of the nanoparticles is that of an antireflection coating. The antireflection effect of the metal nanoparticles differs from conventional antireflection coatings by its resonant character. Because it is based on the PP resonance, it has a pronounced spectral dependence. In particular, this is a problem in the short wavelength range, at optical frequencies above the resonance frequency. Here, the light scattered by the plasmonic electron oscillation has an optical phase different from that of unscattered light. This results in a Fano-like destructive interference between light waves scattered in forward direction and unscattered light waves, giving rise to an increased light reflection rather than increased transmission [48].

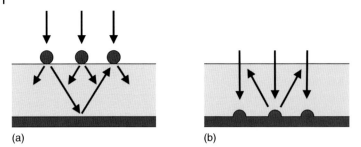

Figure 8.3 Schematic representations of plasmonic nanostructures for thin-film solar cells. (a) Light scattering off an array of metal nanoparticles at the front surface of the solar cell. (b) Light scattering/diffraction off an array of protrusions at the back contact of the solar cell. In both geometries, scattering/diffraction increases the optical path length in the absorber layer. For sufficiently high scattering angles, light can be trapped in the layer by multiple reflections/scattering events.

A further important problem for the use of metal nanoparticles as an antireflection coating is the absorption by the nanoparticles. Small gold nanoparticles absorb almost all of the incident light and only a small fraction is scattered. One approach to solve this problem is to superradiantly increase the scattering quantum efficiency by increasing the nanoparticle volume, as described in Section 2.2.4. For example, the scattering quantum efficiency of a spherical gold nanoparticle in glass at the spectral position of the dipolar PP resonance can be increased from 12 to 81% by increasing the particle radius from 20 to 50 nm. An appropriate selection of the metal used may bring a further improvement. For instance, a spherical aluminum nanoparticle in glass has a calculated scattering quantum efficiency of 90% at the spectral position of the dipolar PP resonance; for silver, it is even 95%. One reason for the superiority of aluminum and silver over gold is that interband damping of the dipolar PP resonance in these metals is weaker or even absent, respectively. As shown in Ref. [48], aluminum is superior to silver in the short wavelength range above a photon energy of approximately 2 eV, because the PP resonance (and hence destructive interference effects and parasitic absorption) lie at higher energies for aluminum.

To solve the problem of low absorption in thin-film solar cells, one can attempt to improve the light trapping by making use of the fact that nanoparticles scatter light not only exactly in forward direction but in an angle distribution around that direction [40]. For perpendicularly incident light, every deviation of the direction of scattered waves from the surface normal results in a lengthening of the optical path length in the absorber layer and thus an increased absorption. This increase of path length becomes particularly large when the scattering angle lies outside the escape cone of the absorber layer and is thus sufficient to guide the scattered light inside the layer due to total internal reflection. The scattering angle distribution can be influenced by interferences between adjacent metal nanoparticles. A special case of such scattering is the diffraction off a periodic array of nanoparticles, where the directions of scattered light group together into diffraction orders [41].

In a majority of studies on metal nanostructures on front sides, the nanostructures were randomly arranged metal nanoparticles [8, 27, 29, 38–40, 46, 47], which were, for example, deposited by evaporation. In these studies, the focus was mostly on antireflection aspects. The plasmon-induced increase of transmission through the surface was investigated and the resulting photocurrent increases were measured. For example, increases of the external quantum efficiency (EQE) and short-circuit current density of more than 8% relative to a flat reference cell were observed in Ref. [38]. Usually, the model of independent nanoparticles was used for an understanding of these increases. In addition to randomly arranged metal nanoparticles, periodic arrays of nanoparticles on the front side have also been studied [39, 41, 42, 44, 48]. For example, an increase of 22% in the spectrally integrated EQE with respect to a flat reference structure was achieved in Ref. [48]. Another important result was that a decrease of front-side transmission with increasing wavelength can be compensated by a simultaneous increase of light trapping efficiency that results from the increase of diffraction angle with increased wavelength [46].

8.5
Rear-Side Plasmonic Nanostructures

In addition to front-side structuring, it is also possible to provide the rear side of thin film solar cells with a plasmonic nanostructure (Figure 8.3b). Similarly to its front-side counterpart, one purpose of this kind of nanostructure is to scatter incident light at sufficiently large angles as to increase the optical path length in the absorber layer and improve light trapping [46, 49–55]. In contrast to the front-side nanostructure, it has to scatter the light in backward direction. The demands on the spectral bandwidth are less comprehensive: It only needs to scatter light that has not been absorbed already during the first passage through the absorber layer. In typical thin-film silicon solar cells, only light of wavelengths above 500 nm reaches the rear side through the amorphous or microcrystalline silicon layer, respectively, and needs to be scattered; the structure thus needs not be optimized for shorter wavelengths [46, 52].

In a study on rear-side nanoparticle arrays, the geometrical parameters relevant for a high light trapping efficiency were systematically studied [50]. The following conditions were identified as important: first, a high scattering cross-section of the individual nanoparticles to achieve a high diffraction efficiency of the array, implying a metallic fill factor of the rear surface of at least ~20%. Second, the period of the array needs to be large enough (e.g., minimally 300 nm in Si for a wavelength of 1000 nm) to give rise to at least one diffracted order in the absorber layer at long wavelengths.

One way to maximize the fraction of backscattered light is to integrate the metal nanostructure into the metallic back contact of the solar cell [51–53]. If the latter is thick enough, scattering in forward direction is impossible, and all light that reaches the structured metal back contact and is not absorbed there is scattered back into the absorber layer. Similarly to front-side nanostructures, it is important to minimize absorption losses in the metal and make the scattering quantum efficiency as large as possible. In analogy to nanoparticles at the front side, this can be achieved by making the scattering structures at the back contact large enough. For example, in Ref. [52] semi-ellipsoidal silver protrusions with a radius of 110 nm and height of 80 nm on a silver layer resulted in an absorption in the metallic back contact below 12%. The spatial field distributions of the localized SPP modes of these protrusions look similar to those of PP modes in similarly sized silver nanodisks without a silver substrate [53].

Similarly to the case of front-side random nanoparticle arrangements, the metal nanoparticles or protrusions of the back-side structure may be randomly arranged. Improved light trapping due to diffuse scattering from distributions of protrusions has been observed in terms of photocurrent enhancements with respect to flat reference structures without plasmonic nanostructures [46]. Good control over the scattering-angle distributions can be achieved with periodic arrays of protrusions. For example, the silver protrusions in Ref. [52] were arranged in a quadratic array with a period of 500 nm. For wavelengths greater than 500 nm, the scattering angle of the discrete diffraction orders was large enough as to guide the diffracted light in the absorbing layer by total internal reflection at the front surface and metallic reflection at the silver back contact. The resulting increase of the optical path length was revealed in increases of the EQE and short-circuit current (relative short-circuit current increase: 19%) with respect to a flat reference structure. These values even slightly exceeded those of a state-of-the-art solar cell with conventional dielectric light-trapping texture [52]. A similar improvement over a conventional light-trapping texture has been observed for a nanostructured plasmonic back contact in an ultrathin a-Si:H solar cell [51]. A still larger improvement of EQE and short-circuit current in Ref. [52] is prevented by the fact that a fraction of the guided light is coupled out of the structure by renewed diffraction off the array [52]. Furthermore, the achieved improvement shows a clear dependence on the period of the array. This dependence is caused by two effects: first, the diffraction angles of the light scattered off the reflective array become smaller with increasing period, which results in less efficient light trapping. Second, the surface coverage of the protrusions decreases with increasing period, resulting in less efficient diffraction [52].

8.6
Further Concepts

The front-side and rear-side plasmonic nanostructures discussed in Sections 8.4 and 8.5 have been the subject of a great number of studies. In particular, rear-side structures have been very successful. However, some other concepts also deserve mentioning. An interesting extension of both the front-side and rear-side concepts are nanostructured metallic intermediate reflectors for tandem thin-film solar cells [56, 57]. The reflector in such cells has the task of reflecting short-wavelength light back into the top cell, which has the larger band gap. However, the topic of intermediate reflectors for tandem cells is covered elsewhere in this volume, and will not be pursued here. Another interesting concept is based on the strong local-field enhancement generated by PPs. If photoexcited metal nanoparticles are in direct contact with a material with high absorption rates, the enhanced local field may be able to transfer the PP energy more rapidly to the surrounding material than nonradiative relaxation in the metal can transform it into heat in the metal. In this way, even small metal nanoparticles can be used as "receiver antennas" for light. For example, this concept has been shown to lead to enhanced efficiencies of ultrathin-film organic solar cells [58, 59].

Another field in which plasmonic nanostructures may be of potential use for photovoltaics are fluorescence concentrators and frequency downconverters. Fluorescence concentrators are based on the absorption of sunlight in large-area layers in which it is transformed, via fluorescence emission, into light of smaller photon energies, which is guided to a solar cell at the edge of the layer [60]. Another concept based on the conversion into radiation of lower photon energy are frequency downconverters. Here, a photon absorbed in the layer is transformed, through fluorescence, into one or two photons of lower energy, which can be absorbed in a solar cell that has a better efficiency in the low-energy spectral region than in the high-energy spectral region [61].

The suitability of both kinds of fluorescent layers is deteriorated if the radiative quantum efficiency of the fluorescent species (e.g., organic fluorophores, semiconductor quantum dots, or rare earth irons) is small. This can be caused by nonradiative relaxation processes, in particular electron–phonon interaction and concentration quenching [61, 62]. To solve this problem, it has been suggested to increase the quantum efficiency of poorly radiating emitters in fluorescence concentrators by coupling them electromagnetically to metallic nanoparticles [63]. Through this coupling, the optical nearfield of the emitter excites PPs in the nanoparticle; due to the much higher dipole moment of the coupled system, the emitter radiates away its energy much faster than it could on its own. The coupling to such an metallic nanoantenna thus results in a shortening of the radiative lifetime of the emitter, reducing the effect of the nonradiative relaxation, and increasing the radiative quantum efficiency of the fluorescent transitions.

Such emission enhancements due to antenna effects, which are based on the coupling of atomic, ionic, and molecular emitters to metallic nanoparticles, have been studied in a great number of recent experimental and theoretical works [41, 64–76]. It has been shown that the observability of the emission enhancement depends on the orientation and position of the emitting dipole relative to the nanoparticle surface. If the dipole is not oriented perpendicular to the surface but parallel to it, no increase but a decrease of the emission rate occurs due to a reduced dipole moment of the coupled system. Furthermore, a certain fraction of the excitation energy is lost to nonradiative relaxation processes in the metal. For too small distances between emitter and nanoparticle, this contribution even becomes dominant so that the radiative quantum efficiency decreases (*fluorescence quenching*) [66, 77].

Conversely to the described coupling, it is also possible that the excitation of a metal nanoparticle transfers energy to a nearby emitter via the near field of the PP. The local-field enhancement generated by the PP can thus cause, in an emitter located in the near field, an increase of the power absorbed from the light field. This increase results in enhanced fluorescence intensity, similarly to the intensity enhancement caused by an increase of the quantum efficiency.

An enhancement of fluorescence intensity through absorption enhancement could be important in situations in which the fluorescent layer (e.g., due to material costs) needs to be so thin that not all incident light is absorbed without the help of nanoantennas. In contrast, an improvement of the radiative quantum efficiency could be important if the fluorescence yield is reduced by nonradiative relaxation processes.

Many of the recent theoretical works on fluorescence enhancement using nanoantennas restrict themselves to situations in which the emitter is located in specific positions and with a specific dipole orientation relative to the exciting field [73]. However, closer to applications are geometries in which emitters are distributed homogeneously in a spatial region around a nanoantenna, as it is the case with metal nanoantenna arrays deposited on fluorescent layers. For such geometries, we have studied, in particular, photoluminescence enhancement of rare earth ions such as Sm^{3+} and Er^{3+}. For a simple geometry in which the ions homogeneously surround a single spherical nanoparticle, the results of our model calculations, in which we used extended Mie theory [78], showed that, due to detrimental effects (light absorption by the nanoparticle, luminescence quenching of the emitters by the metal, absence of ions within the metal volume, destructive interference effects at emission frequencies above the plasmon resonance frequency), no enhancement of photoluminescence intensity nor yield may be expected.

Better results can be achieved for geometries which allow one to make use of the particular spatial distribution of optical near and far fields. For example, a geometry looks interesting in which circular metal nanodisks couple to a flat layer of ions. In order to predict and optimize the absorption and emission behavior of such a system, we have calculated the changes of the excitation and decay rates in the vicinity of a circular aluminum nanodisk (Figure 8.4) [79]. The results of the calculations show that the expected enhancement of the photoluminescence in the relevant locations below the nanoparticle can be several times higher in the case of the disk than in that of the spherical nanoparticle. While this result is encouraging, it is not clear what results are to be expected for a coupling to an extended nanoparticle array, in which mutual electromagnetic interaction between adjacent nanoparticles and collective emission effects can occur. For the excitation process, array and fields can be described using periodic boundary conditions; the same, however, does not apply to a spontaneous emitter located at a single position. As yet, there has been no method for a treatment of this situation. We have therefore developed a theoretical method in which the dyadic Green's function

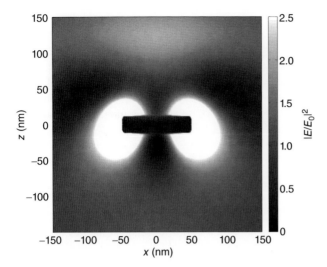

Figure 8.4 Example of calculated relative excitation rate of Sm^{3+} ions in the vicinity of an aluminum nanodisk embedded in glass. A plane light wave at a photon energy of 2.30 eV is incident from above. Shown in a gray-scale representation is the ratio of the excitation rate with and without the nanodisk, in a plane through the nanodisk center. The light wave is polarized in the plane. The relative excitation rate was calculated as $|E/E_0|^2$, where E and E_0 are the electric-field amplitudes of the optical field with and without the nanodisk, respectively. Bright areas indicate enhanced excitation rates due to plasmon-induced local-field enhancement. The calculation was carried out solving Maxwell's equations using a finite-element method.

of the problem is computed for a given position of the emitter. Calculations of the emission changes brought about by the coupling to an extended nanoparticle array are presently in preparation.

Two specific features of rare-earth ions make an exact calculation of the photoluminescence enhancement more challenging. First, Förster energy-transfer processes are important in ensembles of rare earth ions at high ion densities. Being electromagnetic processes, their rates can potentially be changed by coupling to metallic nanoparticles. Using the theoretical approach of Ref. [80] and finite-element calculations, we were able to show that the effect of metallic nanoparticles on the Förster transfer rates is negligibly small. The reason is the small Förster radius of rare earth ions, which allows for a sizable ion-ion coupling through a metallic nanoparticle only at distances on the order of ∼1 nm and smaller from the particle surface; at such small distances, however, photoluminescence quenching by the nanoparticle is dominant. Second, some optical transitions in rare-earth ions (e.g., the 1523 nm transition in Er^{3+}) have a partially magnetic character and can therefore not be described using the conventional description of the emitter as an electrical point dipole. In a first step toward the treatment of transitions that are not purely electric, we have carried out calculations of the plasmon-induced emission changes of magnetic dipole emitters. The emitters were modeled as magnetic point dipoles in a finite-element calculation.

The nanoparticles considered were metallic nanospheres, rings, and disks. The most important result of these calculations is that the electric field induced by the magnetic dipoles induces emission rate changes of a quantitative size similar to those from electric dipoles. As a consequence, if transitions in rare earth ions such as Er^{3+} have a sizable magnetic component, the contribution of these components to the plasmon-induced emission changes should not be neglected.

8.7
Summary

The purpose of this contribution was to provide an overview of several concepts for linear optical applications of plasmonic metal nanostructures in thin-film solar cells. It started off by describing the optical properties of metal nanostructures, with a special focus on metal nanoparticles and the unique qualities of PPs. It went on to discuss various nanoparticle parameters that can be used to control their linear optical properties with the goal of enhancing solar-cell performance. We discussed the concepts of front-side and rear-side plasmonic metal nanostructures. Both concepts seek to improve light trapping in the solar cell; in addition, front-side nanostructures are also able to reduce reflection losses over part of the optical spectrum. Both concepts have shown enhanced spectral response as compared to flat reference structures. However, they both also suffer from absorptive losses in the metal, necessitating careful optimization of the nanostructures with respect to scattering quantum efficiencies and other properties. As a result of such optimization, thin-film solar cells with plasmonic rear-side metal nanostructures have advanced to a point where their EQEs and short-circuit current densities now match or just slightly exceed those of randomly textured thin-film solar cells. In contrast, the concept of plasmonically enhanced down-converters, which is still in the stage of fundamental research on nanoantenna effects, will require substantial further advances to prove its potential benefits for photovoltaics.

Plasmonic metal nanostructures offer the advantages of very compact dimensions, large scattering/diffraction efficiencies, spectral tunability, and potentially useful optical near-field effects. On the downside, plasmons also cause parasitic optical absorption and undesired destructive interference effects at and above the plasmon resonance energy. The future potential of plasmonic metal nanostructures for enhancing thin-film solar-cell performance will hinge on how well these adverse effects can be minimized and the overall properties of the metal nanostructures optimized, whether new ideas for advanced plasmonic structures can be brought to fruition for photovoltaics, and ultimately, how cost-intensive the required nanostructuring will be in large-area fabrication.

Acknowledgments

We thank U. W. Paetzold, R. Carius, E. A. Moulin, K. Bittkau, J. C. Goldschmidt, S. Fischer, R. Wehrspohn, D. Schneevoigt, C. Rockstuhl, R. Filter, J. Qi, D. Lehr, M.

Zilk, and T. Pertsch for discussions. The authors acknowledge financial support from the Deutsche Forschungsgemeinschaft (DFG) through Project PI 261/5-1.

References

1. Kreibig, U. and Vollmer, M. (1995) *Optical Properties of Metal Clusters*, Springer-Verlag.
2. Alivisatos, P. (2003) The use of nanocrystals in biological detection. *Nat. Biotechnol.*, **22**, 47–52.
3. Langhammer, C., Schwind, M., Kasemo, B., and Zoric, I. (2008) Localized surface plasmon resonances in aluminum nanodisks. *Nano Lett.*, **8**, 1461–1471.
4. Dulkeith, E. et al (2004) Plasmon emission in photoexcited gold nanoparticles. *Phys. Rev. B*, **70**, 205424.
5. Sönnichsen, C. et al (2002) Drastic reduction of plasmon damping in gold nanorods. *Phys. Rev. Lett.*, **88**, 0774021–0774024.
6. Wokaun, A., Gordon, J.P., and Liao, P.F. (1982) Radiation damping in surface-enhanced Raman scattering. *Phys. Rev. Lett.*, **48**, 1574.
7. Kelly, K., Coronado, E., Zhao, L., and Schatz, G. (2003) The optical properties of metal nanoparticles: the influence of size, shape, and dielectric environment. *J. Phys. Chem. B*, **107**, 668–677.
8. Schaadt, D., Feng, B., and Yu, E. (2005) Enhanced semiconductor optical absorption via surface plasmon excitation in metal nanoparticles. *Appl. Phys. Lett.*, **86**, 063106.
9. Liao, P.F. and Wokaun, A. (1982) Lightning rod effect in surface enhanced Raman-scattering. *J. Chem. Phys.*, **76**, 751–752.
10. Liebsch, A. (1993) Surface-plasmon dispersion and size dependence of Mie resonance: silver versus simple metals. *Phys. Rev. B: Condens. Matter*, **48**, 11317–11328.
11. Johnson, P.B. and Christy, R.W. (1972) Optical constants of noble metals. *Phys. Rev. B*, **6**, 4370–4379.
12. Templeton, A.C., Pietron, J.J., Murray, R.W., and Mulvaney, P. (2000) Solvent refractive index and core charge influences on the surface plasmon absorbance of alkanethiolate monolayer-protected gold clusters. *J. Phys. Chem. B*, **104**, 564–570.
13. Krenn, J.R. et al (2000) Design of multipolar plasmon excitations in silver nanoparticles. *Appl. Phys. Lett.*, **77**, 3379–3381.
14. Meier, M. and Wokaun, A. (1983) Enhanced fields on large metal particles – Dynamic depolarization. *Opt. Lett.*, **8**, 581–583.
15. Perez-Juste, J., Pastoriza-Santos, I., Liz-Marzan, L.M., and Mulvaney, P. (2005) Gold nanorods: synthesis, characterization and applications. *Coord. Chem. Rev.*, **249**, 1870–1901.
16. Ditlbacher, H. et al (2005) Silver nanowires as surface plasmon resonators. *Phys. Rev. Lett.*, **95**, 257403.
17. Jin, R. et al (2001) Photoinduced conversion of silver nanospheres to nanoprisms. *Science*, **294**, 1901–1903.
18. Mock, J.J., Barbic, M., Smith, D.R., Schultz, D.A., and Schultz, S. (2002) Shape effects in plasmon resonance of individual colloidal silver nanoparticles. *J. Chem. Phys.*, **116**, 6755–6759.
19. Kottmann, J.P., Martin, O., Smith, D.R., and Schultz, S. (2000) Spectral response of plasmon resonant nanoparticles with a non-regular shape. *Opt. Express*, **6**, 213–219.
20. Nordlander, P. (2009) The ring: a leitmotif in plasmonics. *ACS Nano*, **3**, 488–492.
21. Wang, H., Brandl, D.W., Le, F., Nordlander, P., and Halas, N.J. (2006) Nanorice: a hybrid plasmonic nanostructure. *Nano Lett.*, **6**, 827–832.
22. Nordlander, P., Oubre, C., Prodan, E., Li, K., and Stockman, M.I. (2004) Plasmon hybridization in nanoparticle dimers. *Nano Lett.*, **4**, 899–903.
23. Jensen, T., Kelly, L., Lazarides, A., and Schatz, G. (1999) Electrodynamics of noble metal nanoparticles and nanoparticle clusters. *J. Cluster Sci.*, **10**, 295–317.

24. Quinten, M. and Kreibig, U. (1993) Absorption and elastic-scattering of light by particle aggregates. *Appl. Opt.*, **32**, 6173–6182.
25. Dahmen, C., Schmidt, B., and von Plessen, G. (2007) Radiation damping in metal nanoparticle pairs. *Nano Lett.*, **7**, 318–322.
26. Lamprecht, B. *et al* (2000) Metal nanoparticle gratings: influence of dipolar particle interaction on the plasmon resonance. *Phys. Rev. Lett.*, **84**, 4721–4724.
27. Stuart, H. and Hall, D. (1998) Island size effects in nanoparticle-enhanced photodetectors. *Appl. Phys. Lett.*, **73**, 3815–3817.
28. Lukosz, W. and Kunz, R.E. (1977) Light emission by magnetic and electric dipoles close to a plane dielectric interface. II. Radiation patterns of perpendicular oriented dipoles. *J. Opt. Soc. Am.*, **67**, 1615–1619.
29. Pillai, S., Catchpole, K.R., Trupke, T., and Green, M.A. (2007) Surface plasmon enhanced silicon solar cells. *J. Appl. Phys.*, **101**, 093105.
30. Aizpurua, J. *et al* (2005) Optical properties of coupled metallic nanorods for field-enhanced spectroscopy. *Phys. Rev. B*, **71**, 235420.
31. Raether, H. (1988) *Surface Plasmons on Smooth and Rough Surfaces and on Gratings*, Springer.
32. Schroter, U. and Heitmann, D. (1999) Grating couplers for surface plasmons excited on thin metal films in the Kretschmann-Raether configuration. *Phys. Rev. B*, **60**, 4992–4999.
33. Barnes, W.L., Dereux, A., and Ebbesen, T.W. (2003) Surface plasmon subwavelength optics. *Nature*, **424**, 824–830.
34. Wedge, S., Wasey, J., Barnes, W.L., and Sage, I. (2004) Coupled surface plasmon-polariton mediated photoluminescence from a top-emitting organic light-emitting structure. *Appl. Phys. Lett.*, **85**, 182–184.
35. Atwater, H.A. and Polman, A. (2010) Plasmonics for improved photovoltaic devices. *Nat. Mater.*, **9**, 205–213.
36. Green, M.A. (2012) Harnessing plasmonics for solar cells. *Nat. Photonics*, **6**, 130–132.
37. Gu, M. *et al* (2012) Nanoplasmonics: a frontier of photovoltaic solar cells. *Nanophotonics*, **1**, 1.
38. Derkacs, D., Lim, S.H., Matheu, P., Mar, W., and Yu, E.T. (2006) Improved performance of amorphous silicon solar cells via scattering from surface plasmon polaritons in nearby metallic nanoparticles. *Appl. Phys. Lett.*, **89**, 093103.
39. Hägglund, C., Zäch, M., Petersson, G., and Kasemo, B. (2008) Electromagnetic coupling of light into a silicon solar cell by nanodisk plasmons. *Appl. Phys. Lett.*, **92**, 053110–053110-3.
40. Catchpole, K.R. and Polman, A. (2008) Design principles for particle plasmon enhanced solar cells. *Appl. Phys. Lett.*, **93**, 191113.
41. Hallermann, F. *et al* (2008) On the use of localized plasmon polaritons in solar cells. *Phys. Status Solidi A*, **205**, 2844–2861.
42. Rockstuhl, C. and Lederer, F. (2009) Photon management by metallic nanodiscs in thin film solar cells. *Appl. Phys. Lett.*, **94**, 213102–213102-3.
43. Akimov, Y.A., Ostrikov, K., and Li, E.P. (2009) Surface plasmon enhancement of optical absorption in thin-film silicon solar cells. *Plasmonics*, **4**, 107–113.
44. Akimov, Y.A., Koh, W.S., Sian, S.Y., and Ren, S. (2010) Nanoparticle-enhanced thin film solar cells: metallic or dielectric nanoparticles? *Appl. Phys. Lett.*, **96**, 073111.
45. Akimov, Y.A. and Koh, W.S. (2010) Resonant and nonresonant plasmonic nanoparticle enhancement for thin-film silicon solar cells. *Nanotechnology*, **21**, 235201.
46. Pillai, S., Beck, F.J., Catchpole, K.R., Ouyang, Z., and Green, M.A. (2011) The effect of dielectric spacer thickness on surface plasmon enhanced solar cells for front and rear side depositions. *J. Appl. Phys.*, **109**, 073105.
47. Cai, B., Jia, B., Shi, Z., and Gu, M. (2013) Near-field light concentration of ultra-small metallic nanoparticles for absorption enhancement in a-Si solar cells. *Appl. Phys. Lett.*, **102**, 093107.
48. Hylton, N.P. *et al* (2012) Loss mitigation in plasmonic solar cells: aluminium

nanoparticles for broadband photocurrent enhancements in GaAs photodiodes. *Sci. Rep.*, **3**, 2874.
49. Haug, F.J., Söderström, T., Cubero, O., Terrazzoni-Daudrix, V., and Ballif, C. (2008) Plasmonic absorption in textured silver back reflectors of thin film solar cells. *J. Appl. Phys.*, **104**, 064509.
50. Mokkapati, S., Beck, F.J., Polman, A., and Catchpole, K.R. (2009) Designing periodic arrays of metal nanoparticles for light-trapping applications in solar cells. *Appl. Phys. Lett.*, **95**, 53115.
51. Ferry, V.E., Verhagen, E., Schropp, R.E.I., Atwater, H.A., and Polman, A. (2010) Light trapping in ultrathin plasmonic solar cells. *Opt. Express*, **18** (Suppl. 2), A237–A245.
52. Paetzold, U.W. et al (2011) Plasmonic reflection grating back contacts for microcrystalline silicon solar cells. *Appl. Phys. Lett.*, **99**, 181105.
53. Paetzold, U.W., Moulin, E., Pieters, B.E., Carius, R., and Rau, U. (2011) Design of nanostructured plasmonic back contacts for thin-film silicon solar cells. *Opt. Express*, **19**, A1219–A1230.
54. Palanchoke, U. et al (2012) Plasmonic effects in amorphous silicon thin film solar cells with metal back contacts. *Opt. Express*, **20**, 6340–6347.
55. Palanchoke, U. et al (2013) Influence of back contact roughness on light trapping and plasmonic losses of randomly textured amorphous silicon thin film solar cells. *Appl. Phys. Lett.*, **102**, 083501.
56. Fahr, S., Rockstuhl, C., and Lederer, F. (2009) Metallic nanoparticles as intermediate reflectors in tandem solar cells. *Appl. Phys. Lett.*, **95**, 121105.
57. Fahr, S., Rockstuhl, C., and Lederer, F. (2010) The interplay of intermediate reflectors and randomly textured surfaces in tandem solar cells. *Appl. Phys. Lett.*, **97**, 173510.
58. Westphalen, M., Kreibig, U., Rostalski, J., Lüth, H., and Meissner, D. (2000) Metal cluster enhanced organic solar cells. *Sol. Energy Mater. Sol. Cells*, **61**, 97–105.
59. Rand, B., Peumans, P., and Forrest, S. (2004) Long-range absorption enhancement in organic tandem thin-film solar cells containing silver nanoclusters. *J. Appl. Phys.*, **96**, 7519–7526.
60. Debije, M.G. and Verbunt, P.P.C. (2012) Thirty years of luminescent solar concentrator research: solar energy for the built environment. *Adv. Energy Mater.*, **2**, 12–35.
61. Badescu, V. and De Vos, A. (2007) Influence of some design parameters on the efficiency of solar cells with down-conversion and down shifting of high-energy photons. *J. Appl. Phys.*, **102**, 073102.
62. Rowan, B.C., Wilson, L.R., and Richards, B.S. (2008) Advanced material concepts for luminescent solar concentrators. *IEEE J. Sel. Top. Quantum Electron.*, **14**, 1312–1322.
63. Wilson, H. (1987) Fluorescent Dyes interacting with small silver particles – A system extending the spectral range of fluorescent solar concentrators. *Sol. Energy Mater.*, **16**, 223–234.
64. Hayakawa, T., Selvan, S., and Nogami, M. (1999) Field enhancement effect of small Ag particles on the fluorescence from $Eu3+$-doped SiO_2 glass. *Appl. Phys. Lett.*, **74**, 1513–1515.
65. Lakowicz, J. (2001) Radiative decay engineering: biophysical and biomedical applications. *Anal. Biochem.*, **298**, 1–24.
66. Dulkeith, E. et al (2002) Fluorescence quenching of dye molecules near gold nanoparticles: radiative and nonradiative effects. *Phys. Rev. Lett.*, **89**, 203002.
67. Dulkeith, E. et al (2005) Gold nanoparticles quench fluorescence by phase induced radiative rate suppression. *Nano Lett.*, **5**, 585–589.
68. Biteen, J., Lewis, N., Atwater, H., Mertens, H., and Polman, A. (2006) Spectral tuning of plasmon-enhanced silicon quantum dot luminescence. *Appl. Phys. Lett.*, **88**, 131109.
69. Mertens, H. and Polman, A. (2006) Plasmon-enhanced erbium luminescence. *Appl. Phys. Lett.*, **89**, 211107.
70. Pan, S., Wang, Z., and Rothberg, L.J. (2006) Enhancement of adsorbed dye monolayer fluorescence by a silver nanoparticle overlayer. *J. Phys. Chem. B*, **110**, 17383–17387.
71. Fort, E. and Grésillon, S. (2007) Surface enhanced fluorescence. *J. Phys. D Appl. Phys.*, **41**, 013001.

72. Härtling, T.T., Reichenbach, P.P., and Eng, L.M.L. (2007) Near-field coupling of a single fluorescent molecule and a spherical gold nanoparticle. *Opt. Express*, **15**, 12806–12817.
73. Kaminski, F., Sandoghdar, V., and Agio, M. (2007) Finite-difference time-domain modeling of decay rates in the near field of metal nanostructures. *J. Comput. Theor. Nanosci.*, **4**, 635–643.
74. Mertens, H., Koenderink, A.F., and Polman, A. (2007) Plasmon-enhanced luminescence near noble-metal nanospheres: comparison of exact theory and an improved Gersten and Nitzan model. *Phys. Rev. B*, **76**, 115123.
75. Taminiau, T.H., Moerland, R.J., Segerink, F.B., Kuipers, L., and Van Hulst, N.F. (2007) Lambda/4 resonance of an optical monopole antenna probed by single molecule fluorescence. *Nano Lett.*, **7**, 28–33.
76. Bakker, R.M. *et al* (2008) Enhanced localized fluorescence in plasmonic nanoantennae. *Appl. Phys. Lett.*, **92**, 043101.
77. Gueroui, Z. and Libchaber, A. (2004) Single-molecule measurements of gold-quenched quantum dots. *Phys. Rev. Lett.*, **93**, 166108.
78. Kim, Y.S., Leung, P.T., and George, T.F. (1988) Classical decay-rates for molecules in the presence of a spherical surface – a complete treatment. *Surf. Sci.*, **195**, 1–14.
79. Sprafke, A. (2014) *Optische Nahfeld-Wechselwirkungen von Plasmonen mit ihrer Umgebung*, RWTH Aachen University.
80. Govorov, A.O., Lee, J., and Kotov, N.A. (2007) Theory of plasmon-enhanced Forster energy transfer in optically excited semiconductor and metal nanoparticles. *Phys. Rev. B*, **76**, 125308.

9
Up-conversion Materials for Enhanced Efficiency of Solar Cells

Jan Christoph Goldschmidt, Stefan Fischer, Heiko Steinkemper, Barbara Herter, Sebastian Wolf, Florian Hallermann, Gero von Plessen, Jacqueline Anne Johnson, Bernd Ahrens, Paul-Tiberiu Miclea, and Stefan Schweizer

9.1
Introduction

Owing to the discrete band-gap energy of the employed semiconductor material, solar cells generally do not utilize sub-band-gap photons. Silicon solar cells lose about 20% of the energy incident from the sun, because sub-band-gap photons are transmitted. Up-conversion (UC) aims to reduce these losses by converting two sub-band-gap photons into one photon with an energy above the band-gap [1, 2]. UC enhances the radiative efficiency limit of silicon solar cells from close to 30% [3] up to more than 40% [4].

The simple UC process entails the subsequent absorption of two photons via an intermediate energy level to populate a higher energy level in an up-converter system. From this highly excited energy level a photon with more energy than each of the two absorbed ones can be emitted spontaneously. A more efficient, but also more complex process is energy transfer up-conversion (ETU). For an energy transfer process, one excited species, namely the donor, needs to be in the vicinity of at least one other active ion, namely, the acceptor. The energy of the donor ion is transferred to the acceptor to excite higher energy levels.

An early realization of a photovoltaic device with UC was presented by Gibart *et al.* [5]. The up-converter was attached to the rear of a GaAs solar cell. For experimental realizations of silicon solar cells with UC, hexagonal sodium yttrium tetrafluoride (β-NaYF$_4$) doped with trivalent erbium (Er^{3+}) [6], specifically with a doping ratio of one erbium ion to four yttrium ions (β-NaEr$_{0.2}$Y$_{0.8}$F$_4$), has shown comparatively high values of the external quantum efficiency (EQE) in the near infrared (NIR). An estimated luminescence up-conversion quantum yield (UCQY) of 16.7% was determined by Richards and Shalav [7] with an up-converter silicon solar cell device with laser illumination at 1523 nm and an irradiance of 24 000 W m^{-2}, corresponding to a photon flux density of 1.84×10^{23} s^{-1} m^{-2}. Owing to the non-linearity of the UC processes the UC quantum yield increases with increasing irradiance. The irradiance from the sun in the spectral region

Photon Management in Solar Cells, First Edition. Edited by Ralf B. Wehrspohn, Uwe Rau, and Andreas Gombert.
© 2015 Wiley-VCH Verlag GmbH & Co. KGaA. Published 2015 by Wiley-VCH Verlag GmbH & Co. KGaA.

from the band-gap of silicon (~1100 nm) to the lowest energy transition, the ground state absorption (GSA), from the energy level $^4I_{15/2}$ to $^4I_{13/2}$ in Er^{3+} (up to ~1580 nm) is roughly 120 W m^{-2} (7.41 × 10^{23} s^{-1} m^{-2}). This means that more than 200× solar concentration is needed to reach the irradiance [7] used in the experiment described. For an equivalent solar concentration of only 4 suns an UCQY of roughly 1.1% was determined [8] under monochromatic irradiation. For a complete system of silicon solar cell with an attached up-converter, measured under broad spectrum illumination with a suns concentration of 730×, an average external quantum of 1.1% was observed in the spectral region from 1460 to 1600 nm [9].

In this chapter, we investigate up-converter materials based on lanthanides. The trivalent lanthanide ions have a rich energy spectrum from the NIR to the ultra violet spectral region which makes them prominent for many optical applications, like spectral conversion. Trivalent erbium (Er^{3+}) especially, shows conveniently spaced energy levels for UC of photons around 1520 nm. In Section 9.2 we describe the potential of transparent erbium-doped fluorozirconate (FZ) glasses as attractive UC systems. Another material is trivalent erbium doped hexagonal sodium yttrium fluoride (β-$NaYF_4$:20% Er^{3+}) as described in Section 9.3. Both materials are very suitable up-converters that show high UC quantum yield and emission/absorption characteristics that match the spectral characteristics of silicon solar cells [7–10]. In Section 9.4 the UC efficiencies are simulated on the basis of a rate equation model. As we will see, the results indicate that additional means are necessary to increase efficiency. Ways to increase this efficiency are discussed in Section 9.5.

9.2
Up-Conversion in Er^{3+}-Doped ZBLAN Glasses

9.2.1
Samples

The Er^{3+}-doped FZ glasses investigated are based on the well-known ZBLAN composition [11]. The nominal compositions of the glasses are given in Table 9.1. For the 2 and 5 mol% doped samples, ErF_3 doping was compensated by ZrF_4, while for 9.1 and 13 mol% doping, ErF_3 is compensated proportionally by all the other components. The constituent chemicals were weighed in a platinum crucible in an inert atmosphere of nitrogen and melted at 750 °C for 1 h. The melt was poured

Table 9.1 Nominal composition (in mol%) of the Er^{3+}-doped FZ glasses.

No.	ZrF_4	BaF_2	NaF	LaF_3	AlF_3	InF_3	ErF_3
1	51	20	20	3.5	3	0.5	2
2	48	20	20	3.5	3	0.5	5
3	48.2	18.2	18.2	3.2	2.7	0.5	9.1
4	46.1	17.4	17.4	3	2.6	0.4	13

into a brass mold that was at a temperature of 200 °C, that is, below the glass transition temperature of 260 °C for an FZ-based glass [11], before being slowly cooled to room temperature. The visible appearance of the 2, 5, and 9.1 mol% Er-doped glasses is clear but they have a pink tinge due to the erbium starting material. The glass doped with 13 mol% ErF_3 appears milky in addition to the pink tinge; this indicates that there is partial crystallization in the glass; probably caused by the high Er concentration. The samples were cut into approximately. 7 mm × 7 mm × 2 mm plates and polished.

9.2.2
Optical Absorption

Figure 9.1 shows the optical density of a 5 mol% Er^{3+}-doped FZ glass sample. As already described in [12], the glass has strong Er^{3+}-related absorption bands at 254, 377, and 520 nm and very weak absorption at 229, 274, and 800 nm. At other wavelengths the absorption strengths are intermediate. In addition to the observed Er^{3+} absorption bands, the material shows some background absorption below 300 nm.

The Er^{3+} absorption band at approximately 1530 nm is shown in more detail in Figure 9.2a, for FZ glasses doped with 2, 5, 9.1, and 13 mol%. The spectra were corrected for the reflectivity of the glasses. All spectra show a double band structure which is due to the crystal field splitting of the Er^{3+} ground state [14, 15]. The height of the absorption band depends significantly on the Er-doping level, that is, the maximum absorption for an approximately 2 mm thick sample is approximately 34% for a doping level of 2 mol%, 65% for 5 mol%, 86% for

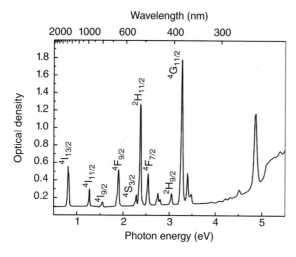

Figure 9.1 Optical density of the 5 mol% Er^{3+}-doped FZ glass sample. The sample thickness was 2.2 mm. The labeled transitions start from the $^4I_{15/2}$ ground state level and end on the levels indicated. *Source:* From Ref. [13]. © IOP Publishing. Reproduced by permission of IOP Publishing.

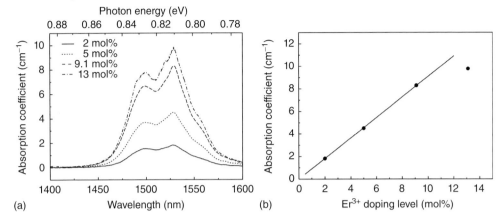

Figure 9.2 (a) Absorption spectra of Er^{3+} doped FZ glasses in the spectral range from 1400 to 1600 nm. The spectra are shown for the 2 mol% (solid curve), 5 mol% (dotted), 9.1 mol% (dashed), and 13 mol% (dash-dotted) sample. All spectra were corrected for their background, which is caused by reflection losses at the glass surface. (b) Absorption coefficient at 1530 nm versus the Er^{3+} doping level. *Source:* From Ref. [16]. ©IOP Publishing. Reproduced by permission of IOP Publishing.

9.1 mol%, and 89% for 13 mol%. The Er^{3+} absorption coefficient at 1530 nm shows a linear behavior versus the doping concentration (see Figure 9.2b). For the 13 mol% glass, however, the absorption coefficient is lower than expected. This is probably caused by the fact that it is not possible to embed that much Er in the glass; ErF_3 does not completely enter the FZ glass during the melting process. Preliminary investigations on even higher doped samples indicate that there is saturation in the Er doping level between 10 and 15 mol%.

9.2.3
Up-Conversion

For intense up-converted fluorescence it is important to absorb as many infrared photons as possible. It is shown in Figure 9.2 that the absorption coefficient increases linearly by appropriate increase in the Er-doping level. However, an increase in the doping level affects the glass quality and it may also lead to a quenching of the up-converted fluorescence.

The UC experiments were carried out with an excitation wavelength of 1540 nm in resonance with a transition from the $^4I_{15/2}$ ground state to the $^4I_{13/2}$ excited state. Note, that the high Er^{3+} concentration and the sample thickness of 2 mm are responsible for the relatively high optical density. Self-absorption of the emitted light by Er^{3+} will probably affect the shape of the up-converted fluorescence spectrum. Figure 9.3 shows the energy level diagram of Er^{3+} in FZ glasses [14, 15]. Possible UC routes and assignments of the main up-converted emission bands are also shown.

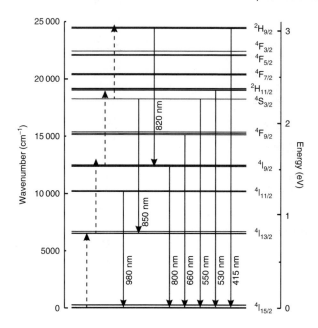

Figure 9.3 Energy level diagram of Er^{3+} in FZ glasses. Possible up-conversion routes (dashed arrows) and assignments of the main up-converted emissions (solid arrows) are indicated. *Source:* From Ref. [13, 14, 15]. ©IOP Publishing. Reproduced by permission of IOP Publishing.

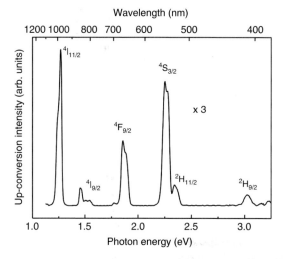

Figure 9.4 Up-converted fluorescence spectra of the 9.1 mol% Er^{3+}-doped FZ glass sample. Excitation is carried out with a cw laser diode operating at 1540 nm with 17.2 mW. The range from 900 down to 400 nm was blown up by a factor of 3. The labeled transitions start from the levels indicated and end on the $^4I_{15/2}$ ground state level. The NIR emission at 850 nm is caused by a transition from the $^4S_{3/2}$ to the $^4I_{13/2}$ state. *Source:* From Ref. [13]. ©IOP Publishing. Reproduced by permission of IOP Publishing.

The UC spectrum for the 9.1 mol% Er-doped sample in the 400–1100 nm spectral range can be seen in Figure 9.4. Besides the 980 nm emission in the NIR spectral range, the most intense bands in the visible are located in the green (530 and 550 nm) and red spectral range (660 nm). The NIR emission at 980 nm can be attributed to a transition from the $^4I_{11/2}$ excited state to the $^4I_{15/2}$ ground state. The weak 800 nm band arises from the $^4I_{9/2}$ state. The $^4I_{9/2}$ and $^4I_{11/2}$ levels are accessible with two 1540 nm photons; the $^4I_{11/2}$ level via a subsequent non-radiative relaxation from the $^4I_{9/2}$ level. The NIR emissions at 850 and 820 nm are caused by a transition from the $^4S_{3/2}$ to the $^4I_{13/2}$ state and a transition from the $^2H_{9/2}$ to the $^4I_{9/2}$ state, respectively [15, 17]. The visible emissions can be attributed to transitions from the $^2H_{11/2}$ (530 nm), $^4S_{3/2}$ (550 nm), and $^4F_{9/2}$ (660 nm) excited states to the $^4I_{15/2}$ ground state. These levels are energetically accessible with three 1540 nm photons. Four-photon up-converted fluorescence can be observed at 415 nm, which arises from a transition from the $^2H_{9/2}$ state to the ground state. All transitions to the $^4I_{15/2}$ ground state are split by about 0.03 eV which is caused by the crystal field splitting of the ground state.

Figure 9.5 shows the relative contribution of the different up-converted emissions for the 9.1 mol% Er^{3+}-doped FZ glass. The main contribution to the overall up-converted fluorescence intensity comes from the two-photon (980 nm) and from the visible three-photon up-converted emissions (530, 550, and 660 nm). All the other emissions play only a minor role; their contribution is summarized as "others." It can be seen that the contribution from the two-photon UC decreases significantly upon increasing the excitation power while the contribution of the three-photon up-converted emissions, in particular if the 550 nm band increases:

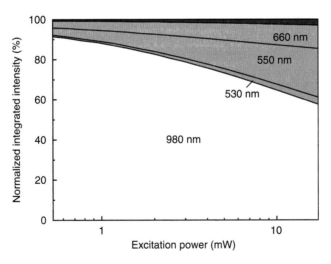

Figure 9.5 Relative contribution of the two-photon up-converted emission at 980 nm (white) and the three-photon up-converted emissions at 530, 550, and 660 nm (light gray) in the 9.1 mol% sample; all other up-converted emissions play only a minor role for the overall up-conversion efficiency and are summarized as "others" (dark gray). Source: From Ref. [13]. ©IOP Publishing. Reproduced by permission of IOP Publishing.

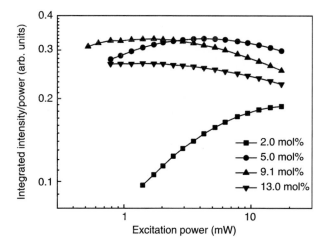

Figure 9.6 Power dependence of the integrated up-converted fluorescence intensity divided by the excitation power of Er^{3+}-doped FZ glasses. Excitation is carried out with a cw laser diode operating at 1540 nm. Source: From Ref. [13]. ©IOP Publishing. Reproduced by permission of IOP Publishing.

For an excitation power of approximately 0.5 mW, the contribution of the main emissions to the total intensity is 0.7% (530 nm), 3.5% (550 nm), 3.5% (660 nm), and 91.4% (980 nm) while it is 3.6% (530 nm), 24.2% (550 nm), 11.8% (660 nm), and 57.6% (980 nm) for an excitation power of 17 mW. A similar behavior can be observed for the 2, 5, and 13 mol% Er^{3+}-doped glass samples (not shown).

Figure 9.6 shows the power dependence of the integrated up-converted fluorescence intensity normalized to the excitation power. Fluorescence spectra were recorded over 2 orders of magnitude of excitation power, for example, from approx. 17 mW (maximal laser diode output power) down to a few tenths of a milliwatts. On a double-logarithmic scale, for each of the four curves a "saturation" of the UC intensity can be observed. The point of saturation depends critically on the ratio between excitation power, intermediate energy level lifetime, and the relative contribution of excited state absorption (ESA) and ETU processes to the overall up-converted fluorescence. The maximum EQE is at more than 10 mW excitation power for the 2 mol% doped sample, at 4–5 mW for a concentration of 5 mol%, approximately 1–2 mW for the 9.1 mol% sample, and about 0.8 mW (with a trend to even lower values) for the 13 mol% Er^{3+}-doped glass.

9.3
Up-Conversion in Er^{3+}-Doped β-NaYF$_4$

As an alternative up-converter material, β-NaYF$_4$ doped with Er^{3+} can be used. The results reviewed in this paper were obtained using 20% doping of Er^{3+} [8, 9], which was produced in the group of Karl Krämer at University of Bern [10].

This material is a microcrystalline powder. Therefore, the β-NaYF$_4$:20% Er^{3+} was filled in a powder cell with an optical window on the front for photoluminescence measurements. The thickness of the powder layer is more than 3 mm in order to achieve complete absorption of the incident photons. The up-converter was illuminated through the optical window with an IR-Laser ECL-210 from Santec, with an output wavelength λ_{inc}, tunable from 1430 to 1630 nm. The luminescence spectra of the up-converter were measured with a grating monochromator H25 from Jobin Yvon, a silicon photodiode detector from OEC, and a lock-in-amplifier 7265 from Signal Recovery. The detector is thermo-electrically cooled to minimize thermal noise.

A calibration of the setup allowed the determination of the calibrated efficiency of the UC on an absolute scale. Overall, an error of 8% is estimated for the uncertainty of the effective efficiency of the system of the powder cell with the up-converter powder as determined with the photoluminescence measurement.

Throughout this chapter, we define the spectral optical UC efficiency, $\eta_{UC,spectral}(\lambda_{inc}, \lambda_{UC}, I)$, at a certain luminescence wavelength, λ_{UC}, under the excitation with a wavelength, λ_{inc}, and an irradiance, I, as

$$\eta_{UC,spectral}(\lambda_{inc}, \lambda_{UC}, I) = \frac{\Phi_{UC}(\lambda_{UC}, I)}{\Phi_{inc}(\lambda_{inc}, I)}, \quad (9.1)$$

where $\Phi_{UC}(\lambda_{UC}, I)$ is the photon flux of the up-converted photons and $\Phi_{inc}(\lambda_{inc}, I)$ the total incident flux of photons with a wavelength, λ_{inc}, onto the powder cell.

With this definition, the maximum optical UC efficiency that can be reached is 50%, because at least two sub-band-gap photons must be absorbed to generate one up-converted photon. The maximum optical UC efficiency for UC involving three incoming photons is correspondingly lower.

By integrating over the luminescence wavelength, λ_{UC}, the integrated optical UCQY of the up-converter UCQY(λ_{inc}, I) at a specific excitation wavelength, λ_{inc}, was calculated:

$$UCQY(\lambda_{inc}, I) = \int \eta_{UC,spectral}(\lambda_{UC}, \lambda_{inc}, I) \times d\lambda_{UC} \quad (9.2)$$

In Figure 9.7a, the integrated optical UCQY in dependence on the excitation wavelength is plotted. The UCQY(λ_{inc}, I) peaks at a wavelength of 1523 nm, reaching 3.0% for a constant irradiance of 880 W m^{-2}. In the spectral range from 1492 to 1547 nm the efficiency of the UC is higher than 1.5%.

This excitation spectrum features distinctive peaks. These peaks are a result of the interaction of the 4f valence electrons of the erbium, which are responsible for the optical transitions, and the crystal field of the host material [14]. The UC luminescence and especially the UC efficiency not only depend on the excitation wavelength, but also on the irradiance of the sample.

To measure the dependence of the UC luminescence on the excitation power, the NaYF$_4$:20% Er^{3+} was excited at its most efficient wavelength of 1523 nm with different laser powers. The laser power was varied from 0.1 mW to the maximum stable output power of 7.9 mW. In the setup used, this corresponds to an irradiance of 17–1370 W m^{-2}. Figure 9.7b shows the dependence of the integrated UCQY on

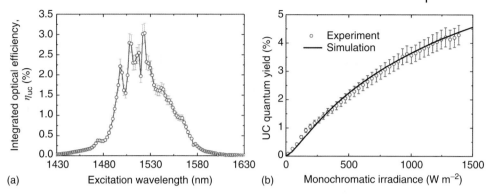

Figure 9.7 Excitation spectrum of β-NaYF$_4$:20% Er^{3+} under monochromatic excitation with an irradiance of 880 W m^{-2} (a). One can see that photons with a wavelength of 1523 nm are most efficiently up-converted. In the spectral region from 1492 to 1547 nm, photons are up-converted with quantum efficiencies exceeding 1.5%. The shape of the excitation spectrum is formed by the sub-energy levels of the Er^{3+}. (b) The dependence of the up-conversion quantum yield from the irradiance is shown for an excitation wavelength of 1523 nm. One can see that the UC quantum yield increase with the irradiance. Furthermore, the results of a simulation of this dependence are shown, the model is described in Section 9.4. The agreement between model and measurement is very good.

the irradiance of the excitation. It can be seen how the UC efficiency increases with the irradiance. For higher irradiance values the slope decreases, making saturation effects visible. The UCQY(λ_{inc}, I) reached in the presented setup was 4.3% for an irradiance of 1370 W m^{-2}.

9.3.1 Device Measurements

A powder of the UC material β-NaYF$_4$:20% Er^{3+} was mixed with the binding agent zapon varnish 79550 – Zapon Lacquer and attached to the grid-free side of a bifacial, back junction silicon solar cell with an active area of 4.5 mm × 4.5 mm. The placement of the up-converter at the grid-free side of the solar cell avoids problems with the contacts of the solar cell, but results in shading losses. Hence, this is not a configuration to reach the highest efficiencies, but provides a convenient test device. Under one-sun AM1.5G illumination on the grid-covered side, the solar cell exhibits 16.7% efficiency [9].

For the measurement of the EQE the solar cell was placed on a cooled gold-coated measurement chuck, which has a cavity for the up-converter. The solar cell UC system was illuminated with the Santec ECL-210 laser from the grid-covered side. The highest EQE was measured at a wavelength of 1523 nm, reaching 0.64% for an irradiance of 2305 W m^{-2}. Extrapolations of these results agree very well with other studies [7], on the same material β-NaYF$_4$:20% Er^{3+}, and constitute the highest UC efficiencies achieved in the context of silicon solar cells, when the relatively low irradiance values are taken into account [8, 9].

The application of UC to harvest solar energy implies that the up-converter operates under broad spectrum illumination and not under monochromatic laser excitation. Therefore we investigated the described UC solar cell device also under the light of a Xe-lamp concentrated by lenses onto the device [9]. A polished Si wafer served as a long pass filter to increase the relative impact of the up-converter. By comparison of the short-circuit current measured with UC material attached to the solar cell, to the case where a polytetrafluoroethylene (PTFE) reflector was attached instead, it was ensured that any measured effect was due to UC. We observed a significant current due to UC, which corresponds to an average UC efficiency of $1.07 \pm 0.13\%$ in the spectral range from 1460 to 1600 nm. The photon flux in this spectral range corresponded to an effective concentration of 732 ± 17 suns. It is interesting to note that the observed average EQE is higher than the peak values of the EQE determined under monochromatic illumination with comparable irradiance levels. This is attributed to the resonant excitation of the two most important optical transitions under broad spectrum illumination, which is not the case under monochromatic illumination. The measured quantum efficiency corresponds to a relative efficiency increase of 0.014% for the used bifacial silicon solar cell with its 16.70% overall efficiency. This increase is too small to make UC relevant in photovoltaics. Therefore, additional means of increasing UC efficiency are necessary, for example, the use of plasmon resonance in metal nanoparticles.

9.4
Simulating Up-Conversion with a Rate Equation Model

The task to increase UC efficiencies beyond the experimentally observed levels requires a profound theoretical understanding. Therefore, we developed a simulation model that describes the UC dynamics. The UC dynamics of β-NaYF$_4$:20% Er^{3+} can be modeled by rate equations [18]. The model presented in this section considers ground and ESA, spontaneous and stimulated emission, energy transfer, and multi phonon relaxation (MPR) [19]. The considered transitions and energy levels are shown in Figure 9.8. The most important input parameters of the model are the Einstein coefficients of the involved radiative transitions. They were derived from the absorption coefficient data of the β-NaYF$_4$:20% Er^{3+} powder using the theory of Judd and Ofelt [20, 21]. Other parameters, such as the values of the overlap integrals describing the energy transfer were estimated based on literature data [22]. The parameters describing MPR were estimated from literature values of similar materials [23] and adjusted to achieve good agreement between simulation and measurement. More details on the rate equation model can be found in Refs. [18, 19].

The occupation of the energy levels is determined in the steady state for a constant irradiance at a wavelength of 1520 nm. For every transition, the luminescence rate L_{if}, of a certain transition from the initial energy level, i, to the final energy level, f, is calculated by the occupation and the Einstein coefficient for spontaneous emission. The total UCQY from the simulation, UCQY$_{tot,Sim}$, can be

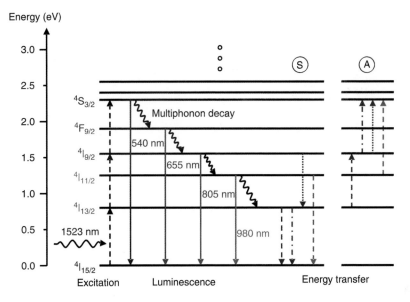

Figure 9.8 Energy levels of Er^{3+} in the host crystal β-NaYF$_4$ with corresponding luminescence wavelengths of the photoluminescence transitions from the excited states to the ground state (solid arrows). In the rate equation model excitation with 1523 nm is assumed. Higher energy levels are populated by ground and excited state absorption (black dashed arrows). Additionally, energy transfer occurs, where energy is transferred from the sensitizer ion (S), to the activator ion (A) (dashed and pointed arrows). Furthermore, multi-phonon relaxation between neighboring energy levels is considered (waved arrows).

determined from the relation of the luminescence rate of all transitions that result in the emission of photons usable by a silicon solar cell and the absorption rate, Abs,

$$\text{UCQY}_{if,\text{Sim}} = \frac{L_{if}}{\text{Abs}} \quad \text{UCQY}_{\text{tot,Sim}} = \frac{\sum_i L_{if}}{\text{Abs}}, \tag{9.3}$$

where UCQY$_{if,\text{Sim}}$ is the simulated UCQY for the transition from energy level i to f. With this definition the maximum UCQY is 50%, because at least two photons are needed to populate the energy level $^4I_{11/2}$ with a corresponding luminescence wavelength of 980 nm. This transition is responsible for more than 99% of the UCQY depending on the irradiance [8]. Figure 9.7b shows that the developed model describes the dependence of the UC quantum yield on the irradiance very well. At this point we should note that this model with all parameters was specifically adapted to experimental UCQY measurements of β-NaY$_{0.8}$Er$_{0.2}$F$_4$ and the significance for other materials and doping concentration, for instance, are limited and must be interpreted with care. The electric field of the crystal and consequently the Einstein coefficients are modified for different materials which is expressed in a shift of the energy gap between the energy levels, for example, [24]. Nevertheless, this UC model is very sophisticated and we think that fundamental trends are represented correctly.

9.5
Increasing Up-Conversion Efficiencies

While the presented experimental results can be considered as a successful integration of UC into photovoltaics, the achieved efficiencies are still too low, to make UC attractive for application in commercial solar cells. UC of sub-band-gap photons faces two major challenges:

1) The spectral width of the absorption from the Er^{3+} is very narrow compared to the broad solar spectrum, which means that only a small fraction of the solar photons can be absorbed by the up-converter. Consequently, the amount of effectively used photons is not particularly large and the density of the photons in the absorption range of the up-converter is small.
2) A sufficiently high photon flux in the absorption range of the up-converter is necessary, to achieve even the modest UCQYs presented above.

In the following text, we present an overview of different possibilities to enhance the UCQY. The concepts can be divided into three groups. The first group comprises internal concepts, for example, the host material itself, the size of the up-converter material, dopant concentration, and so on. In this work, we will focus on phonon energies of the host crystal and the dopant concentration and how these parameters affect the UCQY. The second group consists of external methods which change the physical environment around the up-converter. This can be achieved by plasmon resonances in metal nanoparticles, or by dielectric photonic structures. The third possibility is to change the excitation spectrum incident on the up-converter in order to use a larger part of the solar spectrum. This can be achieved by spectral concentration with infrared luminescent materials like nanocrystalline quantum dots. We assess the impact of the different approaches, based on a sophisticated rate equation model presented in the previous section [19].

9.5.1
The Up-Converter Material

9.5.1.1 Phonon Energy
The probability of multi-phonon relaxation (MPR) $W_{MPR,if}$, depends on the number of phonons needed to bridge the energy gap, ΔE_{if}, from the initial energy level i to the final energy level f:

$$W_{MPR,if} = W_{MPR} \cdot e^{-\kappa \cdot \Delta E_{if}}. \tag{9.4}$$

where W_{MPR} and κ are material constants. This relationship is known as energy gap law [25]. Trivalent erbium ions have been incorporated into many materials with different phonon energies. Fluoride hosts, like β-$NaYF_4$ or LaF_3, are one of

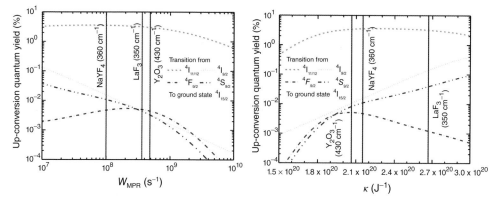

Figure 9.9 The simulations show that the parameters for MPR change the potential UCQY of the various transitions, UCQY$_{if,Sim}$. For large phonon energies, for example, large W_{MPR} and small κ, the total UCQY is decreased by several orders of magnitude. Some prominent host materials with their average phonon energies are shown. These host materials are on a broad plateau of the UCQY of the transition from $^4I_{11/2}$ to $^4I_{15/2}$, which is the most significant one for the total UCQY. Consequently all these host materials are suitable candidates for large values of the UCQY [46].

the most prominent host materials, because the phonon energies are very low and they are chemically stable. While oxide hosts, like Y_2O_3, are also well suited even if the phonon energies are slightly higher, other materials like chlorides, bromides, or iodides with lower phonon energies than fluorides are less suited, because they are often hydroscopic and thus not stable under ambient conditions [24].

In our UC model, the parameters of W_{MPR} and κ for β-NaY$_{0.8}$Er$_{0.2}$F$_4$ have been estimated from the average phonon energy of 360 cm^{-1} and a fit to the data of LaF$_3$ [10, 26]. Afterwards, the parameters were adapted to the irradiance dependent UCQY measurements [8]. The influence of these parameters on the UCQY of various transitions, UCQY$_{if,Sim}$, is shown in Figure 9.9. Some values from prominent host materials have been added into the graph [19, 26, 27]. With MPR the contribution of the energy levels on the total UCQY can be modified. For the most dominant transition from $^4I_{11/2}$ to the ground state $^4I_{15/2}$ a broad maximum is determined which means that for many host materials with slightly different phonon energies the expected UCQY does not change significantly. We conclude that low phonon energies are beneficial for the UCQY, but too low phonon energies, for example, low W_{MPR} and high κ, decrease the potential of the UCQY. Most likely this is caused by a decreased occupation of the energy level, $^4I_{11/2}$, which is mainly populated by MPR.

9.5.1.2 Doping Concentration

In principle, the doping concentration only affects the distance, d, between the lanthanide ions and the only process depending on the distance is energy transfer (ET). ET is very sensitive to the distance between the donator and the acceptor ion (probability for ET $\sim d^{-6}$) and increases with a shorter distance between the

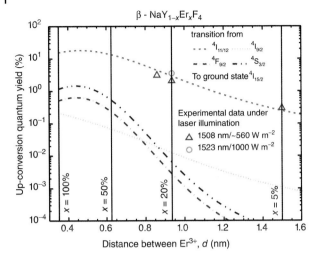

Figure 9.10 The simulated $UCQY_{if,Sim}$ determined for an irradiance of 1000 W m^{-2} and an excitation wavelength of 1523 nm. The $UCQY_{if,Sim}$ increases for shorter distances between the Er^{3+}. For a distance of 0.46 nm (equivalent to an Er^{3+} doping of ~80%) the largest UCQY is determined. For shorter distance values down to the minimum distance of ~0.35 nm the UCQY decreases slightly. The trend of the simulation is supported by UCQY measurements of β-NaY$_{1-x}$Er$_x$F$_4$ with different erbium doping $x = 5$, 20, and 25% [46].

lanthanide ions. The electric field of the host crystal, however, may change with the doping concentration similar as for different host materials and, consequently, the Einstein coefficients and thus the dynamics of the up-converter could be modified [24].

Neglecting these secondary effects we calculated the dependence of the $UCQY_{if,Sim}$ on the doping concentrations by varying the distance between the Er^{3+} for an irradiance of 1000 W m^{-2} and an excitation wavelength of 1523 nm [8]. The $UCQY_{if,Sim}$ increases with shorter distances between the ions, d, for all considered transitions as presented in Figure 9.10. This means that a larger total UCQY can be achieved with larger doping concentration. There is, however, a maximum at an Er^{3+} concentration of 80%. We expect that cross-relaxation processes are quenching the UCQY for higher Er^{3+} concentrations.

Three samples of β-NaY$_{1-x}$Er$_x$F$_4$ were synthesized with erbium concentrations of 5, 20, and 25%. These values are equivalent to distances of approximately 1.50, 0.93, and 0.86 nm, respectively. For these doping concentrations the UCQY was measured as described in Ref. [8]. The determined UCQY values for an irradiance of roughly 560 W m^{-2} are plotted in Figure 9.10. Our laser, however, was not stable at 1523 nm and the excitation wavelength was shifted to a more stable operation point at 1508 nm, where another peak in the UCQY is located [8]. Therefore, slightly lower UCQY values are expected compared to an excitation wavelength of 1523 nm and an irradiance of 1000 W m^{-2}, as used for the simulations. Nevertheless, the trend to larger UCQY values for shorter distance is clearly visible.

Additionally, a data point for an excitation wavelength of 1523 nm and an irradiance of 1000 W m^{-2} is plotted in Figure 9.10, from previously published work [8].

9.5.2
The Environment around the Up-Converter

9.5.2.1 Plasmon Enhanced Up-Conversion

Plasmon resonance in metal nanoparticles can lead to strongly enhanced fields locally. These higher local field intensities can positively influence UC efficiency because of the non-linear nature of UC. Additionally, the metal nanoparticles also influence the transition probabilities of the luminescent system [28–31]. Such plasmonic effects were investigated intensively [32–36]. For Er^{3+}, Mertens and Polman [33] demonstrated an increase of the photoluminescence by more than a factor of two, when Er^{3+} were located in the proximity of an array of elongated Ag nanoparticles. Schietinger et al. [35] demonstrated a 3.8-fold increase in UC luminescence, when a NaYF$_4$ nanocrystal co-doped with Yb^{3+}/Er^{3+} was placed at an optimized position near a gold nanosphere with 60 nm diameter.

To investigate the interaction between plasmon resonance and UC, we linked the model of the up-converter with calculations on plasmon resonance. The effects were simulated for a cubic volume with one spherical gold nanoparticle in the center. The edge length of the simulation volume was six times the diameter of the nanoparticle. The resolution of the simulation was 5 nm. Details on the simulation approach can be found in [19, 37].

Because of the high computational cost of a complete analysis, only the change of the intensity of the electric field was considered in a first run. We calculated the intensity distribution around gold nanoparticles with diameters from 40 to 480 nm for illumination with 1523 nm using Mie-theory [38]. The resulting different levels of irradiance were used as input parameters for the UC rate equation model. It was found that for a particle diameter of 200 nm the increase in UC luminescence due to the field enhancement effect was the largest [37].

In a second step, we calculated how the radiative and non-radiative transition probabilities within the up-converter are changed in the proximity of a particle with a diameter of 200 nm. We used the exact electrodynamic theory [39–41] for this purpose. The theory is based on Mie theory, which provides solutions to Maxwell's equations for spherical boundary conditions. The theoretical model is implemented in Matlab, and details can be found in Ref. [34]. The relative change of the Einstein coefficients that describe the transitions used in the rate equation model, is calculated by the enhancement factor $\gamma_{tot,if}$ for a certain transition from energy level, i, to energy level f. The $\gamma_{tot,if}$ consists of a radiative factor, $\gamma_{rad,if}$, for emitted photons and a non-radiative factor, $\gamma_{nrad,if}$, for transitions in which the excitation energy is lost due to energy transfer to the metal nanoparticle. Therefore, the $\gamma_{rad,if}$ was used to calculate the luminescence of the up-converter.

In the exact electro-dynamic theory, the up-converter is treated as a dipole emitter. The emitting dipole can be oriented in two different polarization directions with respect to the surface of the gold nanoparticle, either parallel to the surface

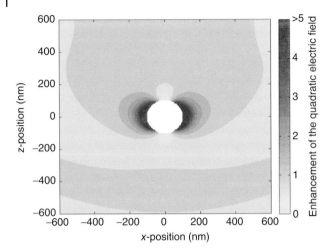

Figure 9.11 Enhancement factor of the quadratic electric field, γ_E, due to a spherical gold nanoparticle with a diameter of 200 nm (white circle in the center) at an illumination with wavelength of 1523 nm. The plot shows a cut through the simulation volume in the x–z-plane at $y = 0$ nm. The incident light propagates in the positive z-direction from the bottom of the graph. The strongest field enhancement with a value of $\gamma_E = 16$ is reached close to the surface of the gold nanoparticle [42].

(PPOL) or perpendicular to the surface (SPOL). Consequently, the different factors must be calculated for every polarization. The changed transition probabilities and the changed irradiance were used to calculate the UC luminescence at every point in the volume around the particle for each polarization. The excitation wavelength was 1523 nm with an irradiance of 1000 W m^{-2}.

Figure 9.11 shows the factor, γ_E, by which the quadratic electric field changes due to a spherical gold nanoparticle with a diameter of 200 nm under an illumination at a wavelength of 1523 nm [42]. The plot shows a cut through the simulation volume in the x–z-plane at $y = 0$ nm. The incident light propagates in the positive z-direction from the bottom of the graph. In the direction perpendicular to the light path, a very high field enhancement occurs, with γ_E values up to 16, close to the surface of the gold nanoparticle. Behind and in front of the nanoparticle, regions occur in which the field is decreased. For most of the simulated volume, the γ_E values are roughly one, indicating a marginal impact of the nanoparticle.

9.5.2.2 Modeling Dielectric Nanostructures

An alternative to metal nanoparticles are dielectric photonic structures. They have the potential to achieve similar or better field enhancement factors, but with lower parasitic absorption. Simulations of the dielectric photonic structures were performed with the finite-difference time-domain (FDTD) method, using Meep, a freely available software package [43]. The structures were simulated in a two dimensional configuration, thus infinitely extended in the third dimension. Additionally, periodic boundary conditions are assumed in the x-direction.

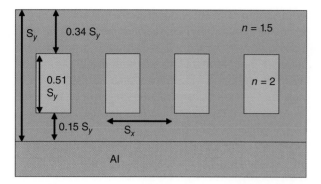

Figure 9.12 Dielectric grating waveguide structure, consisting of a grating like structure with a refractive index $n = 2$ in a material with a refractive index $n = 1.5$. The feature sizes are on the order of magnitude of the wavelength of the incident light of 1523 nm. An aluminum mirror is placed at the rear of the structure [47].

We analyzed a grating waveguide structure, which is shown in Figure 9.12. The waveguide structure is illuminated using a continuous line-source at a wavelength of 1523 nm. This source is situated in vacuum several wavelengths away from the structure. After a steady state is reached in the simulation, the time-averaged magnitude of the Poynting vector was calculated yielding the field intensity at each lattice point. This quantity was divided by the intensity resulting from the same source in vacuum in order to obtain the relative intensity increase, γ_E.

The spatial distribution of the energy density enhancement factor, γ_E, was calculated for different values of the characteristic lengths, S_X, and S_Y, of the waveguide. The spatial distributions of γ_E were than analyzed and histogram data were generated giving the relative frequency $P_i(S_X, S_Y)$ of the different enhancement levels, $\gamma_{E,i}$. It was assumed that the irradiance without any waveguide structure, I_0, would be 250 W m^{-2}. Consequently, the resulting irradiance levels, I_i, could be calculated via $I_i = \gamma_{E,i} I_0$. For each of the irradiance levels, the UC luminescence, Lum(I_i), and the absorption of the up-converter, Abs(I_i), could be calculated using the model described above. With the relative frequencies of the different irradiance levels, the average enhancement of the UC luminescence, rLum(S_X,S_Y), and the absorption by the up-converters, rAbs(S_X,S_Y), could be calculated via

$$\mathrm{rLum}(S_X, S_Y) = \frac{\sum_i \mathrm{Lum}(I_i) P_i(S_X, S_Y)}{\mathrm{Lum}(I_0)} \quad \text{and}$$

$$\mathrm{rAbs}(S_X, S_Y) = \frac{\sum_i \mathrm{Abs}(I_i) P_i(S_X, S_Y)}{\mathrm{Abs}(I_0)}. \tag{9.5}$$

The quantum yield of the up-conversion can be defined as the ratio of photons emitted at a wavelength a silicon solar cell can utilize, divided by the number of photons absorbed by the up-converter. The relative change, rQY(S_X,S_Y), of the

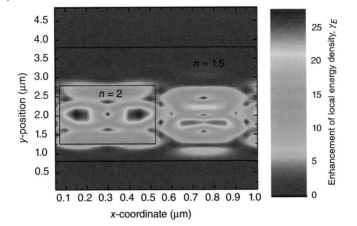

Figure 9.13 Enhancement factor, γ_E, of the local energy density in the grating waveguide structure. The direction of incidence is from the top of this page. The wavelength of the incident light is 1523 nm with an irradiance of 250 W m^{-2}.

quantum yield can then be calculated via

$$rQY(S_X, S_Y) = \frac{rLum(S_X, S_Y)}{rAbs(S_X, S_Y)}. \tag{9.6}$$

Figure 9.13 displays the enhancement factor, γ_E, of the local energy density in the grating waveguide structure. It can be seen that the field is strongly enhanced in the waveguide, especially in the higher refractive index region. The characteristic lengths of the waveguide structure were $S_X = 0.9$ μm and $S_Y = 3$ μm for the presented simulation. Figure 9.14 presents the results of the analysis of how the average enhancement factors for the absorption of the up-converter, the UC luminescence, and the UCQY, change depending on the characteristic lengths, S_X. For this analysis S_Y was fixed to be 2.5 μm. A strong resonance is observed for $S_X = 8$ μm, resulting in a strong increase in UC luminescence by a factor of 7.7. This increase can be attributed to both an increase in the absorption of the up-converter by a factor of 2.3 and an increase of the UCQY by a factor of 3.3.

It has to be noted that the presented results are only preliminary, because only the change of the local energy density has been considered. Within the waveguide, the local density of states for the photons is modified as well. This will influence the transition probabilities of the up-converter and must be considered for a thorough analysis.

9.5.3
Spectral Concentration

The combination of an up-converter with a second luminescent material was proposed by Strümpel *et al.* [44] to use a broader part of the solar spectrum for UC. The concept is visualized in Figure 9.15. An advanced UC solar cell design was proposed by Goldschmidt *et al.* [45] with geometrical and spectral

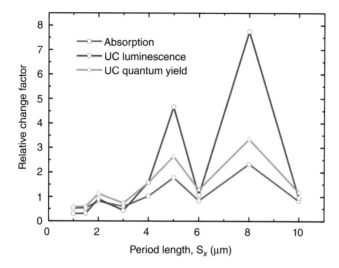

Figure 9.14 Change of the overall absorption, up-conversion luminescence, and the quantum yield in dependence on the period length, S_x. For a period length of 8 μm a strong enhancement is observed. The increase in up-conversion luminescence is partly caused by increased absorption and partly by an increase of the up-conversion quantum yield.

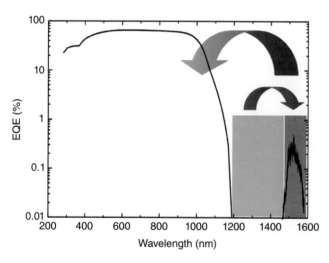

Figure 9.15 Concept of spectral concentration of the photons from the solar spectrum between the band-gap of the solar cell and the absorption range of the up-converter. The blue line shows the EQE of a silicon solar cell with an attached up-converter. The green area indicates the spectral region lost without any additional means. With a second luminescent material the photons in this spectral region can be absorbed and emitted in the active region of the up-converter (red). More photons are used by the up-converter and additionally the photon flux density in the absorption range of the up-converter and thus the up-conversion efficiency itself is increased.

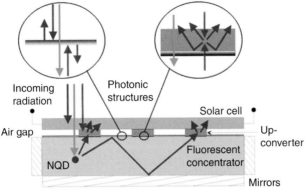

Radiation
→ v_1 can be utilised by the solar cell
→ v_2 can be absorbed by the nanocrystal quantum dot (NQD)
→ v_3 can be upconverted

$v_1 > v_2 > v_3$

Figure 9.16 Potential setup of an advanced up-converter system. The solar cell absorbs photons with energies above the band-gap (v_1). Photons with less energy are transmitted (v_2, v_3). The up-converter transforms especially low energy photons, with energies in the absorption range of the up-converter (v_3), into high energy photons, which can be absorbed by the solar cell (v_1). Photons with energies below the band-gap but above the absorption range of the up-converter (v_2) are absorbed in the fluorescent material, which emits photons in the absorption range of the up-converter (v_3). The emitted radiation is guided by total internal reflection and/or photonic structures to the up-converter. As the up-converter does not cover the whole area, a geometric concentration takes place. Radiation which is emitted from the up-converter toward the fluorescent concentrator is back reflected by a spectrally selective photonic structure.

concentration, which is depicted in Figure 9.16. In this concept, the second luminescent material is embedded in a transparent matrix to form a fluorescence concentrator. Currently, there are no suitable organic dyes for the infrared with an emission wavelength around 1520 nm. Therefore, nanocrystalline quantum dots were investigated for dye sensitized UC.

In the active region of the up-converter, Er^{3+}, from approximately 1460–1580 nm, the solar spectrum contains a photon flux density of only 1.94×10^{20} s^{-1} m^{-2}, which carries about 25 W m^{-2} equivalent to only 2.5% of the total solar irradiation. In contrast, in the spectral range from the band-gap of silicon (~1100 nm) to the absorption range of the up-converter (~1460 nm) the integrated photon flux is roughly 5.47×10^{20} s^{-1} m^{-2} (93 W m^{-2}). Usually, the EQE of silicon solar cells is quite poor for wavelengths larger than 1050 nm, which is caused by the large penetration depth of photons with long wavelengths. If these photons are transmitted through the solar cell, the photon flux density that can be used by UC is about 7.08×10^{20} s^{-1} m^{-2} (~120 W m^{-2}) for the spectral range from 1050 to 1460 nm. By shifting the photons into the absorption range of the up-converter by one absorption/emission event, the photon flux density on the up-converter can be enhanced by a factor of at least 3.5. Hence, spectral concentration is one of the most promising concepts to make UC significant for solar cell applications.

9.6 Conclusion

There are a number of ways to enhance the efficiency of solar cells by UC. This chapter, necessarily, mentions only a few. However, the text provides the reader with a broad knowledge base on which to build. Topics covered include material choices, experimental methods, both fundamental and device oriented, and simulations. In particular the chapter discusses UC in Er^{3+}-doped ZBLAN glasses with increasing doping level. Guidance is given on maximum doping levels and absorption and potential UC mechanisms, namely excited state absorption and energy transfer upconversion and how this affects the EQE. The other UC material discussed was a microcrystalline powder, β-$NaYF_4$ with 20% Er^{3+}-doping. The UCQY in this material is measured and stated, reaching a maximum of 4.3% with the setup described. The dependence of the UC luminescence as a function of excitation power was found. Of course, fundamental science is a prerequisite to device development but ultimately, this must be done to further the solar cell industry. With this in mind, device measurements were made with UC materials coupled to a silicon solar cell. The overall quantum efficiency increase was very small at 0.014%; therefore, additional means of increasing solar cell efficiency were explored.

Simulations are an increasingly important aspect of experimental research, acting as a guide to future experiments and verification of results. In this chapter, UC was simulated with a rate equation model. Parameters considered in the model included ground and excited state absorption, spontaneous and stimulated emission, energy transfer, and multi-phonon relaxation. The UCQY was calculated to be a maximum of 50% for β-$NaYF_4$. The calculations agree well with experiments for this material and provide a useful starting point for other materials.

Finally and most importantly, four methods are described to increase UC efficiencies. The first is phonon energy. While it is well known that low phonon energy materials are desirable hosts for the optically active element, it is less well known and elucidated from this work that too low phonon energies decreases the UCQY. Doping concentration is the second parameter to be discussed. We find here that increasing Er^{3+}-concentration increases the UCQY up to a value of 80%; beyond this, cross relaxation and quenching diminish the UCQY. Thirdly, the environment around the up-converter was considered using plasmon-enhanced UC and by modeling dielectric nanostructures. A 3.8-fold increase in UC luminescence was found when a β-$NaYF_4$ nanocrystal co-doped with Er and Yb was optimally positioned near a gold nanosphere. Dielectric photonic structures have potential for an even higher enhancement of 7.7, shown by calculations, though these results are preliminary. Finally, the most promising method to increase UC efficiency involves the combination of an up-converter with a second luminescent material in order to create spectral concentration.

From the information given in this chapter, the reader can devise further up-converting materials to take solar cell efficiency into the next decade using a multi-pronged approach and subsequent commercialization.

Acknowledgments

The research leading to these results has received funding from the German Federal Ministry of Education and Research (BMBF) within "NanoVolt" (Project No. 03SF0322A), "Infravolt" (Project No. 03SF0401B), and the Centre for Innovation Competence SiLi-nano® (Project No. 03Z2HN11), and from the European Community's Seventh Framework Programme (FP7/2007–2013) in the project NanoSpec under grant agreement n° [246200]. In addition, this work was supported by the FhG Internal Programs under Grant No. Attract 692 034. Stefan Fischer gratefully acknowledges the scholarship support from the Deutsche Bundesstiftung Umwelt DBU. The up-converter material β-NaYF$_4$:20% Er^{3+} has been synthesized in the group of Karl Krämer at the University of Bern, Switzerland. Jacqueline A. Johnson thanks the National Science Foundation, Grant # DMR 1001381.

References

1. Auzel, F. (1990) Upconversion processes in coupled ion systems. *J. Lumin.*, **45**, 341–345.
2. Trupke, T., Green, M.A., and Würfel, P. (2002) Improving solar cell efficiencies by up-conversion of sub-band-gap light. *J. Appl. Phys.*, **92** (7), 4117–4122.
3. Shockley, W. and Queisser, H.J. (1961) Detailed balance limit of efficiency of p-n junction solar cells. *J. Appl. Phys.*, **32** (3), 510–519.
4. Trupke, T. *et al.* (2006) Efficiency enhancement of solar cells by luminescent up-conversion of sunlight. *Sol. Energy Mater. Sol.* , **90**, 3327–3338.
5. Gibart, P. *et al.* (1996) Below band-gap IR response of substrate-free GaAs solar cells using two- photon up-conversion. *Jpn. J. Appl. Phys*, **35** (8), 4401–4402.
6. Krämer, K.W. *et al.* (2004) Hexagonal sodium yttrium fluoride based green and blue emitting upconversion phosphors. *Chem. Mater.*, **16**, 1244–1251.
7. Richards, B.S. and Shalav, A. (2007) Enhancing the near-infrared spectral response of silicon optoelectronic devices via up-conversion. *IEEE Trans. Electron Devices*, **54** (10), 2679–2684.
8. Fischer, S. *et al.* (2010) Enhancement of silicon solar cell efficiency by upconversion: Optical and electrical characterization. *J. Appl. Phys.*, **108**, 044912 1–044912 11.
9. Goldschmidt, J.C. *et al.* (2011) Experimental analysis of upconversion with both coherent monochromatic irradiation and broad spectrum illumination. *Sol. Energy Mater. Sol. Cells*, **95** (7), 1960–1963.
10. Suyver, J.F. *et al.* (2006) Upconversion spectroscopy and properties of NaYF$_4$ doped with Er^{3+}, Tm^{3+} and/or Yb^{3+}. *J. Lumin.*, **117**, 1–12.
11. Aggarwal, I.D. and Lu, G. (1991) *Fluoride Glass Fiber Optics*, Academic Press.
12. Bullock, S.R. *et al.* (1997) Energy upconversion and spectroscopic studies of ZBLAN: Er^{3+}. *Opt. Quantum Electron.*, **29** (1), 83–92.
13. Henke, B., Pientka, F., Johnson, J. A., Ahrens, B., Miclea, P. T., and

Schweizer, S. (2010) Saturation effects in the upconversion efficiency of Er-doped fluorozirconate glasses. *Journal of Physics: Condensed Matter*, **22**, 155107. doi: 10.1088/0953-8984/22/15/155107

14. Dieke, G.H. and Crosswhite, H.M. (1963) The spectra of the doubly and triply ionized rare earths. *Appl. Opt.*, **2** (7), 675–686.
15. Wetenkamp, L., West, G.F., and Többen, H. (1992) Optical properties of rare earth-doped ZBLAN glasses. *J. Non-Cryst. Solids*, **140**, 35–40.
16. Schweizer, S., Henke, B., Ahrens, B., Paßlick, C., Miclea, P. T., Wenzel, J., Reisacher, E., Pfeiffer, W., Johnson, J.A. (2010) Progress on up-and down-converted fluorescence in rare-earth doped fluorozirconate-based glass ceramics for high efficiency solar cells. *Photonics for Solar Energy Systems III*, Proceedings of SPIE, vol. 7725 (eds R.B. Wehrspohn and A. Gombert), SPIE, p. 77250X. doi: 10.1117/12.853943
17. Shinn, M.D. et al. (1983) Optical transitions of Er^{3+} ions in fluorozirconate glass. *Phys. Rev. B*, **27** (11), 6635–6648.
18. Goldschmidt, J.C. (2009) *Novel Solar Cell Concepts*, Verlag Dr. Hut, München, 274 pp.
19. Fischer, S. et al. (2012) Modeling upconversion of erbium doped microcrystals based on experimentally determined Einstein coefficients (vol 111, 013109, 2012). *J. Appl. Phys.*, **111** (8), 134–140.
20. Judd, B.R. (1962) Optical absorption intensities of rare-earth ions. *Phys. Rev.*, **127** (3), 750–761.
21. Ofelt, G.S. (1962) Intensities of crystal spectra of rare-earth ions. *J. Chem. Phys.*, **37** (3), 511–520.
22. Henderson, B. and Imbusch, G.F. (1989) *Optical Spectroscopy of Inorganic Solids (Monographs on the Physics and Chemistry of Materials)*, Clarendon Press, p. 661.
23. Weber, M.J. (1967) Probabilities for radiative and nonradiative decay of Er^{3+} in LaF_3. *Phys. Rev.*, **157** (2), 262–272.
24. Lüthi, S.R. et al. (1999) Near-infrared to visible upconversion in Er^{3+}-doped $Cs_3Lu_2Cl_9$, $Cs_3Lu_2Br_9$, and $Cs_3Y_2I_9$ excited at 1.54 µm. *Phys. Rev. B*, **60** (1), 162–178.
25. Weber, M.J. (1968) Radiative and multiphonon relaxation of rare-earth ions in Y_2O_3. *Phys. Rev.*, **171** (2), 283–291.
26. Weber, M.J. (1979) *Handbook of the Physics and Chemistry of Rare Earths*, North Holland, New York, pp. 275–316.
27. Shalav, A., Richards, B.S., and Green, M.A. (2007) Luminescent layers for enhanced silicon solar cell performance: up-conversion. *Sol. Energy Mater. Sol. Cells*, **91**, 829–842.
28. Gersten, J. and Nitzan, A. (1981) Spectroscopic properties of molecules interacting with small dielectric particles. *J. Chem. Phys.*, **75** (3), 1139–1152.
29. Sun, G., Khurgin, J.B., and Soref, R.A. (2007) Practicable enhancement of spontaneous emission using surface plasmons. *Appl. Phys. Lett.*, **90** (11), 111107-1–111107-3.
30. Muskens, O.L. et al. (2007) Strong enhancement of the radiative decay rate of emitters by single plasmonic nanoantennas. *Nano Lett.*, **7** (9), 2871–2875.
31. van Wijngaarden, J.T. et al. (2011) Enhancement of the decay rate by plasmon coupling for Eu^{3+} in an Au nanoparticle model system. *Europhys. Lett.*, **93** (5), 57005-1–57005-6.
32. Johansson, P., Xu, H., and Käll, M. (2005) Surface-enhanced Raman scattering and fluorescence near metal nanoparticles. *Phys. Rev. B*, **72** (3), 035427-1–035427-17.
33. Mertens, H. and Polman, A. (2006) Plasmon-enhanced erbium luminescence. *Appl. Phys. Lett.*, **89**, 211107-1–211107-3.
34. Hallermann, F. et al. (2008) On the use of localized plasmon polaritons in solar cells. *Phys. Status Solidi A*, **205** (12), 2844–2861.
35. Schietinger, S. et al. (2009) Plasmon-enhanced single photon emission from a nanoassembled metal–diamond hybrid structure at room temperature. *Nano Lett.*, **9** (4), 1694–1698.
36. Verhagen, E., Kuipers, L., and Polman, A. (2009) Field enhancement in metallic subwavelength aperture arrays probed by erbium upconversion luminescence. *Opt. Express*, **17** (17), 14586–14598.
37. Hallermann, F. et al. (2010) Calculation of up-conversion photoluminescence in

Er^{3+} ions near noble-metal nonoparticles. *Proc. SPIE*, **77250Y** 1–8.
38. Bohren, C.F. and Huffman, D.R. (1983) *Absorption and Scattering of Light by Small Particles*, Wiley Science Paperback Series, Wiley-VCH Verlag GmbH, Weinheim, p. 544.
39. Young Sik, K., Leung, P.T., and George, T.F. (1988) Classical decay rates for molecules in the presence of a spherical surface: a complete treatment. *Surf. Sci.*, **195** (1-2), 1–14.
40. Kaminski, F., Sandoghdar, V., and Agio, M. (2007) Finite-difference time-domain modeling of decay rates in the near field of metal nanostructures. *J. Comput. Theor. Nanosci.*, **4** (3), 635–643.
41. Mertens, H., Koenderink, A.F., and Polman, A. (2007) Plasmon-enhanced luminescence near noble-metal nanospheres: Comparison of exact theory and an improved Gersten and Nitzan model. *Phys. Rev. B*, **76** (11), 115123-1–115123-12.
42. Goldschmidt, J.C. *et al.* (2012) Increasing upconversion by plasmon resonance in metal nanoparticles – a combined simulation analysis. *IEEE J. Photovoltaics*, **2** (2), 134–140.
43. Oskooi, A.F. *et al.* (2010) MEEP: a flexible free-software package for electromagnetic simulations by the FDTD method. *Comput. Phys. Commun.*, **181** (3), 687–702.
44. Strümpel, C. *et al.* (2005) Erbium-doped up-converters of silicon solar cells: assessment of the potential. Proceedings of the 20th European Photovoltaic Solar Energy Conference, 2005, Barcelona, Spain.
45. Goldschmidt, J.C. *et al.* (2008) Advanced upconverter systems with spectral and geometric concentration for high upconversion efficiencies. Proceedings IUMRS International Conference on Electronic Materials, 2008, Sydney, Australia.
46. Fischer, S. *et al.* (2012) SPIE EU photonics 2012. Proc. SPIE **8438** Photonics for Solar Energy Systems IV, 843806 (June 1, 2012). doi:10.1117/12.922183
47. Goldschmidt, J.C. *et al. Conference: Proceedings of SPIE - The International Society for Optical Engineering.* Vol. 8256 doi:10.1117/12.910915

10
Down-Conversion in Rare-Earth Doped Glasses and Glass Ceramics

Stefan Schweizer, Christian Paßlick, Franziska Steudel, Bernd Ahrens, Paul-Tiberiu Miclea, Jacqueline Anne Johnson, Katharina Baumgartner, and Reinhard Carius

10.1
Introduction

The efficiency of solar cells in the short wavelength range is limited by the transparency or collection efficiency of the contact layers. Furthermore, the transparency of the glass cover of solar modules will add to these losses. Part of these losses can be compensated by conversion of short wavelength photons into a wavelength range, where the quantum efficiency of the respective solar cell is highest. To use most of the short wavelength spectrum it would be advantageous to replace the solar cover glass by a so-called "down-converter" (Figure 10.1). This sets the most important requirement for the very high transparency of the down-converter system in the whole spectral range of the solar cell response. In particular, where the sun spectrum has its highest intensity, even weak losses there can easily eliminate the gains of the down-conversion process. It should be mentioned here that the AM1.5 standard solar spectrum is used for the evaluation of the efficiency of the down-converter systems in this contribution. This is not sufficient for a final evaluation as for different locations around the world the short wavelength spectrum might have higher or lower intensities and thus make the implementation of such a system more or less valuable. On the other hand, different solar cells with different spectral response may require different down-converting materials in order to optimize the balance of the gain with the losses typically inevitably related with the implementation of additional components into the optical path of a solar cell or module. In this contribution, investigations on different down-converter materials are presented, which have potential to increase the efficiency of different types of thin film solar cells. We restrict ourselves to the rare-earth ions europium (Eu) and

Photon Management in Solar Cells, First Edition. Edited by Ralf B. Wehrspohn, Uwe Rau, and Andreas Gombert.
© 2015 Wiley-VCH Verlag GmbH & Co. KGaA. Published 2015 by Wiley-VCH Verlag GmbH & Co. KGaA.

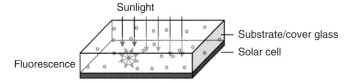

Figure 10.1 Concept of the implementation of a down-converter into solar modules. Rare-earth ions acting as down-converters are embedded in a transparent matrix (glass, glass ceramic, etc.) and attached to the surface or implemented into the cover glass.

samarium (Sm) for down-conversion. It should be noted, however, that results on terbium ions in various glass matrices such as Tb-doped SiO_2 and Al_2O_3 films have been published showing no overall gain in efficiency when integrated into an a-Si:H solar cell [1]. In the following the choice of the material selection is motivated.

The photon-conversion is an umbrella term for two conversion mechanisms, namely the up- and down-conversion. Down-conversion is presented in more detail in the following chapter. Two processes are observed, namely down-shifting and quantum cutting. Down-shifting or also known as photoluminescence (PL) describes the conversion of a high energetic photon into one lower energetic photon. Therefore, the high energetic photon is absorbed from a luminescent center, for example, a Eu or Sm ion, which hence is excited from the ground state into a higher energetic state. Relaxing back to the ground state, the ion releases its energy by emitting a photon with the same or less energy when the remaining energy is absorbed by the crystal lattice and phonons, respectively.

Higher photon-conversion efficiency is reached in the case of quantum cutting where a high energetic photon is converted into two or more lower energetic photons. Here, several conversion routes over different intermediate states and also different ions which are located close enough to each other for transferring energy between them are possible. Figure 10.2 summarizes the principle mechanisms: The easiest way is illustrated in (a) where an optically active center like an ion is for example excited with a high energetic ultraviolet (UV) photon to state E_2. Then it emits two lower energetic photons hv_1 and hv_2 by first relaxing in the intermediate state E_1 and then back into the ground state E_0. In (b) a second ion gets excited by a partial energy transfer from the excited ion I which is relaxing from excited state E_3 to state E_2. Both ions then drop back into the ground state by emitting a photon in each case. (c) The excited ion I relaxes from excited state E_3 to state E_2 by emitting a photon with energy hv_1 the residual energy from the relaxation to the ground state is transferred to ion II, which itself drops back into ground state by emitting a second photon hv_2. In (d) the excited ion I relaxes back to ground state by transferring its whole energy to two neighboring ions II and III, which both then relax to their ground states by emitting a photon.

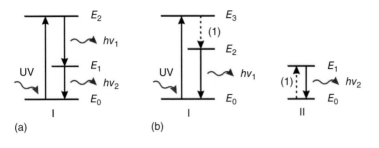
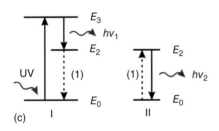
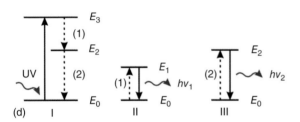

Figure 10.2 Scheme of different quantum-cutting mechanisms. (a) The excited ion I emits two lower energetic photons by first relaxing in an intermediate state E_1 and then into the ground state E_0. (b) The excited ion I relaxes from its excited state E_3 to state E_2 by a partial energy transfer (1) to ion II. Both excited ions then drop back into their ground state by emitting a photon in each case. (c) The excited ion I relaxes to state E_2 by emitting a photon and after that transfers its energy to ion II for relaxation to the ground state. Ion II drops back into its ground state by emitting a photon again. (d) The energy for relaxation of ion I from the excited state to the ground state becomes transferred to ion II and ion III. Both now can drop back to state E_0 by emitting a photon.

10.2 Physical Background

10.2.1 Rare-Earth Ions

Scandium, yttrium, and the lanthanides of the periodic table of elements are called rare-earths (REs). The elements are not as rare as their name suggests; for example, thulium, the rarest RE element, is more abundant than gold. A characteristic of RE ions is the incompletely filled 4f shell of the ion, which is shielded by the surrounding filled $5s^2$ and $5p^6$ orbitals with larger radial extension [2]. Therefore, the host lattice or phonons have a relatively weak influence on the optical transitions inside the $4f^n$ configuration.

The parity selection rule forbids optical absorption transitions, but can be partially abolished by uneven components of the crystal field due to an occupation of a crystallographic site without inversion symmetry by an RE ion. The wave functions of the 5d and 4f shells can then be mixed and the nominal forbidden 4f-4f transition shows spectral intensity.

Two transition mechanisms with broad absorption bands become partially allowed:

- Charge-transfer transitions ($4f^n \rightarrow 4f^{n+1}L^{-1}$, L = ligand) are found for tetravalent RE ions like Ce^{4+}, Pr^{4+}, and Tb^{4+}), which like to be reduced [3]. Trivalent ions with a tendency to become divalent like Sm^{3+}, Eu^{3+}, and Yb^{3+} show charge-transfer transition absorption bands in the UV.
- $4f^n \rightarrow 4f^{n-1}5d$ transitions are found for divalent ions like Sm^{2+} in the visible region and for Eu^{2+} and Yb^{2+} in the long wavelength UV, which like to be oxidized. Here, trivalent ions with a tendency to become tetravalent like Ce^{3+}, Pr^{3+}, and Tb^{3+} have got 4f-5d transition absorption bands in the UV.

This chapter will focus on trivalent samarium (Sm^{3+}) and europium, which can occur in divalent (Eu^{2+}) and trivalent (Eu^{3+}) valence state. Eu is one of the most reactive lanthanides with atomic number 63 (period 6, block f in the periodic table of the elements). Fast oxidation of Eu^{2+} to Eu^{3+} in ambient atmosphere makes a nitrogen atmosphere during the glass preparation essential. Also, water has to be kept away in which Eu would react with hydroxide ions. When excited with ionizing radiation like X-rays or UV light, Eu^{2+} is luminescent in a wide optical spectral range from the UV to the red depending on the host lattice structure. Eu^{3+} shows less intense luminescence and thus it is important to have the maximum amount of Eu^{2+} inside the glass.

Sm is the lanthanide with the atomic number 62 (period 6, block f, too) and is most stable at the +3 oxidation state, as all lanthanides. It reacts with water to form samarium hydroxide. Trivalent samarium has a strong luminescence in the red spectral range upon excitation in the UV/blue.

10.2.2
Glass Systems

Once having chosen as a luminescent RE for a desired application, the next important step is the selection of the used host material. In this section the used glass hosts, their advantages, and their preparation methods are briefly described.

10.2.2.1 Phonons

One of the most important properties of an RE host material is its maximum phonon frequency or energy ($h\nu_{max}$). Phonons, also known as *quantized thermal energy* or *lattice vibrations*, with low energies reduce non-radiative losses of fluorescent ions and allow radiative emission from energy levels that would normally be quenched in high phonon energy hosts [4]. Multiphonon relaxation (MPR), the non-radiative relaxation from an excited ion back into a lower or ground energy level via energy transfer into one or more phonons, can be reduced by large energy gaps between the excited and lower states as well as by decreasing the maximum phonon energy. With lower phonon energies, more phonons are required to bridge the energy gap, making this process less probable and increasing the lifetime and the luminescence efficiency of the excited state.

For low RE doping concentrations, the energy transfer between RE ions can be neglected and the total transition probability between two levels is given by the reciprocal of the fluorescence lifetime

$$\tau^{-1} = \sum A + W_{MPR}$$

with the total radiative transition probability, A, and the multiphonon relaxation rate, W_{MPR}, which is described by the so-called energy gap law

$$W_{MPR} = C \cdot \exp(-\alpha \Delta E)$$

with the host-dependent constants, C and α, the latter corresponding to the electron–phonon coupling, and ΔE being the energy gap between the emission level and the next lower level [5]. Maximum phonon energies and the host-dependent parameters of the MPR for various glasses are listed in Table 10.1. Low-phonon energies are indispensable to increase the total fluorescence efficiency of the RE ions. Thus, a fluoride-based glass system was chosen as host matrix for this work on which the following subsection will give more insights.

10.2.2.2 ZBLAN Glasses

In 1974 Poulain and Lucas discovered heavy metal fluoride glasses (HMFGs) at the University of Rennes in France [6]. ZBLAN is an abbreviation for glasses that are made of a mixture of zirconium, barium, lanthanum, aluminum, and sodium fluorides. The standard ZBLAN composition introduced by Ohsawa and Shibata [7] is

Table 10.1 Host-dependent parameters of MPR in various glasses [5].

Host	Notation	$h\nu_{max}$ (cm^{-1})	C (s^{-1})	α (10^{-3} cm)
Borate	B_2O_3	1400	2.9×10^{12}	3.8
Phosphate	P_2O_5	1200	5.4×10^{12}	4.7
Silicate	SiO_2	1000	1.4×10^{12}	4.7
Germanate	GeO_2	900	3.4×10^{10}	4.9
Tellurite	TeO_2	700	6.3×10^{10}	4.7
ZrF_4-based	ZrF_4	500	1.9×10^{10}	5.8
$ZnCl_2$-based	$ZnCl_2$	300	5.0×10^{7}	4.1

known for its lowest critical cooling rate among the fluorozirconate-based glasses [8]. Compared with silicate glasses, ZBLAN has extended transparency into the infrared (IR) wavelengths and a multiphonon edge shift to longer wavelengths due to heavier ions so that it is used for ultra-low loss optical fibers [9].

10.2.2.3 Borate Glasses

Borate glasses are known for a long time. A deeper understanding of these glass systems started in 1932 by Zachariasen with the introduction of the "random network theory" [10]. Zachariasen described the possibility of forming three-dimensional glass networks consisting of oxygen and cations such as boron. He concluded that boron oxide has the highest ability to form a glass because of the trivalency of boron. In 1938 Biscoe and Warren [11] found that boron also partly becomes tetravalent depending on the amount of network modifier (sodium) content. Three-coordinated boron becomes four-coordinated during the melt by reducing the network modifier. The borate glass network (see Figure 10.3) consists of BO_3 and BO_4 groups. The four-coordinated boron bonds with metal bound oxygen to form BO-tetrahedra. The reduced metal network modifier stays as an ion in the glass network. The metal ions (network modifiers) prevent crystallization of boron oxide. For a boron oxide to metal bound oxygen ratio of two to one, the amount of four-coordinated boron is at a maximum [13].

10.3
Down-Conversion in ZBLAN Glasses and Glass Ceramics

ZBLAN glasses are additionally doped with chlorine ions by adding barium chloride or barium fluoride and sodium chloride. This enables precipitation of barium

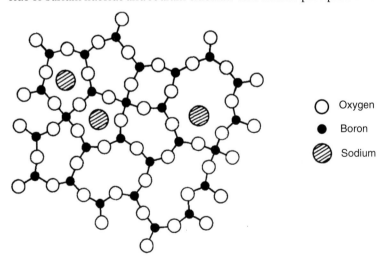

Figure 10.3 Schematic representation of the structure of a sodium borate glass with low sodium content [11, 12].

chloride nanocrystals within the glass upon appropriate annealing. Depending on the structural phase of the barium chloride, the glass-ceramic material can act either as a scintillator (able to convert ionizing radiation to visible light) [14] or as a storage phosphor (able to convert the radiation into stable electron–hole pairs, which can be released at a later time with a scanning laser beam) [15].

10.3.1
Samples

The nominal composition of the ZBLAN glasses was $48ZrF_4 - 10BaF_2 - 10BaCl_2 - 3.5LaF_3 - 3AlF_3 - 20NaCl - 0.5InF_3 - 5EuF_2$ (values in mol%). The glasses are prepared under an argon atmosphere inside a glove box by a two-step melting process. The first step involves mixing and melting of all fluorides at 800 °C for 60 min in a platinum crucible. After being cooled to room temperature chlorides are added to the melt and the whole composition is remelted for 60 min at 745 °C. The last step involves pouring into a 200 °C hot brass mold, where it stays for 60 min before being cooled to room temperature at 50 °C per h.

The additional barium chloride doping enables precipitation of $BaCl_2$ nanocrystals within the glasses upon appropriate thermal treatment. These nanocrystals offer even lower maximum phonon energies (around 200 cm^{-1} [16]) compared to the glass matrix host (590 cm^{-1} for an FZ glass [17]) as well as the above mentioned storage phosphor properties. Depending on the annealing conditions two barium chloride phases are obtained in the glass ceramics. Firstly, at temperatures around 250–270 °C, nucleation and growth of the metastable hexagonal $BaCl_2$ phase occur. With higher temperatures the crystals grow and around 280–290 °C a phase transition to orthorhombic crystal structure occur based on the fact that the thermodynamic system will go over in its state of lowest energy (see Figure 10.13 in Section 10.3.3.2). Note that the given temperatures depend on glass composition and preparation method.

10.3.2
Glass-Ceramic Cover Glasses for High Efficiency Solar Cells

A conventional mono-bandgap solar cell is not able to use the whole solar spectrum. The UV and IR wavelength regions are particularly problematic: The IR light cannot be used, because its energy is lower than the bandgap energy of the semiconductor solar cell material, while in the UV spectral range photon energy is "wasted" by thermalization and surface recombination of the charge carriers. Frequency "down-shifters" like the above mentioned fluorescent glasses or glass ceramics doped with RE ions such as Eu^{2+} are able to convert the high-energy part of the solar spectrum into photons in the visible spectral range, which can be more efficiently absorbed by a solar cell.

Eu^{2+}-doped fluorozirconate-based glass ceramics have suitable optical properties for this purpose, that is, they can be excited in the UV spectral range leading to an emission in the blue spectral range. Here, the conversion efficiency depends mainly on the phonon energy (lattice vibrations) of the host material. FZ-based glasses are well known for their low-phonon energies and therefore are desirable hosts as "down-shifters".

The additional doping with chlorine ions enables the growth of $BaCl_2$ nanoparticles upon thermal annealing in which the Eu^{2+} is incorporated. In the case of ZBLAN glasses with the nominal composition given in Section 10.3.1, annealing for 20 min at temperatures between 260 and 280 °C leads to a phase transformation from hexagonal to orthorhombic $BaCl_2$; at a temperature of 290 °C only orthorhombic phase $BaCl_2$ is observed. The size of the $BaCl_2$ nanoparticles is between 10 and 90 nm (see Figure 10.4).

The previous sections have shown that Eu can be incorporated in the glass in its divalent and trivalent state. Interestingly, Eu^{2+} does not fluoresce in the fluorozirconate base glass although it is clear from electron paramagnetic resonance (EPR) [19] and Mössbauer studies [20] that it is in the glass. However, Eu^{3+} shows its typical emissions in the red spectral range. Upon annealing, some of the Eu^{2+} ions are incorporated into the $BaCl_2$ nanocrystals leading to an intense Eu^{2+}-related and $BaCl_2$ phase dependent fluorescence under UV excitation (see inset of Figure 10.4).

10.3.2.1 Absorption

For photovoltaic applications, the metastable hexagonal phase is preferred because of its larger integral fluorescence intensity, whereas the orthorhombic

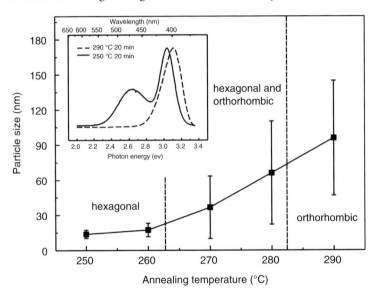

Figure 10.4 Barium chloride ($BaCl_2$) nanoparticle size versus annealing temperature for a 5 mol% Eu-doped glass ceramic. The inset shows the Eu^{2+}-related fluorescence spectra for glass ceramics containing hexagonal (solid curve) and orthorhombic phase (dashed curve) $BaCl_2$ nanocrystals under UV excitation (280 nm) [18].

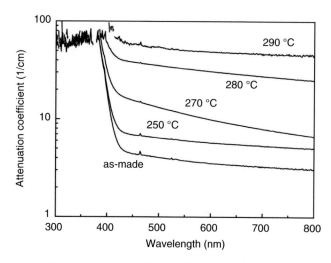

Figure 10.5 Attenuation coefficient of 5 mol% Eu-doped samples. Presented is the sample as-made (bottom) and the samples annealed for 20 min at 250, 270, 280, and 290 °C (from bottom to top) [22].

phase offers a storage phosphor effect which is attractive for medical X-ray imaging [21]. However, nucleation and growth of nanocrystals result in Rayleigh scattering effects if the crystal dimensions reach the size of the visible wavelengths. The aim is to minimize these scattering losses, that is, more visible light can pass the cover glass and reach the solar cell enhancing its efficiency. Figure 10.5 shows the attenuation coefficient for 5 mol% Eu-doped samples with a Cl/(F + Cl) ratio of 10.6% (standard ZBLAN composition as described in Section 10.3.1, but $15BaF_2 - 5BaCl_2$ instead of $10BaF_2 - 10BaCl_2$) and different annealing temperatures (for details on the measurement see Ref. [22]). In general, the transmission is nearly zero for the UV spectral region due to the broad absorption bands of Eu^{2+} and Eu^{3+} up to 380 nm. Thus, the attenuation coefficient falls down rapidly between 380 and 400 nm by more than one order of magnitude for the as-made glass.

For all samples, an upward trend to higher attenuation coefficient with increased annealing temperature is observed due to a higher number and larger nanocrystals inside the glasses. The as-made sample is most transparent at low wavelengths but does not contain any nanocrystals, whereas the fully ceramized sample annealed at 290 °C shows a high scattering up to 1500 nm and therefore is not useful for a solar cell cover glass. It can already be argued here that an exact and constant annealing temperature during the fabrication process is very important for receiving the right crystal sizes; otherwise, the scattering effects rapidly increase and reduce the solar cell efficiency.

Furthermore, all samples show a weak absorption band at 464 nm which can be assigned to the $^7F_0 \rightarrow {}^5D_2$ absorption of Eu^{3+} [23]. The small step at 820 nm found for the curves of the as-made and the 250 °C annealed sample is caused by the detector changeover from photomultiplier to InGaAs detector.

10.3.2.2 Short-Circuit Current

To demonstrate the effect and to show the potential of the Eu-doped FZ-based glass ceramics as down-converters, the short circuit current of a commercial monocrystalline silicon solar cell (*www.hupra.com*) with the glasses on top was measured and compared to the corresponding short circuit current to that of a cell without a cover glass. The samples were not optically coupled to the solar cell. The short circuit current was measured with a source-meter (Keithley Model 2400). The excitation source was a xenon lamp (450 W) in connection with an excitation monochromator of a spectrophotometer (Horiba Jobin Yvon FluoroLog-3).

Figure 10.6 shows this calculated short circuit current ratio for different annealing temperatures. Although the short circuit current has been measured for the 14.1% series, all curves are qualitatively consistent to the courses of the measured absorption of the 10.6% samples, that is, the opaque 290 °C sample does not lead to any measurable short circuit current within the measured wavelength range. The other samples show a short circuit current peak at around 360 nm caused by the Eu^{2+} absorption band leading to an enhancement of solar cell efficiency for the samples annealed at 270 and 280 °C in the ultraviolet spectral region. For the other samples and wavelengths longer than 400 nm, the short circuit ratio drops below "1" meaning a decrease in the solar cell efficiency. This decrease is caused by the high base glass absorption below 400 nm and the reflection and scattering effects of the incident light at the glass ceramic surface, since the short circuit

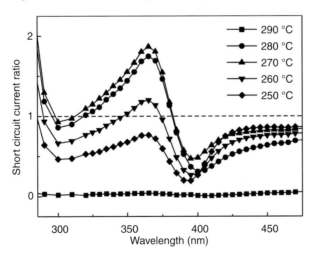

Figure 10.6 Short-circuit current ratio between a conventional silicon solar cell with a glass-ceramic cover to one without one. The Eu-doping level of the glass ceramic is 5 mol% and the Cl/(F + Cl) ratio is 14.1%. The samples were annealed for 20 min at 250 to 290 °C in steps of 10 °C [22].

current of the cell with down-converter is compared to the cell without a cover glass. However, it should be mentioned that in daily use solar cells are protected by a cover glass from weather, water, dust, and other influences leading also to a decrease in the cells' efficiencies. The sample annealed at 270 °C for 20 min is the best possible cover glass from this series.

10.3.2.3 Internal Conversion Efficiency

The internal conversion efficiency, η, of a 5 mol% Eu-doped FCZ glass after annealing at 270 °C, which reveals the highest short circuit current ratio, is investigated. The nominal composition of the glass is given in Section 10.3.1. The absorption coefficient of this glass has been determined by reflection and transmission spectroscopy, α_{RT}, and by photothermal deflection spectroscopy (PDS), α_{PDS} (see Figure 10.7). In PDS only absorbed light converted into thermal energy (parasitic absorption, α_{par}) is measured, down-converted light (α_{down}) does not contribute to the absorption coefficient [24].

From the absorption coefficients the internal conversion efficiency, η, is calculated by

$$\eta = \frac{\alpha_{down}}{\alpha_{par} + \alpha_{down}} = \left(1 - \frac{\alpha_{PDS}}{\alpha_{RT}}\right) \cdot \frac{1}{1 - (E_2 - E_1)/E_1}$$

with E_2 the excitation energy and E_1 the Eu^{2+} emission at 490 nm. The result is depicted in Figure 10.7, blue curve. The internal conversion efficiency has a maximum of 52 % at 393 nm.

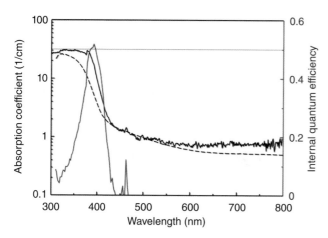

Figure 10.7 Internal conversion efficiency of a 5 mol% Eu-doped FCZ glass tempered at 270 °C (blue curve) calculated from the absorption coefficient determined by reflectance and transmittance measurements (solid black curve) and by PDS (dashed black curve) [24]. The grey curve indicates an internal quantum efficiency value of 50 %.

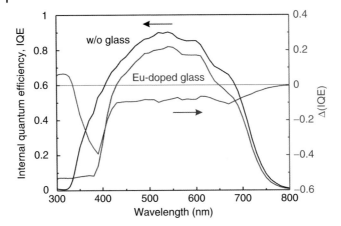

Figure 10.8 Internal quantum efficiencies of the amorphous silicon solar cell without (black curve) and with a 5 mol% Eu-doped FCZ glass tempered at 270 °C (red curve) and difference of both IQEs (blue curve) [24].

10.3.2.4 Quantum Efficiency Increase

The external quantum efficiency (EQE) of an amorphous silicon solar cell (a-Si p-i-n cell on Corning glass with ZnO made by IEK-5 – FZ Jülich) with a 5 mol% Eu-doped FCZ glass placed on the sunny side was measured. The FCZ glass was coupled optically with silicon oil to the Corning cover glass of the solar cell. For comparing the efficiency of the solar cell with and without a down-converter, the EQE was corrected for the reflection losses caused by the glasses. The calculated internal quantum efficiency (IQE) of the solar cell without (black curve) and with (red curve) the down-converter is shown in Figure 10.8. The IQE of the solar cell with the down-converter is significantly lower in the spectral range between 335 and 800 nm due to absorption, reflection, and scattering losses in the FCZ glass. But for wavelengths smaller than 335 nm the efficiency of the solar cell is enhanced by up to 7 %. The difference between both IQE curves shows a drastic loss for the solar cell with down-converter in the spectral range between 335 and 400 nm. This is caused by the base glass absorption of the FCZ, which is at higher wavelengths than for the Corning glass of the solar cell, and demonstrates that the down-converter only provides a benefit in the spectral range where the EQE of a solar cell is low.

A simulation of an IQE for an amorphous silicon solar cell covered with an optimal Eu-doped FCZ glass was performed assuming 100 % internal conversion efficiency of the Eu-ions and no absorption by the glass matrix, so that all the absorbed light is down-converted. The result is shown in Figure 10.9. In the spectral range between 300 and 400 nm the IQE of the simulated solar cell (red curve) is significantly higher than the measured one (black curve). The difference between both IQE is up to 65 % (blue curve).

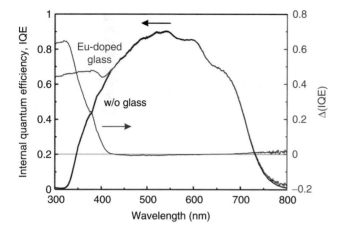

Figure 10.9 Simulation of internal quantum efficiencies of the amorphous silicon solar cell without (black curve) and with a Eu-doped FCZ (red curve) and difference of both IQEs (blue curve) [24].

10.3.3
Influence of Multivalent Europium-Doping

This section will focus on the most important and expensive ingredient of the glasses and glass ceramics described above: divalent europium as optical activator. The reason for the high price in the world market is the serious shortage of RE minerals mostly caused by China's export embargoes and price policy in 2010 and 2011. Doubled or tripled costs of RE materials are first signs of an upcoming crisis which is predicted to reach a high in 2014 and 2015 [25]. Therefore, the Eu^{2+} portion of the image plate is the most expensive part. An attempt to overcome this problem is an *in situ* reduction of the relatively inexpensive $EuCl_3$ to introduce divalent europium ions into the glass.

Several routes are under investigation. MacFarlane *et al.* [20] reduced trivalent europium by using metal hydrides and adding hydrogen to the process atmosphere. Hydrogen reacts with the $EuCl_3$ to produce gaseous hydrogen chloride and $EuCl_2$. Unfortunately, melting in a hydrogen-bearing atmosphere can partially reduce ZrF_4 and form the already mentioned dark gray precipitates in the glass. Phebus *et al.* [26] showed that while trivalent europium was practically all reduced in a 4 % hydrogen in nitrogen atmosphere, the resulting glass was of poor quality. Coey *et al.* [27] showed that fluoride glasses made using 10–20 mol% EuF_2 contained a significant proportion of EuF_3 after processing. Mössbauer spectra had a broad peak centered at -14 mm s^{-1} indicating spin relaxation of the Eu^{2+} ions. The Eu^{3+} produced a much narrower peak environment at 0 mm s^{-1}. Ball *et al.* [28] also used Mössbauer spectroscopy to investigate thermal reduction of $EuCl_3$ processed in a vacuum furnace. They discovered that a mixture of $EuCl_3$, a chlorine deficient phase $EuCl_{2.8}$, and $EuCl_2$ was formed by treatment at 300 °C.

Glassner [29] compiled thermodynamic data for a variety of metal halides including the components of ZBLAN glass, but not for europium dihalides. The data for Zr show that at all temperatures, ZrF_3 is the most stable fluoride. The thermodynamic stability of ZrF_3 in the glasses has not been confirmed. Broer [30] observed black precipitates in ZBLAN glasses and attributed them to partially reduced zirconium halides. Rard [31] reviewed the properties and thermodynamics of RE fluorides and reported reduction of Eu^{3+} on heating. Massot et al. [32] studied reduction of Eu^{3+} in molten salts by using cyclic voltammetry. They reported an equilibrium constant of 0.811 for $[Eu^{2+}]/[Eu^{3+}]$ at 800 °C. This finding is consistent with Rard's assessment and suggests that thermal reduction of $EuCl_3$ is a feasible route for synthesizing the divalent salt.

In the following section, the focus is on the amount of divalent and trivalent Eu chloride and fluoride additives. The nominal multivalent doping of the FCZ glasses investigated varies between $xEuCl_2-(2-x)EuCl_3$ for the chloride series, and between $xEuF_2-(2-x)EuF_3$ for the fluoride series ($x = 0, 0.4, 0.8, 1.6, 2.0$; values in mol%). The effect of multivalent Eu doping on the $BaCl_2$ crystallization is discussed. X-ray absorption near edge structure (XANES) and X-ray diffraction (XRD) investigations were performed to evaluate the accurate Eu^{2+}-to-Eu^{3+} ratios, as well as the upcoming corresponding crystal phases.

10.3.3.1 X-Ray Absorption near Edge Structure

Figure 10.10 shows the normalized Eu L_3 XANES spectrum of the as-poured, nominal 100 % $EuCl_3$-doped FCZ glass. The two resonances located at approximately 6975 and 6983 eV are caused by a dipole-allowed transition from a 2p3/2 core level into an empty 5d state and are associated to the characteristic white lines (WLs) of Eu^{2+} and Eu^{3+}, respectively [33]. The Eu^{2+}-to-Eu^{3+} ratios can be determined by building the relative intensity ratios of the two WLs via fitting each WL by a pseudo-Voigt and arctangent function (see Figure 10.10; for further information about the fitting process see Ref. [34]).

Figure 10.11 shows the measured Eu^{2+} fractions for different ratios of $EuCl_2$-to-$EuCl_3$ in FCZ glasses versus annealing temperature. The actual Eu^{2+} content differs from the nominal Eu^{2+} doping and also shows a slight decrease when the annealing temperature is increased within each doping series, in contrast to previous series. In this case, the heat treatment was done in an inert-argon atmosphere, whereas it was previously done in nitrogen. Another difference is that the studied samples here were exposed to air for several days before the heat treatment. Weber et al. [35] evaluated Mössbauer measurements on a $Eu^{2+}-Eu^{3+}$ mixed powder and attributed the decrease in Eu^{2+}, after 10 weeks exposure in air, to oxidation. The observed behavior here is, therefore, likely caused by a small partial and diffusion-limited oxidation of Eu^{2+} on and near the oxygen-contaminated glass surfaces during the annealing step.

The diffusion coefficient in solids at different temperatures is well known to be proportional to $\exp(-E_A/RT)$, with the activation energy for diffusion E_A, the universal gas constant R, and the temperature T of the solid. Therefore, higher annealing temperatures lead to slightly larger diffusion distances of oxygen and

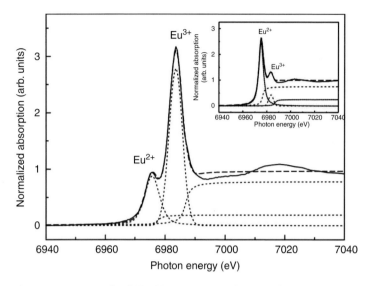

Figure 10.10 Normalized XANES spectrum of the as-made, nominal 100 % EuCl$_3$-doped FCZ glass. A pseudo-Voigt function and an arctan function (dashed curves) were used to fit each of the characteristic white lines of Eu^{2+} and Eu^{3+} and their absorption edges, respectively. The inset shows the normalized XANES spectrum of the as-made, nominal 100 % EuCl$_2$-doped FCZ glass for comparison [34].

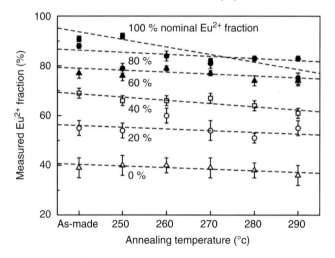

Figure 10.11 Measured Eu^{2+} fraction for differently EuCl$_2$-to-EuCl$_3$-doped FCZ glasses versus annealing temperature. The lines are a guide to the eye [34].

moisture for the same annealing time, leading to a decrease in the actual Eu^{2+} content. The melting process is even more influential. When doped with pure EuCl$_3$, about 40 % of the Eu^{3+} is reduced to Eu^{2+}. Looking at the thermodynamics and at the primary-phase equilibrium for changing the Eu oxidation state, this result fits well to the theory. The reaction equation during heat treatment above

570 K is given by Laptev et al. [36]: $EuCl_3 = EuCl_2 + 1/2\, Cl_2$ and the corresponding equilibrium constant is given by:

$$K = \exp\left(-\frac{\Delta G_T^\ominus}{RT}\right) = \frac{a[EuCl_2] \cdot p^{1/2}[Cl_2]}{a[EuCl_3]}$$

with the free energy of the reaction $\Delta G_T^\ominus$ at process temperature T and the activities, a, of the components and the pressure, p, at which half of the Cl is ligated.

As already mentioned earlier, Massot et al. [32] calculated K to be 0.811 for an operating temperature of 800 °C via an electrochemical study. For pure $EuCl_3$ doping, this would result in an equilibrium consisting of 45 Eu^{2+} and 55 % Eu^{3+}, confirming the measured data for the ZBLAN:$EuCl_3$ melt, that is, the sample with nominal 0 % Eu^{2+} fraction in the as-made state.

However, doping with pure $EuCl_2$ shows the possible Eu^{2+} doping limit for the above explained melting process to be slightly more than 90 % for the as-made glass (see Figure 10.11). It was shown that the InF_3 additive acts as a mild oxidizer preventing a reduction from Zr^{4+} to Zr^{3+} for reasons of glass quality, but at the same time also oxidizes some Eu^{2+} to Eu^{3+}. Moreover, even minimal oxygen contamination in the used raw materials and the above-mentioned long air exposure could lead to this oxidation process and the 90 % limit, respectively. Unlike all other series, the nominal pure 2 mol% $EuCl_2$-doped series shows a significant downward trend to lower Eu^{2+} fractions with higher annealing temperatures (78 % for 20 min annealing at 290 °C) synonymous with an increased conversion from Eu^{2+} into Eu^{3+}. The reason is not completely understood yet, but could be related to a higher Eu fraction near the surface caused by a not fully homogeneously blended melt and glass, respectively, making an interaction with oxygen during air exposure more probable. Looking at the measured Eu^{2+} fractions and their corresponding theoretical Eu^{2+} fractions, it is noticeable that the difference between them is not linear in all regions.

Figure 10.12 shows the measured Eu^{2+} fraction for Eu^{2+}-to-Eu^{3+} chloride- (squares) and fluoride- (triangles) doped FCZ glasses. During melting, more Eu^{2+} is formed when the glasses are doped with $EuCl_3$ as opposed to EuF_3. Thus, for the chloride series and from nominal Eu^{2+} fractions between 0 and 80 %, an even higher Eu^{2+} amount is found because of the reduction of the additional Eu^{3+} dopant. Saturation is observed at nominal Eu^{2+} fractions of more than 80% because of the above-mentioned effects of InF_3 and oxygen contamination. These effects are confirmed when looking at the fluoride series, where the measured Eu^{2+} fraction also saturates at almost the same mean value of about 83 % (Eu^{3+} fraction of about 17 %) for a nominal pure EuF_2-doped glass.

10.3.3.2 X-Ray Diffraction

Figure 10.13 shows a typical XRD pattern for a 2 mol% $EuCl_3$-doped FCZ glass. The as-poured glass does not show any crystalline phases, whereas annealing at 250 °C for 20 min leads to the formation of hexagonal phase $BaCl_2$ nanocrystals

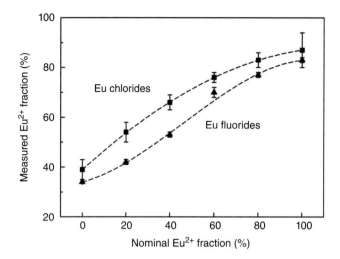

Figure 10.12 Measured Eu^{2+} fraction for Eu^{2+}-to-Eu^{3+} chloride- (squares) and fluoride- (triangles) doped FCZ glasses versus their nominal Eu^{2+} fraction. The lines are a guide to the eye [34].

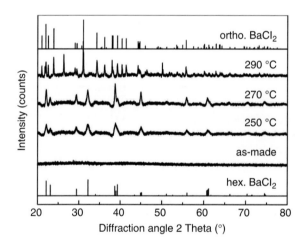

Figure 10.13 XRD patterns of a pure 2 mol% $EuCl_3$-doped FCZ glass as-made and annealed at temperatures from 250 to 290 °C for 20 min. The XRD patterns of hexagonal (bottom, PDF 45-1313) and orthorhombic $BaCl_2$ (top, PDF 24-0094) are shown for comparison [34].

(space group $P\bar{6}2m$ (space group no. 189), see bottom bar pattern). Only diffraction peaks related to hexagonal $BaCl_2$ are detectable up to an annealing temperature of 270 °C, whereas a phase transition to the orthorhombic $BaCl_2$ structure (space group *Pnma* (space group no. 62), see top bar pattern) can be observed upon annealing at 290 °C. In addition, two reflections arise at 26.3° and 50.1°, which are not yet identified.

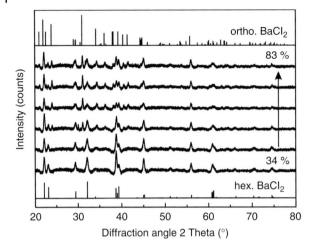

Figure 10.14 XRD patterns of an Eu^{2+}–Eu^{3+} fluoride FCZ glass series annealed at 270 °C for 20 min. The percentages correspond to the measured Eu^{2+} fraction of the total Eu doping. The XRD patterns of hexagonal (bottom, PDF45-1313) and orthorhombic BaCl$_2$ (top, PDF 24-0094) are shown for comparison [34].

Figure 10.14 shows XRD patterns of an Eu^{2+}–Eu^{3+} fluoride FCZ glass series annealed at 270 °C for 20 min. Compared to the nominal 100 % EuCl$_2$-doped sample, which already shows a completely orthorhombic BaCl$_2$ phase at 270 °C, the nominal 100 % EuF$_2$-doped glass still contains a clear measurable amount of both phases. Furthermore, within the fluoride series, a phase mixture at 270 °C is not observed prior to 40 % Eu^{2+} being present, whereas it is observed at 20 % within the chloride series. Because the base-glass compositions (except Eu^{2+} and Eu^{3+} chlorides and fluorides) remain constant, there are two possible reasons for this behavior: First, because of the slightly decreased chlorine content in the Eu fluoride series, higher energies and temperatures are needed to initiate nucleation and phase transition, respectively. Second, the XANES measurements show that, compared to the chloride series, the Eu^{3+} in the fluoride series is more stable for the same melting process. From the previous results on chlorides it is known that a lower amount of Eu^{2+} leads to a more stable hexagonal BaCl$_2$ phase, explaining the later phase transition at higher temperatures for the fluoride series.

10.3.3.3 Photoluminescence

BaCl$_2$ nanocrystal structure sensitive PL spectra of differently annealed and differently Eu^{2+}/Eu^{3+}-doped FCZ samples are plotted in Figure 10.15. The samples were annealed for 20 min at temperatures between 250 and 290 °C. Again the PL was excited in the UV spectral range at 280 nm and all spectra have been normalized to their most intense emission. All samples annealed at 250 °C (dashed-dotted curves) show the typical 5d–4f transition PL peak of Eu^{2+} incorporated

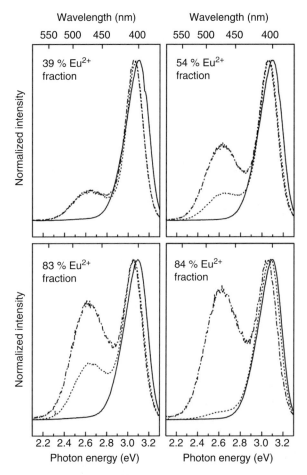

Figure 10.15 Normalized PL spectra for different Eu^{2+} fractions (measured with XANES) in FCZ glass ceramics and annealed at 250 (dashed-dotted), 270 (dotted), and 290 °C (solid curve) for 20 min. The PL was excited at 280 nm [37].

into hexagonal phase $BaCl_2$ particles and the less intense emission at 475 nm, as well as the nominally 100 %-doped Eu^{3+} sample, which confirms the already discussed conversion mechanism from Eu^{3+} to Eu^{2+} during melting. When annealing up to 290 °C, the 475 nm band disappears for all different doping fractions as a result of the phase transformation from hexagonal to orthorhombic phase $BaCl_2$. However, there are differences in between these temperatures such as peak position of the main PL peak and intensity of the second peak, which are in good correlation to the obtained nanocrystal phases.

The spectra of the nominally 100 %-doped Eu^{3+} sample with measured 39 % Eu^{2+} fraction (top left) show no significant differences between 250 and 270 °C annealing; both show the typical Eu^{2+} PL of the hexagonal nanocrystal phase. XRD has shown that annealing at 290 °C leads to the formation of orthorhombic phase $BaCl_2$. Here, the 405 nm band for 250 °C is shifted to 399 nm, the typical orthorhombic bulk $BaCl_2$ PL, while the 475 nm band disappears. At 394 nm, a small dip in the Eu^{2+} emission band is observed, which is assigned to an Eu^{3+} absorption from the ground state $^7F_0$ to state $^5L_6$ [23]. Increasing the divalent doping fraction, this absorption decreases due to less Eu^{3+} in the glass. Furthermore, a higher Eu^{2+} fraction leads to a decrease in the peak intensity ratio $I(475\,nm)/I(405\,nm)$. The 475 nm peak of the 250 °C spectra rises from 39 % Eu^{2+} to 84 % Eu^{2+} fraction. This confirms the theory of the impurity associated Eu^{2+} sites or clusters, which are more probable to form with a higher Eu^{2+} amount.

The PL peak shift of about 5 nm for the phase transition is also observed for the 84 % Eu^{2+} sample (bottom right). However, in contrast to the lower Eu^{2+}-doped glass ceramics, here the 475 nm band is almost gone for 270 °C annealing meaning that the hexagonal phase undergoes the phase transition at lower temperatures. This is also verified by the already discussed XRD results showing a higher amount of divalent Eu leads to a less stable hexagonal $BaCl_2$ phase. The higher Eu^{2+} doping fraction probably results in a higher Eu^{2+} concentration in the nanocrystals causing an earlier hexagonal to orthorhombic phase transformation.

At the same time, the larger number of impurity associated Eu^{2+} sites or clusters disturbs the crystal transformation and is a possible explanation for the increase of the activation energy measured via differential scanning calorimetry (DSC) investigations.

10.3.4
Conclusion

Similarly to the down-conversion efficiency, the transparency of the down-converting cover glass also plays an important role in the solar cell efficiency enhancement. The results showed that the absorption is very sensitive to changes in the annealing procedure. A few degrees higher annealing temperature is sufficient to change the nanocrystal growth conditions and thus the glass transparency. Higher temperatures lead to more and larger nanocrystals and thus to a decreased transparency. These influences also occur in the short circuit current of a solar cell when placing a down-converter on its top. The best enhancement is achieved when the nanocrystals are large in number but their size as small as possible.

XANES measurements show that, during the glass-melting process, Eu^{3+} is partially reduced to Eu^{2+}. Because the measured Eu^{2+} fraction in the glass saturates with higher doping ratio (only 13 % difference between nominal 60 % and 100 % Eu^{2+} fraction), usage of the cheaper $EuCl_3$ for this region would reduce costs without lowering the Eu^{2+} luminescence significantly. These results also show that europium chloride converts more Eu^{3+} into the divalent valence state than their fluoride counterparts and, therefore, are the preferred choice for this method. Furthermore, a slight oxidation of Eu^{2+} during annealing was observed. It is recommended to heat treat the samples soon after the cutting and polishing steps.

The different chemical equilibria of the multivalent Eu-doped FCZ glass melts significantly influence the luminescence output, which is important for $BaCl_2$

crystallization. XRD and PL measurements show that a higher amount of divalent Eu leads to an earlier (in terms of temperature) phase transition from hexagonal to orthorhombic $BaCl_2$. As XANES results suggest, the phase transition takes place at higher temperatures for Eu–fluoride compounds than for their chloride equivalents.

In conclusion, and in terms of a price reduction for image plates for medical diagnostics and down-conversion layers for solar cells, multivalent Eu doping of ZBLAN glasses and glass ceramics could be a technically and financially feasible route.

Concerning the scope of application, a down-converter only provides a benefit to the spectral range where the EQE of a solar cell is low. The maximum internal conversion efficiency of a 5 mol% Eu-doped glass annealed at 270 °C for 20 min was calculated to 52 % at 393 nm excitation using the absorption coefficient of the glass. The calculated IQE of the solar cell with down-converter showed an efficiency enhancement up to 7 % for wavelengths smaller than 335 nm. In the spectral range between 335 and 400 nm the absorption edge of the down-converting FCZ glass resulted in an IQE decrease.

10.4
Down-Conversion in Sm-Doped Borate Glasses for High-Efficiency CdTe Solar Cells

Borate glasses are known for their high optical transparency; they are a robust and inexpensive matrix material for fluorescent ions. These characteristics make them attractive for many applications; the usage as spectral down-conversion glass for photovoltaic applications is a recent example [38]. Samarium-doped borate glasses, for instance, convert the incident violet and the blue part of the solar spectrum to visible red light.

Photovoltaic devices based on CdTe absorbers feature a poor response in the blue and near ultraviolet spectral range due to absorption in the CdS buffer layer. Borate glasses, which are additionally doped with fluorescent RE ions such as trivalent samarium, provide optical properties that perfectly complement the limitations in quantum efficiency of the CdTe solar cell. In the following, barium borate glasses, which were additionally doped with Sm^{3+} ions, are investigated for their potential as a superstrate for CdTe thin film solar cells.

10.4.1
Samples

Borate glasses using barium oxide as a network modifier were prepared. A ratio of 2 mol of boron oxide (B_2O_3) and 1 mol of barium oxide (BaO) was used. The glasses were additionally doped with samarium oxide (Sm_2O_3). The chemicals were weighed in a platinum/gold alloy crucible and melted at 1100 °C for approximately 3 h. The melt was then poured onto a brass block at 550 °C below the glass transition temperature. The glass was kept at this temperature for 3 h to eliminate residual mechanical and thermal stresses before being slowly cooled to room temperature.

10.4.2
External Quantum Efficiency of CdTe Solar Cells

Figure 10.16 illustrates the photon flux density for the AM1.5 solar spectrum (black curves) [39] and the EQE of a CdTe solar cell (red curve) with a CdS buffer layer of 300 nm [40]. The area under the blue curves yields the total number of photons leading to electron–hole pairs, that is, to the maximal short circuit current density, which can be calculated from

$$J_{SC} = q_e \int_{300\ nm}^{900\ nm} \text{AM1.5}(\lambda) \cdot \text{EQE}_{CdTe}(\lambda) d\lambda = q_e \cdot 1.119 \times 10^{21}\ \text{photons}/(s \cdot m^2)$$
$$= 17.94\ \text{mA}/\text{cm}^2$$

with q_e the elementary charge.

10.4.3
Optical Absorption and Fluorescence Emission

Figure 10.17 shows the transmission spectra of barium borate glasses, undoped and 1 mol% Sm_2O_3-doped with a thickness of 3.2 mm each. The spectrum of the doped sample shows the typical Sm^{3+}-related absorption bands at 340, 360, 375, 400, and 470 nm resulting from transitions from the ground state level $^6H_{5/2}$ to the

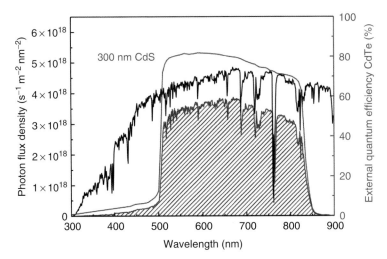

Figure 10.16 Photon flux for the AM1.5 spectrum (black curve) [39], external quantum efficiencies of CdTe solar cells with a CdS buffer layer of 300 nm (red curve) [40], and the total number of photons leading to electron–hole pairs (area under blue curve) [38].

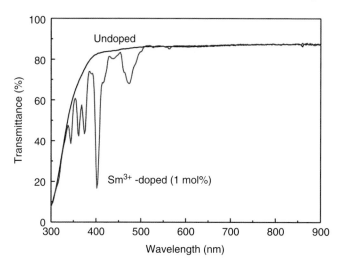

Figure 10.17 Transmission spectra of barium borate glasses, undoped (black curve) and doped with 1 mol% Sm_2O_3 (blue curve). For both samples, the thickness was 3.2 mm [38].

excited states $^4K_{15/2}, ^4D_{3/2}, ^4D_{1/2}, ^6P_{5/2}$, and $^4G_{7/2}$, respectively. Figure 10.18 shows the energy level diagram of Sm^{3+} [41]. Possible absorption and emission routes are indicated.

The number of solar photons being absorbed by a 1 mol% Sm^{3+}-doped barium borate glass with a thickness of 3.2 mm can be calculated from

$$\int_{300\,nm}^{900\,nm} AM1.5(\lambda) \cdot A_{Sm^{3+}}(\lambda) d\lambda = 0.044 \times 10^{21} \text{ photons}/(s \cdot m^2).$$

The solar photons being absorbed by Sm^{3+} are partially converted to photons in the red spectral range where the EQE of the CdTe solar cell is at maximum.

Figure 10.19 shows the normalized fluorescence excitation and emission spectra of a 1 mol% Sm^{3+}-doped barium borate glass. Upon excitation at 402 nm ($^6H_{5/2}$ to $^6P_{5/2}$ transition) the glass shows the typical Sm^{3+} emissions in the red spectral range, which are caused by transitions from the $^4G_{5/2}$ state to the ground state levels $^6H_{5/2}$ (560 nm), $^6H_{7/2}$ (600 nm), $^6H_{9/2}$ (645 nm), and $^6H_{11/2}$ (705 nm). In addition to these visible emissions, there are further emissions in the near infrared spectral (NIR) range, namely the $^4G_{5/2}$ to $^6H_{15/2}$ (910 nm) and the $^4G_{5/2}$ to $^6F_{5/2}$ (955 nm) transitions. The NIR emissions are more than 2 orders of magnitude less intense and thus not considered any further. The fluorescence excitation spectrum agrees well with the Sm^{3+} absorption spectrum shown in Figure 10.17.

10.4.4
Efficiency Increase

Preliminary estimates have shown that 80 % of the emitted photons reach the CdTe absorber; the escape cone losses are approximately 20 %. In the red spectral range

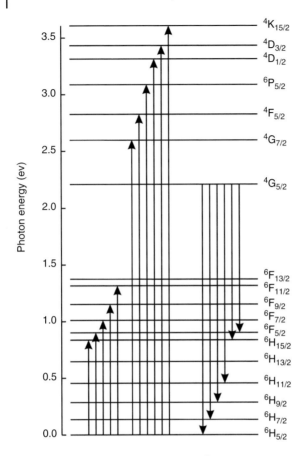

Figure 10.18 Energy level diagram of Sm^{3+}. Possible absorption (left arrows) and emission routes (right arrows) are indicated [12, 41].

of the Sm^{3+} emission, the quantum efficiency of the CdTe solar cell is approximately 80%, that is, the majority of the converted photons generate additional electron–hole pairs leading to an enhanced short circuit current density. Assuming an efficiency of 50 % for the conversion of photons in the violet and blue to photons in the red spectral range the total enhancement achieved by a 1 mol% Sm^{3+}-doped cover glass amounts to

$$q_e \cdot 0.044 \times 10^{21} \text{ photons}/(s \cdot m^2) \cdot \underbrace{50\ \%}_{\substack{\text{conversion efficiency} \\ \text{Sm}^{3+}}}$$

$$\cdot \underbrace{80\ \%}_{\text{coupling efficiency}} \cdot \underbrace{80\ \%}_{\substack{\text{external quantum efficiency} \\ \text{CdTe solar cell}}} = q_e \cdot 0.014 \times 10^{21} \text{ photons}/(s \cdot m^2)$$

$$= 0.22 \text{ mA cm}^{-2}$$

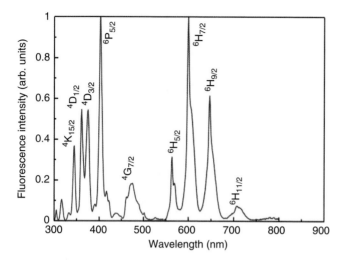

Figure 10.19 Normalized fluorescence excitation (blue) and emission (red) spectra of Sm^{3+}-doped barium borate glasses. The emission spectrum was excited at 402 nm; the excitation spectrum detected at 600 nm. The Sm^{3+} excitation bands are caused by transitions from the $^6H_{5/2}$ ground state to the levels indicated; the emissions start from the $^4G_{5/2}$ excited state and end on the ground state levels indicated [38].

However, since the CdTe solar cell has also a quantum efficiency in the ultraviolet/blue spectral range of the Sm^{3+} absorption bands, the maximal short circuit current density of a Sm^{3+}-doped glass superstrate-based cell has to be corrected for the Sm^{3+} absorptance, that is,

$$q_e \cdot \int_{300\,nm}^{900\,nm} AM1.5(\lambda) \cdot A_{Sm^{3+}}(\lambda) \cdot EQE_{CdTe}(\lambda) d\lambda$$
$$= q_e \cdot 0.003 \times 10^{21}\,photons/(s \cdot m^2) = 0.05\,mA\,cm^{-2}$$

In total, the maximal short circuit current density amounts to

$$J_{SC}^{Sm^{3+}} = J_{SC} + 0.22\,mA\,cm^{-2} - 0.05\,mA\,cm^{-2} = J_{SC} + 0.17\,mA\,cm^{-2}$$
$$= 18.11\,mA\,cm^{-2}$$

meaning an increase of +0.9%. These calculations have been repeated with respect to the Sm_2O_3 doping level and the Sm^{3+} conversion efficiency. The result is depicted in Figure 10.20. Please see [38] for further information.

10.4.5
Conclusion

The influence of the Sm^{3+} conversion efficiency and the Sm_2O_3 doping level on the short circuit current density has been estimated. Sm^{3+}-doped barium borate glasses have potential as a superstrate for CdTe solar cells. To achieve an increase, a Sm^{3+} conversion efficiency of more than 14 % is needed for CdTe solar cells with

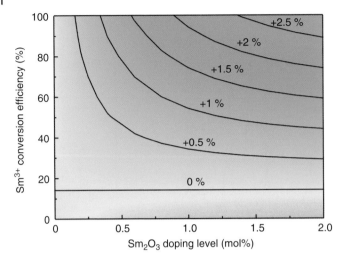

Figure 10.20 Relative changes in the CdTe solar cell efficiency with a CdS buffer layer of 300 nm with respect to the Sm_2O_3 doping level and the Sm^{3+} conversion efficiency [38].

a CdS buffer layer of 300 nm. The performance of CdTe solar cells with poor quantum efficiency in the ultraviolet and blue spectral range can thus be significantly increased by the use of frequency converting cover glasses such as samarium-doped borate glasses.

10.5
Summary

In this work, down-converters based on RE ions in ZBLAN, glass ceramics and borate glasses have been investigated. The down-conversion efficiency of europium and samarium has been optimized for application in different types of solar cells. An improvement of the quantum efficiency of the solar cells is demonstrated in the spectral region where the EQE of the cells is low. For silicon-based solar cells, this was achieved by Eu ions which absorb in the near UV and emit in the deep blue. However, as the quantum efficiency of the silicon is high even at wavelengths below 400 nm the gain by the enhanced EQE in the UV was spoiled by extra absorption from the glass. For CdTe solar cells, Sm was chosen as it matches favorably the low EQE at wavelengths below 500 nm. As the photon flux of the sun light is low in the spectral region where down-converters are appropriate it turns out to be extremely important to avoid all spurious losses in the remaining part of the spectrum. Even additional absorption or reflection losses of 1 % can easily abolish any gain achieved by an optimized down-converter. Yet, there is still plenty of room for finding solutions to deal with these challenges.

Acknowledgment

This work was supported by the FhG Internal Programs under Grant No. Attract 692034. In addition, the authors would like to thank the German Federal Ministry for Education and Research (BMBF) for the financial support within the Centre for Innovation Competence SiLi-nano® (Project No. 03Z2HN11) and the "Verbundvorhaben: NanoVolt – Optische Nanostrukturen für die Photovoltaik" (Project No. 03SF0322A). We would like to thank J. Klomfaß for PDS measurements and M. Hülsbeck for technical assistance.

References

1. Baumgartner, K., Ahrens, B., Angelov, O., Sendova-Vassileva, M., Dimova-Malinovska, D., Holländer, B., Schweizer, S., and Carius, R. (2010) Proceeding of the 25th European Photovoltaic Solar Energy Conference and Exhibition/5th World Conference on Photovoltaic Energy Conversion, September 6–10, 2010, Valencia, Spain, p. 245.
2. Blasse, G. and Grabmaier, B.C. (1994) *Luminescent Materials*, Springer-Verlag, Berlin, Heidelberg.
3. Hoefdraad, H.E. (1975) *J. Inorg. Nucl. Chem.*, **37**, 1917–1921.
4. Miyakawa, T. and Dexter, D.L. (1970) *Phys. Rev. B*, **1** (7), 2961–2969.
5. Shojiya, M., Takahashi, M., Kanno, R., Kawamoto, Y., and Kadono, K. (1997) *J. Appl. Phys.*, **82** (12), 6259.
6. Poulain, M., Poulain, M., Lucas, J., and Brun, P. (1975) *Mater. Res. Bull.*, **10** (4), 243.
7. Ohsawa, K. and Shibata, T. (1984) *J. Lightwave Technol.*, **LT-2** (5), 602.
8. Aggarwal, I.D. and Lu, G. (1991) *Fluoride Glass Fiber Optics*, Academic Press, London.
9. Wetenkamp, L.G., West, F., and Többen, H. (1992) *J. Non-Cryst. Solids*, **140**, 35–40.
10. Zachariasen, W.H. (1932) *J. Am. Chem. Soc.*, **54**, 3841–3851.
11. Biscoe, J. and Warren, B.E. (1938) *J. Am. Ceram. Soc.*, **21**, 287.
12. Dyrba, M., Miclea, P.T., and Schweizer, S. (2010) Spectral down-conversion in Sm-doped borate glasses for photovoltaic applications Photonics for Solar Energy Systems III, In Ralf B. Wehrspohn and Andreas Gombert (eds) *Proc. of SPIE*, Vol. 7725, 77251D. DOI: 10.1117/12.853942
13. Müller-Warmuth, W. and Eckert, H. (1982) *Phys. Rep.*, **88**, 91.
14. Johnson, J.A., Schweizer, S., Henke, B., Chen, G., Woodford, J., Newman, P.J., and MacFarlane, D.R. (2006) *J. Appl. Phys.*, **100**, 034701.
15. Johnson, J.A., Schweizer, S., and Lubinsky, A.R. (2007) *J. Am. Ceram. Soc.*, **90** (3), 693–698.
16. Sadoc, A. and Guillo, A. (1971) *C. R. Acad. Sci. Ser., B*, **273**, 203.
17. Aasland, S., Einarsrud, M.-A., and Grande, T. (1996) *J. Phys. Chem.*, **100** (13), 5457–5463.
18. Paßlick, C., Henke, B., Ahrens, B., Miclea, P.-T., Wenzel J., Reisacher, E., Pfeiffer, W., Johnson, J. A., and Schweizer, S. (2010) Glass-ceramic covers for highly efficient solar cells. SPIE Newsroom, 25 May 2010. http://dx.doi.org/10.1117/2.1201005.002938 (accessed 4 November 2014).
19. Furniss, D., Harris, E.A., and Hollis, D.B. (1987) *J. Phys. C: Solid State Phys.*, **20**, L147.
20. MacFarlane, D.R., Newman, P.J., Cashion, J., and Edgar, A. (1999) *J. Non-Cryst. Solids*, **53**, 256–257.
21. Johnson, J.A., Weber, R., Schweizer, S., MacFarlane, D.R., Newman, P., and Chen, G. (2007) *Multi-Radiation Large Area Detector*, The University of Chicago.
22. Paßlick, C., Henke, B., Császára, I., Ahrens, B., Miclea, P.-T., Johnson, J.A.,

and Schweizer, S. (2010) *Proc. SPIE*, **2**, 7772–7779.
23. Samek, L., Wasylak, J., and Marczuck, K. (1992) *J. Non-Cryst. Solids*, **140**, 243–248.
24. Baumgartner, K. (2011) Charakterisierung von Materialien für die Konversion von Sonnenlicht zur Effizienzsteigerung von Solarzellen. PhD-Thesis, RWTH Aachen.
25. UPI Information about the Cost of Rare-Earth Minerals in Recent Times, http://www.upi.com/Business_News/Energy-Resources/2011/06/23/rare-earth-minerals-prices-skyrocket/UPI-76601308849756/ (accessed 23 June 2011).
26. Phebus, B., Getman, B., Kiley, S., Rauser, D., Plesha, M., and Terrill, R.H. (2005) *Solid State Ionics*, **176**, 2631–2638.
27. Coey, J.M.D., McEvoy, A., and Shafer, M.W. (1981) *J. Non-Cryst. Solids*, **43**, 387–92.
28. Ball, J., Jenden, C.M., Lyle, S.J., and Westall, W.A. (1983) *J. Less-Common Met.*, **95**, 161–70.
29. Glassner, A. (1953) *A survey of the free energies of formation of the fluorides, chlorides, and oxides of the elements to 2500 K*. University of Michigan Library, Argonne, IL.
30. Broer, M.M. (1989) *J. Am. Ceram. Soc.*, **72** (3), 492–495.
31. Rard, J.A. (1985) *Chem. Rev.*, **85**, 555–582.
32. Massot, L., Chamelot, P., Cassayre, L., and Taxil, P. (2009) *Electrochim. Acta*, **54**, 6361–6366.
33. Rohler, J. (1987) *Handbook on the Physics and Chemistry of Rare Earths*, North-Holland, Amsterdam.
34. Paßlick, C., Müller, O., Lützenkirchen-Hecht, D., Frahm, R., Johnson, J.A., and Schweizer, S. (2011) *J. Appl. Phys.*, **110**, 113527.
35. Weber, J.K.R., Vu, M., Paßlick, C., Schweizer, S., Brown, D.E., Johnson, C.E., and Johnson, J.A. (2011) *J. Phys. Condens. Matter*, **23**, 495402.
36. Laptev, D.M., Kiseleva, T.V., Kulagin, N.M., Goryushkin, V.F., and Voronkov, E.S. (1986) *Russ. J. Inorg. Chem.*, **31**, 1965.
37. Paßlick, C. (2013) Optical and Structural Analysis of Luminescent Nanoparticles in Glasses for Medical Diagnostics, PhD Thesis, Martin Luther University of Halle-Wittenberg.
38. Steudel, F., Dyrba, M., and Schweizer, S. (2012) *Proc. SPIE*, **8438**, 843803.
39. Reference Solar Spectral Irradiance: Air Mass 1.5, http://rredc.nrel.gov/solar/spectra/am1.5/ (accessed 28 June 2011).
40. Hädrich, M., Kraft, C., Metzner, H., Reislöhner, U., Löffler, C., and Witthuhn, W. (2009) *Phys. Status Solidi C*, **6** (5), 1257.
41. Binnemans, K. (1998) *J. Non-Cryst. Solids*, **238**, 11.

11
Fluorescent Concentrators for Photovoltaic Applications

Jan Christoph Goldschmidt, Liv Prönneke, Andreas Büchtemann, Johannes Gutmann, Lorenz Steidl, Marcel Dyrba, Marie-Christin Wiegand, Bernd Ahrens, Armin Wedel, Stefan Schweizer, Benedikt Bläsi, Rudolf Zentel, and Uwe Rau

11.1
Introduction

In a luminescent collector, a luminescent material embedded in a transparent matrix absorbs sunlight and subsequently emits photons with a different wavelength. Total internal reflection traps most of the emitted photons and guides them to the edges of the collector (Figure 11.1). This principle was first used in scintillation counters [1, 2]. When solar cells are optically coupled to the collector they can convert the guided light into electricity. A geometric concentration is achieved, when the area from which light is collected is larger than the area of the attached solar cells. This application to concentrate solar radiation onto a solar cell was proposed in the late 1970s [3, 4]. Probably the most outstanding feature of luminescent solar concentrators is their ability to concentrate both direct and diffuse radiation. In consequence, luminescent concentrators do not require tracking systems that follow the path of the sun, in contrast to concentrator systems that use lenses or mirrors. These features, alongside their optical appearance, make luminescent concentrators interesting for application such as in building integrated photovoltaics, or in indoor applications.

Luminescent concentrators were investigated intensively in the early 1980s, for example in Refs. [5, 6]. Research at that time aimed at cutting costs by using the concentrator to reduce the need for expensive solar cells. After more than 20 years of progress in the development of solar cells and luminescent materials, and with new concepts, several groups such as those of Refs. [7–25] are currently reinvestigating the potential of luminescent concentrators. High efficiencies have been achieved [19, 25, 26] and there has been also considerable progress in the understanding and theoretical description, for example in Refs. [22, 24, 26–28]. However, efficiencies are still too low and system sizes too small for commercial applications.

The concept of luminescent concentrators has been given many different names, among them are fluorescent collectors [4], quantum dot (QD) solar concentrator

Photon Management in Solar Cells, First Edition. Edited by Ralf B. Wehrspohn, Uwe Rau, and Andreas Gombert.
© 2015 Wiley-VCH Verlag GmbH & Co. KGaA. Published 2015 by Wiley-VCH Verlag GmbH & Co. KGaA.

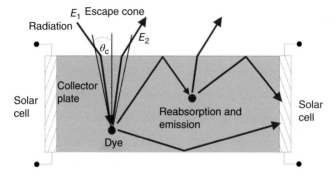

Figure 11.1 Principle of a luminescent concentrator. A luminescent material (dye) in a matrix absorbs incoming sunlight (E_1) and emits radiation with a different energy (E_2). Total internal reflection traps most of the emitted light and guides it to solar cells optically coupled to the edges. Emitted light that impinges on the internal surface with an angle steeper than the critical angle θ_c is lost due to the escape cone of total internal reflection. A part of the emitted light is also reabsorbed and then possibly reemitted [23].

[20–22, 29], and organic solar concentrator [30]. In this work, we will use the term *fluorescent concentrator* for the overall concept, as mostly fluorescent materials are considered for this application. For clarity, *fluorescent collector* will identify the collector plate without attached solar cells and *fluorescent concentrator system* will refer to a system constructed from a collector plate with solar cells attached [31].

Several factors determine the efficiency of a fluorescent concentrator system. The important parameters [32, 33] are the transmission of the front surface with respect to the solar spectrum, the absorption efficiency of the luminescent material due to its absorption spectrum with respect to the transmitted solar spectrum, the quantum efficiency of the luminescent material, the Stokes shift between the wavelength of the absorbed and emitted photons, the fraction of the emitted light that is trapped by total internal reflection, the efficiency of this light guiding, the self-absorption of the emitted light, the losses caused by scattering or absorption in the matrix, the optical coupling of the solar cells and the fluorescent collector, and finally the efficiency, by which the solar cells convert the light they receive from the collector.

In Section 11.2, we present a theoretical description of fluorescent concentrators. We will investigate which fundamental limits exist for the efficiency of fluorescent concentrators and discuss concepts how these limits can be shifted by the addition of photonic structures.

Many of the parameters determining the efficiency of fluorescent concentrators depend on the used luminescent material and the matrix that surrounds it. Therefore, we will present in Section 11.3 an overview of the different organic and inorganic materials, which are currently being investigated for their application in fluorescent concentrators.

Different system configurations have been proposed for luminescent concentrators. Variations include deposition of the luminescent material in a film on a transparent slab [34], stacking of different luminescent collectors [4], cylindrical collectors [35], solar cells coupled to the bottom of the collector [13, 15], and many more. The experimental results for a range of system configurations will be presented in Section 11.4.

11.2
The Theoretical Description of Fluorescent Concentrators

11.2.1
Detailed Balance Considerations

To highlight the different aspects that determine the efficiency of fluorescent concentrators, we model a fluorescent concentrator consisting of an acrylic plate with the refractive index $n_r = 1.5$ and embedded fluorescent dye molecules with a thermodynamic model. Figure 11.2 depicts the absorption/emission behavior of the fluorescent dye used in the following. We assume a stepwise increase of the absorption constant α from zero at energies $E < E_2$ to a value α_2 for $E > E_2$ and a further increase to α_1 for energies $E > E_1$. The emission coefficient e is linked to the absorption coefficient α via Kirchhoff's law

$$e(E) = \alpha(E) n_r \Phi_{bb}(E) \tag{11.1}$$

with the black body spectrum $\Phi_{bb}(E)$.

The absorption/emission dynamics used in the following is given by a two-level scheme as used earlier to describe the detailed balance limit of fluorescent concentrators [13, 14, 36]. The choice of this simple approach ensures a certain generality of our results such that the trends caused by the collector geometries or

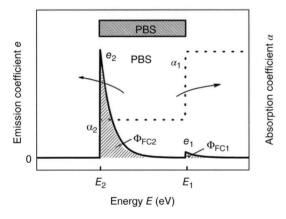

Figure 11.2 Sketch of the absorption and emission behavior as assumed in Section 11.2.3. The dye absorption is given by a step function. Incoming photons have the energy E_1 and a high absorption coefficient α_1. The lower absorption coefficient α_2 holds for the lower energy E_2 and leads with Kirchhoff's law (Eq. (11.1)) to a high emission coefficient e_2. The model also features the possibility of an energy selective photonic band stop (PBS) that keeps the emitted photons in the fluorescent concentrator system [49].

by the introduction of loss mechanisms should be equally found in real systems with more complex spectral absorption/emission properties. In the following, we assume $E_1 = 2.0$ eV, $E_2 = 1.8$ eV, and absorption coefficients $\alpha_1 = 3/d$, $\alpha_2 = 0.03/d$. With a choice of absorption coefficients $\alpha_1 = 100\alpha_2$ we ensure that the probability for emission at E_2 is about 20 times larger than for the emission at E_1.

Figure 11.2 also features an ideal photonic structure which has the characteristic of a band-stop filter (PBS – photonic band stop Filter). Thus, the PBS filter has a reflection $R = 1$ and a transmission $T = 0$ for photon energies $E < E_{th}$. For the other part of the spectrum $R = 0$ and $T = 1$ is assumed. Therefore, E_{th} denotes the upper cut-off energy of the filter. We choose $E_{th} = E_1$ for the simulations in Section 11.2.3.

11.2.2
Photonic Structures to Increase Fluorescent Collector Efficiency

Photonic structures reduce one dominant loss mechanism in fluorescent concentrators which is the escape cone of total internal reflection. All light impinging on the internal surface with an angle smaller than the critical angle $\theta_c(\lambda_{emit})$ leaves the collector and is lost (see also Figure 11.1). The critical angle is given by

$$\theta_c(\lambda) = \arcsin\left(\frac{1}{n(\lambda_{emit})}\right). \quad (11.2)$$

The light which impinges with greater angles is totally internally reflected. Under the assumption of isotropic emission, integration gives a fraction

$$\eta_{trap}(\lambda) = \sqrt{1 - n(\lambda_{emit})^{-2}} \quad (11.3)$$

of the emitted photon flux that is trapped in the collector [37]. For PMMA (polymethylmethacrylate) with $n = 1.5$, this results in a trapped fraction of around 74%, which means that a fraction of 26% is lost after every emission process. The 26% ratio accounts for the losses through both surfaces. An attached mirror does not change this number, as with a mirror the light leaves the collector through the front surface after being reflected.

However, the Stokes shift between absorption and emission opens the opportunity to reduce escape cone losses significantly: a selective reflector, which transmits all the light in the absorption range of the luminescent material and reflects the emitted light, would trap nearly all the emitted light inside the collector [38]. In Ref. [12], hot mirrors were proposed to serve as selective reflectors and in Ref. [13], photonic structures.

The concept is illustrated in Figure 11.3. Incoming photons have the energy E_1. The dye absorbs these photons and emits spatially randomized photons at a lower energy E_2. Emitted photons impinging at the top surface with an incident angle θ higher than the critical angle θ_c for total internal reflection are guided to the collector sides. Without a photonic structure, photons with $\theta < \theta_c$ leave the collector as is has been presented in Figure 11.1. Two- and three-dimensional photonic crystals [39–41] are also promising materials that might be used as omnidirectional PBS in fluorescent concentrator systems. Another possible realization of

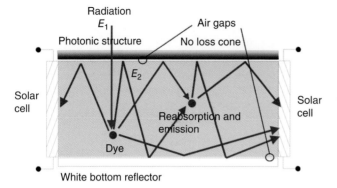

Figure 11.3 A selective reflector, realized as a photonic structure, reduces the escape cone losses. The photonic structure acts as a band stop reflection filter. It allows light in the absorption range of the dyes to enter the collectors, but reflects light in the emission range [23].

such a selective reflector is a so-called Rugate filter. It features a continuously varying refractive index profile that results in a single reflection peak. However, some unwanted side lobes remain. Optimized Rugate filters [42] show only one single reflection peak for a certain wavelength and almost no other reflections. Technological developments have led to dielectric mirrors used as band pass filters with almost rectangular cut-off characteristics for normal incident photons [43, 44]. In Ref. [25] it was shown that overall system efficiencies of luminescent concentrator systems can be increased with the help of such an optimized filter. These Rugate-filters show a high angular dependency by blocking only photons with almost perpendicular incidence. This can lead to the situation wherein some emitted photons with $E \leq E_1$ are neither reflected by the PBS nor subject to total internal reflection and leave the system. We analyze this effect on a fundamental basis in Section 11.2.3.2.

11.2.3
Possible System Configurations – Side-Mounted and Bottom-Mounted Solar Cells

In the classical fluorescent concentrator system presented by Goetzberger and Greubel [4] the solar cells are attached to the edges of the fluorescent collector. Figure 11.4a shows such a classic fluorescent concentrator system with a plate of length l and thickness d doped with a fluorescent dye. The collecting solar cells are mounted at the sides of the plate. The coverage fraction $f = A_{cell}/A_{coll}$ is the ratio between the area $A_{cell} = 4\,dl$ of the solar cells in the system and the illuminated collector area $A_{coll} = l^2$. For the configuration in Figure 11.4a, we have $f = 4d/l$; hence, the coverage fraction depends only on the ratio between the collector thickness d and the side length l. Such systems have been theoretically and experimentally analyzed, for example in Refs. [5, 25, 26, 45] and achieve highest measured system efficiencies [19]. Figure 11.4b features a variant of the side-mounted fluorescent concentrator where only a part of each side is covered with solar cells. Perfect

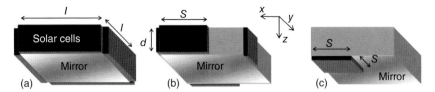

Figure 11.4 Sketch of different fluorescent collector geometries (seen from the bottom). (a) Classical design with the solar cells mounted at each side of the collector with length l and thickness d. (b) Modified classical system where only a fraction of the respective collector sides is covered with a solar cell area $A_{cell} = ds$. (c) Collector with solar cells with an area $A_{cell} = s^2$ mounted at the bottom. For the simulations, the (remaining) rear side is covered with a mirror [49].

mirrors cover the remainders. The coverage fraction for the system in Figure 11.4b is $f = 4ds/l^2$ with the side length $s \leq l$ of the solar cells. Thus, coverage fraction f and collector length l are decoupled and this geometry offers an additional degree of freedom for the collector design. In both system geometries, the fluorescent concentrator rear side is covered by a mirror. The collector design of Figure 11.4c uses a square solar cell with a side length s at the rear side of the fluorescent concentrator. Thus, the solar cells in this bottom-mounted system cover a fraction $f = s^2/l^2$ of the surface. As in the system of Figure 11.4b, the fluorescent concentrator edges and the remaining parts of the rear side are covered with mirrors.

In the following, we will present simulations that compare systems with side- and bottom-mounted solar cells. In order to model a constant coverage fraction f while varying the collector length l, we compare the systems of Figure 11.4b,c.

Keeping a constant coverage fraction f upon variation of the collector length l requires the adjustment of the solar cell side length s to $s/d = f(l/d)^2/4$ for the normalization of all quantities to the collector thickness d. The maximum coverage fraction is $f_{max} = 4d/l$ because in this case the side length s of the solar cell equals the collector length, and the system is equal to the system in Figure 11.4a where an increased collector length causes a decreased coverage fraction.

Figure 11.5a,b compares the collection probabilities of side- and bottom-mounted systems for fixed collector length $l = 10d$ (Figure 11.5a) and $100d$ (Figure 11.5b). Therefore, for the side-mounted system, coverage fractions up to $f_{max} = 0.4$ and 0.04 are modeled.

The results in Figure 11.5a outline that without applied PBS the side-mounted system performs better for low coverage fractions. The application of PBS on top of the collector yields higher photon collection for the bottom-mounted system in this region. In the region of $f > 10^{-2}$ both systems reach collection probabilities close to 100%. Without the application of a PBS, side-mounted solar cells provide slightly higher collection probabilities for $10^{-3} < f < 4 \times 10^{-1}$. The most important feature is that the systems with PBS have a considerably higher collection probability p_c than those without PBS. This is because the PBS decreases the emission of photons through the surface of the collector as shown in Figure 11.3. For systems without PBS, this non-radiative loss occurs whenever a photon falls into the critical angle θ_c of total reflectance. For the system with PBS, the photon

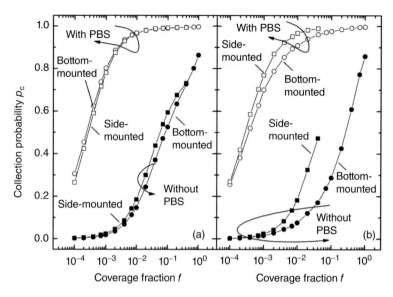

Figure 11.5 Comparison of fluorescent concentrator systems with solar cells at the sides or at the back as sketched in Figure 11.4b,c. (a) Systems at collector length $l = 10d$. Without PBS side-mounted solar cells provide slightly higher collection probabilities for low f. Applying PBS eliminates the difference in collection probability p_c for high coverage fractions. Both systems achieve approximately $p_c = 1$ for coverage fractions $f \approx 1$. For coverage fractions $f < 10^{-2}$ mounting solar cells at the fluorescent concentrator rear side is of slight advantage. (b) Systems at collector length $l = 100d$. Mounting solar cells at collector sides leads to higher p_c for all cases. Compared to the system of the left graph, the side-mounted system obtains the same values. Therefore, this system works still in the small-scale limit, whereas for this collector length the bottom-mounted system is already in the transition regime to the large-scale limit [49].

additionally must be emitted at an energy $E \geq E_1$. This emission probability is low, but non-zero for reasons of detailed balance. For the same reason, systems with PBS obtain the high values of p_c also for lower coverage fractions f, whereas, for the systems without PBS, p_c drops considerably already upon slight decreases of f. Furthermore, with decreased f also the number of photons absorbed by the dye a second or third time increases. Each absorption event leads to θ-randomization of the re-emitted photon and, in consequence, to a certain probability that the photon is lost by emission from the collector surface.

Figure 11.5b shows the comparison of the two systems with $l = 100d$. Compared to Figure 11.5a, the side-mounted system obtains the same values for p_c. As found in Ref. [46], the collection probability for photons p_c for photovoltaic systems with fluorescent concentrator at constant coverage fraction f drastically depends on the collector length l/d. They display asymptotic behavior for high values at small l/d (small-scale limit) as well as for significantly smaller values at large l/d (large-scale limit). The results of Figure 11.5b imply that the side-mounted system still works in the small-scale limit whereas the bottom-mounted system passes into the large-scale region. Therefore, in Figure 11.5b, the fluorescent concentrator system

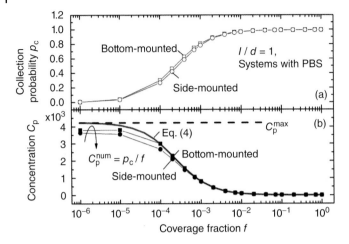

Figure 11.6 (a) Collection probabilities p_c for photons derived with a Monte-Carlo simulation for a side- and a bottom-mounted system in the statistical limit and with applied PBS. (b) Symbol-lines are the values for the concentration c_p derived from p_c in graph (a). Solid lines show the analytically calculated $c_p^{an}(f)$ from Eq. (11.4). Side-mounted systems show less agreement between the analytically and the numerically derived concentration because photons entering the system close to a solar cell are absorbed with a higher probability. This leads to an inhomogeneous chemical potential μ for the incoming photons which violates the assumptions for the calculation. But, the approach $\mu = $ constant is a good approximation in the statistical limit for the bottom-mounted solar cells [48, 49].

with solar cells covering the sides performs better at all coverage fractions with or without PBS.

11.2.3.1 Thermodynamic Efficiency Limits of Fluorescent Concentrators

Showing large- and small-scale limit is a behavior typical of spatially extended inhomogeneous systems. If the characteristic feature length (here the collector length l) is large with respect to the length scale that is characteristic for interactions within the system (here the mean free path of photons), the system can be looked at as a parallel connection of spatially separated subsystems without interaction. The collection probability $p_{c,ls}$ in this *large scale limit* [47] is then the weighted average of a portion f that has a local p_c of a collector with full back coverage, that is, $p_c \approx 1$, for the system with $p_{nr} = 0$ and a portion $(1-f)$ with $p_c = 0$.

In contrast, approaching the *small scale* limit ($l < 1/\alpha_2$), any ray, reflected back and forth within the collector, has the possibility to hit a cell at the collector rear side often. In this situation, the system might be looked at as spatially homogeneous with statistical cell coverage at its back. In fact, the value of $p_{c,ss}$ in the small scale limit is consistent with quasi-one-dimensional computations that simulate the bottom-mounted solar cells by a probability $p_c = f$ for a photon to be collected by cells at the rear side. Therefore, we denote this case also as the *statistical limit*.

Figure 11.6a,b presents an analytical description for the theoretical maximum

$$c_p^{an}(f) = \frac{c_{p,max}^*}{c_{p,max}^* f + 1} \tag{11.4}$$

with the coverage fraction $f = A_{cell}/A_{coll}$ developed by Prönneke et al. [49] for the statistical limit which holds in the radiative case and for systems with applied photonic structure. As discussed by Markvart and Rau [50, 51], using photon fluxes only describes systems with equal chemical potential μ for the incoming photons. Applying a PBS equalizes μ because the absorption coefficient for all incoming photons is now α_1. In contrast, the system without PBS does not provide a spectrally equal absorption for all incoming photons leading to an inhomogeneous μ. Additionally, photons experience a spatial inhomogeneity for systems beyond the statistical limit also leading to an inhomogeneous μ. Considering these limitations, we compare in the following a side- and a bottom-mounted system in the statistical limit ($l/d=1$) with applied PBS.

In Figure 11.6a, the Monte-Carlo simulation excites the system with a monochromatic beam with $E = 2.22$ eV.

In Figure 11.6b we derive the numerical concentration $c_p^{num}(f) = p_c/f$ from the numerical simulated collection probability p_c of Figure 11.6a and compare this with the analytically calculated concentration $c_p^{an}(f)$ from Eq. (11.4). The analytical solutions fit the statistically derived values for the bottom-mounted system, whereas the side-mounted system shows less agreement between the analytically and the numerically derived concentration.

As described above, the system is described in thermal equilibrium and with the same chemical potential μ for all incoming photons. In particular, it holds $qV_{OC} = \mu$ for the cell at the side of the collector. Yet, under short circuit conditions, the voltage V of the cell equals zero and at the solar cell the chemical potential of the photons is $\mu = 0$. Therefore, μ cannot be constant throughout the system, because a net flux of photons requires local differences in the chemical potential. For low coverage fractions f, at which the period length l exceeds the mean free path of the photons as it is the case for the side-mounted system, the results are not valid any more. Photons entering the system close to a solar cell are absorbed with a higher probability, whereas photons in areas with no solar cell are most likely reabsorbed by the dye and, with a higher probability, reemitted from the collector. However, the approach $\mu = $ constant is a good approximation in the statistical limit for the bottom-mounted solar cells [48]. In both cases, the open-circuit condition is achieved by reducing the coverage fraction f of solar cell area to photovoltaic unreasonable low values. The collection probability p_c decreases with decreasing f, and reduces to nearly zero at $f < 10^{-5}$. Concurrently, the concentration approaches the theoretical maximum $c_p^{max} = 4251$. Note, that c_p is reaching the maximum c_p^{max} for coverage fractions f at which p_c is almost zero. The thermodynamic limit of the concentration lies therefore beyond photovoltaic useful collector dimensions.

11.2.3.2 Non-perfect Photonic Structure

In the simulations the application of a PBS as an energy selective filter increases the photon collection in all cases. However, in realistic filters, blocking the photons depends not only on the energy but also on the angle of incidence of the photon. Figure 11.7a,b outlines the influence of smaller reflection cones ($\theta_{pbs} = 10°$ and $20°$) on side- and bottom-mounted systems with coverage fractions $f = 0.01$ and 0.9.

Figure 11.7a depicts the results for the side-mounted system. The application of a non-perfect PBS filter decreases the photon collection at all collector lengths. However, a relatively higher drop occurs for the system with $f = 0.01$ than for the system with $f = 0.9$. Here, a higher re-absorption rate occurs due to longer distances between the cells. Therefore, photons are more often re-emitted with their direction spatially randomized. The frequent randomization carried out also in unfavorable angles contributes to the number of lost photons. The side-mounted solar cells show an interesting behavior for $l/d > 11$ and $f = 0.9$. Instead of descending into the large-scale limit, p_c slightly increases. We explain this effect as follows: The larger l/d, the more collector edge area is covered with solar cells until they are fully covered. Thus, the photon path length is larger for small l/d than for large l/d because the probability to hit solar cell area is smaller. The longer the path length, the higher is the probability for the photon to be subject to loss mechanisms.

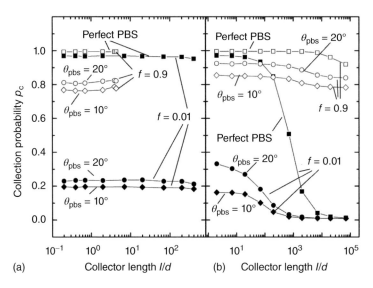

Figure 11.7 Influence of a photonic band stop filter with angular selectivity. (a) Side-mounted system. Coverage fraction $f = 0.9$ achieves higher photon collections p_c than $f = 0.01$ for all θ_{pbs}. (b) Bottom-mounted system. For small scales the system shows a better performance than the side-mounted system because this system benefits from the randomization of photon direction during re-emission [48, 49].

Figure 11.7b presents the calculation results for the bottom-mounted system. At small l/d, this system collects more photons under the application of a non-perfect PBS filter than the side-mounted system. This is due to the effect that the disadvantageous angles for the PBS are very favorable angles for solar cells at the fluorescent concentrator rear side. Therefore, the collection of a photon is more likely than in the side-mounted system.

11.2.3.3 Luminescent Materials in Photonic Structures

Photonic structures on the top of a fluorescent concentrator do not alter the emission characteristic of the dye itself. The guiding due to the enhanced reflection of light emitted in the escape cone, that is, with steep angle with respect to the surface, results in longer paths this light has to travel before it hits the edges. Thus, this escape cone light is heavily subject to path length dependent losses such as scattering or reabsorption.

Therefore, the "NanoFluko" concept was proposed to mitigate the major loss mechanism of escape cone loss and reabsorption by embedding the dye directly into a photonic structure [52].

In contrast to conventional fluorescent collectors, the feature size of such structures is comparable to the wavelength of light. The emission process is therefore subject to photonic effects. It is well known, that photonic structures can influence the emission process as they alter the availability of allowed modes. Already the very first papers on photonic crystals by Bykov [53] and Yablonovitch [39] reported on the influence of such structures on the spontaneous emission process. Depending on the direction, band gaps for photons appear that means that propagation in this direction is forbidden for photons with certain frequencies. The influence on the emission can be theoretically described by Fermi's Golden Rule, which states that the transition rate of an emitter is proportional to the available density of photon states. In photonic band gaps the density of photon states is zero and thus the emission is strongly suppressed at these frequencies. Numerous publications reported on this effect from both theoretical and experimental perspectives [54–60].

The "NanoFluko" concept aims at restricting the emission to in-plane directions, so that light guiding toward the edges is strongly improved through shorter paths and by circumventing the total internal reflection (TIR) escape cone. Further, redistribution of emission due to inhibited emission in a photonic band gap could be used to reduce reabsorption: suppressing the emission in the overlap region of the absorption and emission spectra increases the effective Stokes shift and reduces reabsorption losses. Lastly, the effect of stimulated emission might increase the quantum efficiency of the emitting material.

In the following, we present some simulation results that highlight the potential this concept has, to increase fluorescent concentrator efficiencies. The light guiding was studied using electromagnetic simulations of the one-dimensional photonic structure shown in Figure 11.8. A Bragg stack of 40 pairs of quarter-lambda layers ($n_1 = 1.5$, $n_2 = 2$, and thickness $t_i = \lambda_0/4n_i$) with a dipole source placed in a cavity layer in between was simulated using the finite-difference time-domain

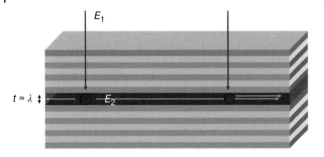

Figure 11.8 In the NanoFluko concept the luminescent material is embedded in a photonic structure. In such structures photonic band gaps occur for the propagation in certain directions, which results in suppression of emission. Spectral and directional redistribution of emitted light can strongly reduce typical losses in fluorescent concentrators [52].

method [61–63] to solve Maxwell's equations [64]. Two major quantities were obtained in the simulations: first, the *relative emission* E_{rel}, defined as the emission spectrum within the "NanoFluko" structure divided by the spectrum of the emitter in a homogeneous medium. It reveals how the surrounding structure influences the emission spectrum due to an altered density of photon states. Second, the light guiding efficiency LGE, defined as the ratio of light emitted at the edges of the structure to total emission. This quantity is a key property for the NanoFluko's application as a light concentrator.

The relative emission E_{rel} is shown in Figure 11.9a. The dashed line ($E_{rel} = 1$) represents undisturbed emission in homogeneous media. Within the "NanoFluko,"

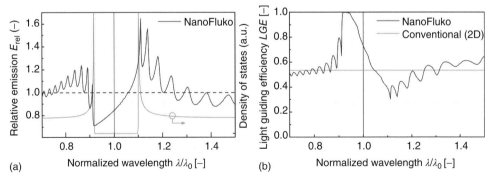

Figure 11.9 (a) The emission spectrum inside the simulated NanoFluko structure is modified compared the spectrum in homogeneous media: emission within the photonic band gap is suppressed, which can be used to reduce the emission at wavelengths that are prone to reabsorption losses. (b) The light guiding efficiency inside the NanoFluko is greatly enhanced in a broad spectral range: up to 99.7% of the emitted photons are guided to the edge faces of the concentrator.

Figure 11.10 Simulated distribution of the energy density (overlay with Nanofluko structure) in the simulation cell for monochromatic emission at $\lambda = \lambda_0$ (design wavelength of Bragg stack). Emission into the escape cone is suppressed and most light is trapped within the NanoFluko structure and thus it is guided to the edges [64].

emission is suppressed around the design wavelength λ_0 of the Bragg stack. In this range the ideal Bragg stack features a photonic band gap (for wave vectors normal to the interfaces) where the density of the photon states is zero. This result shows that the emission process of the dipole can be strongly influenced by the surrounding geometry. Inside the photonic structure, emission is suppressed within the band gap. This effect could be utilized to reduce reabsorption losses by suppressing emission in unwanted spectral regions, specifically at the overlap of absorption and emission spectra of the luminescent material.

The LGE is depicted in Figure 11.9b. In a conventional fluorescent concentrator, the LGE is given by the part of light that is not lost in the escape cone of total internal reflection. As the NanoFluko simulation was treated as a two-dimensional problem for symmetry reasons, it is compared to the two-dimensional LGE of a conventional fluorescent concentrator. This value is given by $1 - \theta_c/90° \approx 53.3\%$ with critical angle $\theta_c = \sin^{-1}(1/1.5)$. In the case of the NanoFluko, strongly enhanced light guiding occurs near the design wavelength λ_0, as shown in Figure 11.10, while for other wavelengths the LGE roughly approaches the TIR limit. Optimum light guiding, however, is obtained not at λ_0, but at slightly smaller wavelengths with efficiencies of up to 99.7% at $\lambda_0 = 0.93\lambda_0$. This shift is a result of the angular dependence of the Bragg stack's reflection as discussed in Ref. [64].

In this study, one-dimensional Bragg stacks were used as a well-known model system to study the fundamental effects of light guiding enhancement and suppression of emission. While these results are very promising, the photonic structures need to be optimized for transmission in the absorption range of the dye. This can be achieved with edge filters such as the one used in Ref. [23]. Additionally, more than one luminescent layer might be necessary to absorb the incoming light, depending on the absorption coefficient and layer thickness. Further, a more complex structure with different luminescent materials has to be considered to cover the whole relevant solar spectrum. In such a stack of NanoFlukos, each photonic structure has to be optimized for the specific emitter, while allowing transmission of incident light to the underlying NanoFlukos.

11.3
Materials for Fluorescent Concentrators

11.3.1
Systems Based on Organic Matrix Materials

11.3.1.1 The Matrix Material

Polymeric matrix materials for fluorescent concentrators must fulfill a number of special requirements. These refer to their optical properties (absorption and scattering losses, refraction index n), easy and effective possibilities of dispersing the luminescent components in the matrix including compatibility between matrix and embedded luminescent molecules or particles; long-term stability of the matrix against solar irradiation and weather influences; physical and chemical properties such as mechanical strength, processability, flame resistance, and others (e.g., non-toxicity); and, last but not least, their price.

In the following some of these properties together with advantages and disadvantages of several polymers as matrices for fluorescent concentrators will be shortly discussed, with the different requirements of compact fluorescent collectors (a plate with embedded luminescing substance) and fluorescent collector layer systems (luminescing layer on a wave-guiding substrate) being taken into account. With regard to the optical properties of the matrix a very small extinction coefficient in the relevant spectral region is the basic requirement. This means not only the absence of absorption bands, but also minimal scattering effects. Intensity losses of the emitted light due to absorption or scattering increase strongly with increasing fluorescent collector size, since the path length of the emitted rays may be a multiple of that size, that is, the rays have to pass the through the matrix for a distance of many decimeters. To give a numerical example, performance gains of about 20% would be theoretically possible, if the absorption coefficient of the polymer matrix could be decreased from $0.3 \, m^{-1}$ (commercial PMMA) to $0.001 \, m^{-1}$ (optical fiber) [65]. Scattering can be caused by, for example, tiny (not visible to the eye) pores or cracks in the material, but also by small density inhomogeneities (revealed by X-ray scattering) of the material formed during the polymerization process. Intensity losses may also be caused by impurities in the matrix, for example, remainings of initiators, or by additives, such as stabilizers. Another point, becoming more important in recent years, refers to absorptions due to overtones of C–H and O–H vibrations of the organic matrix material. These did not play an essential role as long as the fluorescent concentrators were more or less restricted to harvesting visible light using, above all, red luminescence light. However, when the efficiency of the fluorescent concentrators is to be increased by using also the near infrared part of the solar irradiation, those overtone absorptions become more important, in particular for larger fluorescent concentrator sizes. In such cases, it must also be taken into account, that many polymers contain some water with the water content varying in dependence on polymer and environment. In the NIR region, water shows absorptions, for example, around 750, 850, and 970 nm with the extinction coefficient increasing considerably along with wavelength. To give an example, for commercially available PMMA types, a reasonable water content of 0.4% and 1 m path length the matrix-induced intensity loss at 970 nm (water band) would be roughly 40%. All these disturbing absorptions in the NIR could be avoided or at least strongly diminished by the use of halogenized (mostly perfluorated) or deuterated polymers, for example, polyacrylates [66–69]. Another positive effect of halogenization is the reduction of yellowing. However, these polymers may be much more expensive than the basic unmodified polymers. Furthermore, choosing a minimal-loss matrix polymer for a fluorescent concentrator is not always as advantageous as may be expected, since other properties (e.g., outdoor stability) play an important role, too, and additional losses due to other factors (e.g., an absorption tail of the embedded dye overlapping with the emission region) may be comparatively large thus outweighing the positive effect of a low-loss polymer [70–72]. Model calculations showed, that decreasing the polymer matrix-linked loss coefficient beyond a certain value did not result in a further improvement of a given fluorescent concentrator [8].

The index of refraction (n) is another relevant optical property of polymers for fluorescent concentrators with larger n-values resulting in smaller escape cone losses [37], but, on the other side, causing somewhat larger reflection loss of the incident solar light, in particular for non-vertical irradiation.

The two fluorescent collector types (compact plate and thin-film fluorescent collector,) differ to a certain degree with respect to the properties that are required to be met by the matrix polymer, as compact fluorescent collectors are made by a polymerization process, which requires solubility of the used organic dyes in the monomers and stability of the luminescence properties of the dyes against influences due to the chemical reaction. Additionally, since the thickness of the compact fluorescent collector plates is usually in the order of several millimeters, much more substance is needed as for thin-film fluorescent collectors.

A number of transparent polymers has been tested for fluorescent collectors up to now, for example in Refs. [73, 74], the main types being polyacrylates (including polymethacrylates, polyhydroxymethacrylates, and copolymers), polycarbonate, polyurethanes (PURs), epoxy resins, and polysiloxane. Among these, despite not being very cheap, PMMA is widely seen as most suitable as it has very good optical properties, good processability, and mechanical properties as well as outdoor stability. Best results were obtained with PMMA, if the polymer plate was produced by thermal polymerization of freshly distilled monomer [8]. By producing copolymers using different acrylate monomers, the properties of the resulting polymer can be changed. Methylmethacrylate monomers can also be copolymerized with styrene units to make a transparent PMMA/PS (polystyrene)-copolymer, the stability properties of which have been claimed to be improved in comparison with the pure polymer [75]. The refraction index of PMMA/PS-copolymers is, advantageously, between the value of PMMA (1.49) and that of pure PS (1.59). Despite its high refraction index pure PS, which has poor stability to outdoor weathering [75], has not been widely used as the matrix for fluorescent collectors. Tests of commercially available urethanes as well as of an epoxy resin showed the latter to have the best optical properties [73]. However, organic dyes were found to lose their luminescent properties when combined with epoxy resins [73] excluding those combinations for fluorescent collectors. Another transparent polymer studied for fluorescent collectors is polysiloxane, which may be of interest for special cases, in particular, because of its flexibility. But, if polysiloxanes are used for fluorescent collectors containing organic dyes problems concerning solubility of the dyes as well as worsening of luminescing properties occur.

Another way of realizing fluorescent collectors is to have a thin layer that contains the luminescent material on a transparent slab. These layers could be produced by polymerization on the substrate, but mostly it will be strongly advantageous to make them by one of the numerous well-established ways to produce a polymer layer on a substrate without a polymerization step, for example, by casting [8], spin-coating, doctor-blading, spraying, printing, Langmuir–Blodgett techniques [16], or evaporation [30]. Such fluorescent collectors may offer wider possibilities concerning large-area and cheaper devices. Furthermore, the smaller amount of polymer needed for the active layer might allow somewhat more expensive special polymers with better optical properties to be taken into account. In some cases, avoiding a chemical reaction during layer formation may be advantageous also due to the exclusion of harmful interaction of the substances involved with respect to the luminescence properties of the embedded optically active substance. Some limitation for the layer production from solutions of the optically active component in a polymer/solvent mixture may arise from the fact that the dye concentration must be considerably higher than that for compact fluorescent collectors to achieve a comparably light absorption. This means that the dye's solubility in the polymer/solvent mixture has to be large enough. Theoretical considerations [32] as well as experimental measurements [76] show, in principle, equal efficiencies of the compact and of the layer-fluorescent collector type. The refraction index of the substrate should exceed that of the dye-doped polymer layer to reduce losses [30]. Thus, as in most cases, glass plates are used as substrate material, and suitable glass must be selected.

As for compact fluorescent collectors, polyacrylates belong to the most suitable polymers for thin-film fluorescent collector devices, too. Advantageously, the refraction index of the most used polymer of this class, PMMA, is very close to those of common glass types; furthermore, polyacrylates are well soluble, together with organic dyes, in ordinary organic solvents. To achieve sufficient long-term adhesion between the polymer layer and glass substrate the surface of the latter must be carefully cleaned. This can be done by treatment with chemicals or physical methods [77].

Not in all cases the thin luminescent layer has to be produced directly on the substrate, but if thin polymer sheets with incorporated luminescing dyes are available (as, e.g., for polycarbonate) it remains to connect them properly to the substrate. When using polycarbonates suitable types must be selected, as there are polycarbonates that absorb in the deep-blue region [70, 74].

Dyes that are better soluble in water than in organic solvents can be incorporated, for example, in the water-soluble polymer polyvinylalcohol forming very transparent layers. However, sensitivity of such layers to water or very high atmospheric humidity must be taken into account.

Although still on a research level it should be mentioned that efforts are being taken to use conjugated polymers having luminescent properties, which means the "matrix" itself would play an active role. Quantum yields higher than 50% for parts of the visible spectrum have been reported for films made from specially designed materials, but usually the photoluminescence of those polymers is small [78].

The interaction between the two components plays an important role not only for compact polymer-based fluorescent collectors with embedded organic luminescing dyes, but also with layer systems. An example for this is the incorporation of luminescing particles of biological origin, such as phycobilisomes into a polymer matrix to form a fluorescent collector. In this case, the polymer matrix must offer an environment for the particles, which enables them, on the one side, to maintain their luminescence properties, but which has, on the other side, sufficient optical and mechanical properties. To achieve this polyacrylamide hydrogels [79] mixtures consisting of starch and sugars [80] have been used. In the first case, the hydrogel-phycobilisomes layer or even flexible films were produced by polymerization on a polycarbonate substrate, whereas the starch/sugars-phycobilisomes layers were made by doctor-blading and drying solutions on glass.

Sol–gel glasses are another kind of matrix material to incorporate organic dyes in layers, which can be deposited on glass substrates. An example for this are the so-called ORMOCERs, which show mechanical and thermal stability as well as high refraction index [81, 82].

11.3.1.2 The Luminescent Species

Organic Dyes Luminescent materials are the key elements in fluorescence concentrators. They can be divided into two large groups depending on whether they are made of organic or inorganic matter. Organic dyes have been intensely used in the first luminescence concentrator campaign in the 1980s [5, 6, 32, 34, 45, 83–85]. The facile availability of a large number of strongly fluorescent dyes and the possibility to easily incorporate them into transparent host matrices have been beneficial criteria toward cheap and large-area concentrator systems [16, 18, 19, 86, 87]. They were either distributed in a transparent matrix material or applied as a thin layer on a transparent slab of material.

The molecular structure of an organic dye is based on a conjugated π-electron system, which is usually connected to electron-pushing or electron-withdrawing groups. Absorption and emission events are caused by electronic transitions of π-electrons from a ground into an excited state and vice versa. The geometry of the electronic states depends on the molecular structure, that is, the extent of the conjugated π-electron system and the position of the substituents and determines the optical properties of the dye [88]. Thus, the absorptivity, the fluorescence quantum yield, the position, width and shape of the absorption and emission spectra, and the Stokes shift between the absorption and emission maxima, which are all

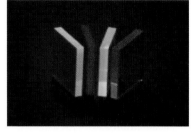

Figure 11.11 The molecular structure of Lumogen Red 300 (a) is based on a perylene imide moiety. The optical and physical properties of the dye are influenced by substituents at the bay region and the imide amines. Examples of luminescence concentrators prepared by doping PMMA plates with dyes from the Lumogen series (b) [5, 6, 32, 45, 83].

critical factors for the applicability of the dyes in fluorescence collectors, can be changed by modifying the molecular architecture. In addition to these properties the photostability and solubility of the dye molecules are important parameters.

The research in the 1980s resulted in luminescent dyes with high luminescent quantum efficiencies above 95% and good stability. Figure 11.11 shows a photograph of a selection of luminescent concentrator materials produced during that time with perylene-based fluorophores in PMMA plates [5, 6, 32, 45, 83]. Today, these luminescent dyes are commercially available and, therefore, also used in recent works [16, 18, 19, 86, 87]. Among these dyes are perylene-based fluorophores, which exhibit favorable characteristics that are sold under the tradename Lumogen®. In Figure 11.11, the molecular structure of Lumogen Red 300 is shown as well. Until now, the highest reported efficiencies [19, 25] were reached with systems based on these organic dyes.

A general problem with most organic dyes are large overlaps between the absorption and emission bands. Light losses during reabsorption and reemission events decrease concentrator efficiencies. Small Stoke shifts also limit the improved guidance of the emitted light, for example, by selective reflection layers [23, 52]. Research efforts to circumvent these problems include the preparation of dyes with larger Stoke shifts [89], the use of phosphorescence instead of fluorescence [30] and the utilization of energy transfer processes [90, 91]. Luminescence concentrators based on rare earth complexes, for example, in which the energy is transferred from absorbing organic ligands to the emitting center ions of the complex have been proposed as systems without self-absorption losses [92]. The luminescence enhancement of dyes by the addition of metal nanoparticles [93, 94] or by optimizing the encapsulation of the dyes in the host material [82, 95] are other approaches to improve the overall efficiency of dye-based fluorescence concentrators. Another limitation is that high quantum efficiencies are only achieved in the visible range of the spectrum, while efficiency remains low in the infrared. In Ref. [32] it is shown that these low quantum efficiencies have fundamental reasons that are difficult to overcome. The main reason is that the energy difference between the excited electronic states moves closer to the energy of vibrational transitions within the molecule, which facilitates non-radiative transitions. Nevertheless, also in recent works, organic dyes are being developed, which extend the efficiently used spectral range to longer wavelength. For example, in Ref. [89] the synthesis of a dye based on a perylene perinone is described, which extends the absorption wavelength range by more than 50 nm in comparison to the perylene-based dye Lumogen Red 305. This extended absorption allows for the collection of potentially 25% more photons at a reasonable luminescent quantum yield and photostability.

Luminescent Nanocrystalline Quantum Dots With the development of nanotechnology, luminescent nanocrystalline QDs have become of interest for luminescent concentrators. QDs are semiconducting nanocrystals that typically have dimensions of several nanometers. QDs have been modeled and experimentally tested for fluorescent concentrators for more than 10 years [29, 65, 81, 96–100]. Additional to QDs having a spherical shape, more recently rod-like semiconducting nanocrystals (nanorods, NRs) have been synthesized and studied, too [101–103]. Owing to their small size and the small number of atoms involved, QDs and NRs represent an intermediate state between molecular structures and macroscopic crystals. Characteristic features are, for example, the formation of discrete energy levels and, upon excitation, the generation of excitons spatially confined to the nanocrystal size. If the latter is smaller than the Bohr radius of the semiconductor material, the exciton energy depends not only on the bandgap of the bulk material, but also on the size of the particles. The properties of QDs and NRs are determined by the bulk material as well as by the surface of the nanoparticles. To tailor these properties and to shield the material against chemical attacks on imperfections and also to passivate dangling bonds, core/shell, and even multishell types of nanocrystals have been developed. Nevertheless, organic ligands are necessary at least for colloidal QDs. Often studied and/or relevant semiconducting nanocrystals consist, for example, of CdS, CdSe, CdSe/ZnS, CdTe, InP, Si, PbS, PbSe, $CuInS_2$, $CuInSe_2$, with some types being commercially available; an example for multishell QDs are nanoparticles consisting of a CdSe core surrounded by successive CdS, Cd0.5Zn0.5S, and ZnS shells [20, 21, 96, 99, 104–108].

QDs and NRs have some properties that make them interesting candidates for fluorescent concentrators: first, they absorb light from UV (where they have highest absorption coefficients) to visible or even (near-)infrared wavelengths. Thus, they can, in principle, harvest a larger part of solar energy than organic dyes, which have rather narrow absorption bands. Second, the bandgap depends on the size of the nanoparticle; therefore, the absorption and emission properties can be tuned by the size and composition of the nanocrystals. Third, luminescence takes place in a single emission band, the width of which is determined by the width of the nanocrystal's size distribution, which can be manipulated by synthesis. Furthermore, the nanocrystals are inherently more stable against light than they are against organic dyes. Finally, the possibility of multiplex exciton generation (if photons have a multiple of the bandgap energy) reported for PbSe QDs [109] could be another property relevant for fluorescent concentrators in future as this would strongly increase the quantum yield of the QDs.

At present there are still some problems with using QDs and NRs in fluorescent concentrators: first, the overlap between absorption and emission region, is generally too large, which results in high reabsorption losses [110]. The extent of this overlap, however, differs for different types of QDs [81]. QDs with larger Stokes shifts are under development in the form of NRs [99], or type II heteronanocrystals [111]). Furthermore, in many cases, the quantum yields are up to now lower than those of organic dyes, although there is encouraging work to increase them with maximum values of >80% already achieved [102, 112, 113]. Finally, many of the luminescent nanocrystals studied are toxic; but there are continued efforts to reduce the toxic potential connected with QDs [114], for example, by using Si QDs [115, 116] or heavy metal-free nanoparticles, for example, InP or $CuInS_2$.

The semiconducting nanocrystals are often very sensitive as long as they are not embedded in a suitable, shielding matrix [117, 118] preventing, for example, contact with oxygen. Embedding QDs while maintaining the nanocrystals as luminescent, single particles was achieved using two types of procedures: *in situ* or *ex situ*. The *in situ* preparation includes synthesis/growth of the nanoparticles in a liquid matrix or solution, being solidified in the process [107, 119, 120]. In the *ex situ* method the particles are made at first, and the ready QDs/NRs are then dispersed into liquid monomers or polymer solutions from which solid bodies or films on substrates are made by various methods (e.g., polymerization). Any agglomeration of the particles (resulting in luminescence quenching) must be strictly avoided. This can be – in principle – achieved by the use of organic ligands having a hydrophilic head (bonding to the anions or kations, respectively, at the surface of the nanocrystal) and a long hydrophobic tail stabilizing the particle/shell complex in a hydrophobic organic environment. However, considerable difficulties arise in case of polar monomers (e.g., MMA) or solvents. Substances often used as organic ligands for stabilization of QDs and NRs are, for example, oleic acid (OA), octadecylamin (ODA), hexadecylamine (HDA), trioctylphosphine (TOP), trioctylphosphine oxide (TOPO), hexylphosphonic acid (HPA), octadecylphosphonic acid (ODPA). Embedding media tested using polymerization of monomer-nanoparticle-dispersions are, for example, PMMA [73, 121], PUR [73, 96], epoxy resin [73], copolymers of laurylmethacrylate (LMA) with other acrylic monomers [99, 103, 122], polybutylacrylate (PBA) [123], and PS [124]. Cellulose triacetate (CTA) is an example for polymers being tested to prepare nanoparticle containing films on substrates without polymerization [103]. Silica produced in sol–gel processes was described as an organic embedding substance [107, 119, 120]. Some QDs/NRs together with embedding preparation will be very shortly described in the following.

CdSe/ZnS-QDs belong to the best studied semiconductor nanocrystals for fluorescent concentrators. Quantum yields up to considerably more than 80% were achieved [125] for this QD type, and very recently considerably increased emissions (by max. 53%) were reached by exploiting plasmonic interaction between QDs and gold nanoparticles [126]. Using two commercial CdSe/ZnS-QD types for embedding tests including MMA/PMMA-solution, some PUR-types, a polyol, and an epoxy resin the last mentioned was found to be best suited since it resulted in the most efficient samples [73]. Successful embedding of QDs of this type capped by TOPO in PMMA was accomplished, if they were dispersed in the high viscous prepolymer, which was then thermally polymerized [121, 127]. CdSe/ZnS as well as CdS-QDs capped originally by TOPO were successfully redispersed using TOP in a mixture of the monomers LMA and ethylene glycol dimethacrylate (EGDM); the mixture was thermally polymerized to small QD-polymer composite rods [122]. CdSe core/multishell QDs capped by OA and ODA [105] were dispersed in mixtures of distilled LMA and EGDM and solidified by UV-polymerization within 15 min, which resulted in transparent fluorescent concentrator plates without agglomerates [99]. Similar multishell QDs may have a luminescence spectrum reaching even into the near-infrared region; in this case a smaller overlap between absorption and emission was found, causing smaller reabsorption and escape cone losses [110]. A very different embedding procedure (but not used till now for fluorescent concentrators) was demonstrated in Ref. [123] including emulsion polymerization of butylacrylate droplets containing TOPO-capped CdSe QDs resulting in QD-containing PBA particles.

Similarly to CdSe/ZnS-QDs also CdSe/CdS core shell NRs, prepared according to [101, 102], and capped by various organic ligands were successfully embedded in poly(lauryl methacrylate)/EGDM copolymers [103]. Initiator concentration proved to be relevant, since initiator radicals may cause luminescence quenching (max. quantum yield achieved: 70%). Owing to their shape, NRs have anisotropic properties, for example, linearly polarized absorption and emission. Thus, NRs-containing polymer composites will also be anisotropic, if the NRs have a preferred orientation, achievable by various known methods, for example, by stretching the polymer matrix or using self-aligning luminescing NRs [114]. NR alignment in the polymer matrix can, for example, reduce escape cone losses [128].

PbS and PbSe QDs are two relevant types of infrared emitting nanocrystals [129–131] considered to be used in fluorescent concentrators [81, 132]. They absorb in a broad region, having relatively high absorption coefficients (though much smaller than that of CdSe/ZnS-QDs), show tunable emission from 850 to 1900 nm with quantum yields up to 80% [113], and have been reported to possess possibilities of multiplex exciton generation [109]. A very important feature for use in fluorescent concentrators is the small overlap between absorption and emission with the Stokes shift of PbS being reported as 122 nm (CdSe/ZnS: 23 nm) [132]. However, commercially available PbS QDs have small quantum yields (about 10%) and bad long-term stability and are expensive [81]. These QDs have been produced by *in situ* processes (e.g., in ethylene-methacrylic acid-copolymer [133] or glass [134]). Recently [118] *ex situ* produced PbS QDs were embedded in PMMA by polymerization according to [121] as well as using the so-called second dispersion (SD) method, where PMMA pellets are solved in toluene, mixed with QD solution, and cured in molds [135]. Embedding caused considerable quantum yield decrease (much stronger for SD-prepared samples) of the QDs incorporated in the full or hollow PMMA cylinders prepared [118].

Nanoparticles on $CuInS_2$ basis are another relevant QD type being of interest for fluorescent concentrators due to their NIR emission, high quantum yield, large Stokes shift, simple synthesis, and absence of heavy metal atoms [136–138]. There are also hybrid approaches, for example, in Ref. [139] Er-doped WO_6 nanocrystals in composite telluride glasses are investigated.

Finally, CdS QDs emitting in the blue and green spectral region are addressed. They are examples for *in situ* produced QDs, which were embedded in silica by a sol–gel process. Solutions containing the precursors for silica and CdS were made and deposited as thin layers on glass by spin-coating [107], or dip-coating [119]. By applying suitable temperature-time regimes (temperatures up to 450 °C) CdS nanoparticles were formed in a glass-like matrix resulting in "stable, highly luminescent QD solar concentrators" [107].

Summarizing, semiconducting nanocrystals have great potential for use as optical active species in fluorescent concentrators, but at present the efficiencies of QD/NR- based fluorescent concentrators are still low due to large reabsorption losses and low quantum yields, whereas embedding problems seem to be solvable. If the optical properties of the nanoparticles will continue to be improved, with prices of commercial types as well as toxicity taken also into account, QDs and NRs can play an important role for efficient fluorescent concentrators.

11.3.2
Completely Inorganic Systems Based on Rare Earths

Because of the instability of organic materials, especially under ultraviolet radiation, completely inorganic material systems have been investigated as well. Promising inorganic materials are glasses and glass ceramics doped with rare earth ions such as Nd^{3+} and Yb^{3+}, or other metal ions such as Cr^{3+} [140–143]. The advantages of these approaches are high stability and a high refraction index of the glasses, which increases the trapped fraction of light. One big disadvantage is the narrow absorption bands of the luminescent materials.

In the following, the optical properties of Nd^{3+}-doped barium borate glasses are investigated (see Figure 11.12). Borate glasses might be good candidates as matrix materials for rare-earth ions since they are robust, inexpensive, and offer a high optical transparency. Please see Ref. [144] for further information.

Figure 11.13 shows the transmission of undoped, 1 and 5 mol% Nd_2O_3-doped barium borate fluorescent collector plates; the sample thickness was 5 mm. In the spectral range from 300 to 950 nm, several Nd^{3+}-related absorption bands can be observed. The most intense band at approximately 585 nm can be assigned to a transition from the $^4I_{9/2}$ ground state to the $^4G_{5/2}$ excited state. Further intense absorption bands are at approximately 680, 745, 805, and 875 nm belonging to transitions from $^4I_{9/2}$ to $^4F_{9/2}$, $^4F_{7/2}$, $^4F_{5/2}$, and $^4F_{3/2}$, respectively [145].

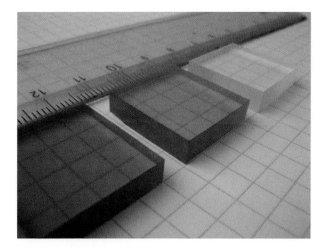

Figure 11.12 Barium borate glasses for use as fluorescent concentrators. The two samples in the front are additionally doped with 5 mol% (left) and 1 mol% (middle) Nd_2O_3. The undoped sample in the back is for reference [114].

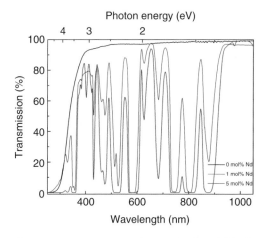

Figure 11.13 Transmission factor of undoped (black), 1 mol% (red), and 5 mol% (green) Nd^{3+}-doped barium borate glasses [114].

To investigate the collection efficiency of the Nd_2O_3-doped barium borate fluorescent concentrator glasses the experimental setup shown in Figure 11.14a, was used. The Nd^{3+} fluorescence was excited with a laser diode working at 795 nm, which is in resonance with the $^4I_{9/2}$ to $^4F_{5/2}$ transition; the fluorescence was detected in the spectral range from 800 to 1200 nm. Figure 11.15 shows the corresponding emission spectra, recorded along the optical axis (indicated as a dashed line in Figure 11.14) for different positions on the 1 mol% Nd^{3+}-doped sample. The intense emissions at 900 and 1060 nm are typical for Nd^{3+} and can be assigned to transitions from $^4F_{3/2}$ to $^4I_{9/2}$ and $^4I_{11/2}$, respectively. There is an

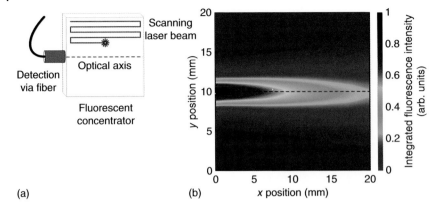

Figure 11.14 (a) Experimental setup to measure the collection efficiency of the Nd^{3+}-doped barium borate glass fluorescent concentrators. (b) Two-dimensional plot for the integrated fluorescence intensity of the 1 mol% Nd^{3+}-doped glass; the fluorescence is excited with a laser diode emitting at 795 nm [114].

additional emission in the near-infrared spectral range at approximately 1340 nm, which has not been detected here; this emission is due to a $^4F_{3/2}$ to $^4I_{13/2}$ transition. In addition, the integral fluorescence intensity, that is, the area under the fluorescence emission spectrum, was calculated and plotted versus the xy position of the scanning laser beam on the sample. The result is shown in Figure 11.14b.

Figure 11.15 shows that the fluorescence intensity drops significantly when the laser spot moves from the 0 mm x position to the 20 mm x position. Apart from these significant changes in intensity, the spectral shape of the $^4F_{3/2}$ to $^4I_{9/2}$ transition changes. The inset shows the fluorescence spectra after normalization to the intense $^4F_{3/2}$ to $^4I_{11/2}$ transition at 1060 nm. The high-energy shoulder of the 900 nm emission decreases significantly upon increasing the distance to the edge of the sample where the fluorescence is detected. This is caused by self-absorption, that is, the absorption and emission band at 900 nm strongly overlap.

In principle, borate glasses are good candidates for fluorescent concentrator applications. The glasses are highly transparent over a broad spectral range and provide light guiding properties over relatively long distances by minimal absorption losses. The use as fluorescent concentrators for the near-infrared spectral range, however, is limited by the relatively poor fluorescence efficiency of doped Nd^{3+} ions and self-absorption effects.

The next step will be to go from borate glasses to fluorozirconate-based glasses or even fluorozirconate-based glass ceramics. These two systems provide lower phonon frequencies enabling reduced non-radiative losses and, therefore, increased fluorescence efficiencies from doped rare-earth ions [146]. To this end, a class of fluorozirconate-based glass ceramics has been developed in which the rare-earth is not only embedded in the glass but also incorporated into $BaCl_2$ nanoparticles formed therein, after appropriate thermal processing [147]. The material costs of fluorozirconate-based glasses and glass ceramics, however,

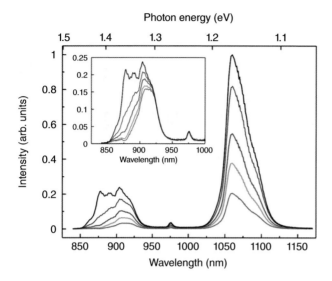

Figure 11.15 Fluorescence spectra of the 1 mol% Nd^{3+}-doped barium borate fluorescent concentrator glass, excited with a 795 nm laser diode and recorded along the optical axis (indicated as a dashed line in Figure 11.14) for different positions on the sample: 0 mm (black), 5 mm (red), 10 mm (blue), 15 mm (green), and 20 mm (magenta). The inset shows the spectral range from 825 to 1000 nm with the spectra normalized to the emission band at 1060 nm [114].

will be significantly higher than for borate glasses. In addition, in the case of glass ceramics, it is essential to develop a method to control the size and size distribution of the fluorescent nanoparticles therein since the nanoparticles have a different refractive index than the base glass thus leading to significant scattering losses for large-sized particles.

11.4
Experimentally Realized Fluorescent Concentrator Systems

11.4.1
Systems with Side-Mounted Solar Cells

One of the most efficient fluorescent concentrator systems [114] produced till now was reported by Slooff *et al.* in 2008 [19]. Based on results of model calculations a PMMA plate (size $5 \times 5 \times 0.5$ cm^3) was prepared by polymerizing a MMA/PMMA mixture containing 0.01 wt% Lumogen F Red305 (BASF) and 0.003 wt% Fluorescence Yellow CRS040 (Radiant Color). After combining the plate with mirrors, a rear diffuse reflector, and varying numbers and types of solar cells (mc-Si, GaAs, InGaP) the efficiencies of the devices were measured. Highest efficiency (7.1%) was found, if four GaAs cells were attached to the sides of the fluorescent collector plate (geometrical concentration (c_g) = 2.5). Comparatively high efficiency values have been presented in Ref. [26], where 6.9% efficiency was reported, with a system of the same geometric dimensions using two types of solar cell, GaInP and GaAs.

It should be noted, however, that, due to size dependence, the direct comparison of the efficiencies of various fluorescent collectors is not possible. To assess fluorescent collectors the relation between efficiency achieved and maximal achievable efficiency (for a given c_g) is relevant. This relation can be deduced using a simple model describing "first generation fluorescent collectors" (randomly oriented absorbing molecules, isotropic emission) [148] showing that for fluorescent collectors with geometrical concentration 2.5 there is still room for an efficiency increase to values several percent larger than the 7.1% obtained.

11.4.2
Systems with Bottom-Mounted Monocrystalline Silicon Solar Module

As discussed in Section 11.2.3, it is also a viable option to attach the solar cells at the bottom of the fluorescent collector. Using fluorescent collectors as cheap and easy to apply low concentrators is especially interesting in photovoltaic technologies where the cell size is limited and the module area is not fully covered with photovoltaic active material. For analyzing the benefits of a fluorescent material in photovoltaic modules, the following experiment is carried out with two parallel connected mono crystalline silicon (c-Si) solar cells in varied distance x. The solar cells are cut out from an industrialized screen-printed c-Si solar cell and placed under acrylic glass, which is either clear or doped with fluorescent dye. Each solar cell area is $A_{cell} = 4 \text{ cm}^2$ plus busbar. The cells are optically coupled to the acrylic glass on top. Here, total internal reflection is out of action in contrast to the surrounding area. The solar cells are attached on top of a clear acrylic glass plate and under this a paperboard is placed, which is either black or white.

Figure 11.16 shows that the fluorescent concentrator is advantageous compared to the clear acrylic glass due to spectrally shifting the absorbed photons into a wavelength range where the quantum efficiency QE of the c-Si cell is higher. Of course, the beneficial shift only occurs for about 80% of the photons with wavelengths $\lambda < 486$ nm where the fluorescent dye absorbs. All other photons are transmitted through the collector. Either they directly hit the solar cell or the surrounding area. If the surrounding cover is black paperboard, it absorbs all transmitted photons, while the dispersive, white material sends part of the transmitted photons back into the collector.

The power output of the system was measured under illumination with an AM1.5G spectrum. The illuminated area was $A_{ill} \approx 256 \text{ cm}^2$. To avoid edge effects the area of the acrylic glass exceeds A_{ill}. The impinging optical power is $P_{opt} = 100 \text{ mW cm}^{-2}$. Figure 11.17 presents the module output power for various distances x of the two solar cells, for the four possible combinations of plate and back materials. For the application of clear acrylic glass on top of the solar cells, white paperboard ($P_{el} \approx 128$ mW) under the set-up is of slight advantage over black paperboard ($P_{el} \approx 125$ mW). Here, the randomized distribution leads

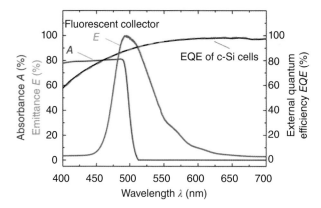

Figure 11.16 The fluorescent collector emits photons in a wavelength range where the quantum efficiency QE of the mono crystalline silicon (c-Si) solar cells is higher than for the absorbed wavelengths. The higher QE leads to more electrons per incoming photons. Additionally to the spatial concentration due to distributing the photons during emission, the fluorescent collector improves the spectral range of incoming photons for the solar cells [153].

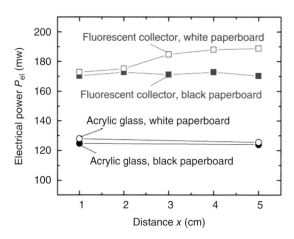

Figure 11.17 Fluorescent collector on top of the solar cells increases the electrical output power significantly from 125 to 170 mW. A white paperboard underneath the solar cell adds further beneficial distribution and increases the efficiency up to 189 mW [153].

impinging photons back into the acrylic glass, giving part of them the chance to be guided to a solar cell. Therefore, the beneficial contribution of a white surrounding lies for solar cells of this size and material at 0.2%.

Replacing the clear acrylic glass with a fluorescent collector results in a power increase from $P_{el} \approx 125$ mW to $P_{el} \approx 170$ mW for all distances if the underlying paperboard is black. Since the black material absorbs impinging photons, this difference is directly ascribed to the spatial distribution and spectral shift in the fluorescent collector. Using underlying white instead of black paperboard, increases the power even further. For small distances ($x=1$ and 2 cm), the power increase is $\Delta P_{el} \approx 45$ mW. But for $x=5$ cm the power increase amounts to $\Delta P_{el} \approx 60$ mW. Thus, increasing the distance x increases the power output under a fluorescent collector but not under a clear acrylic glass. The white paperboard spatially distributes photons and reflects part of them back into the collector. The clear acrylic glass only guides photons entering with proper angles onto a solar cell. The fluorescent collector provides the same mechanism for photons with wavelengths not absorbed by the dye. Additionally, photons in the proper wavelength range experience the benefits of spatial and spectral concentration. The smaller the distance x between the solar cells, the more the solar cells experience the effect that some photons have already been absorbed by the solar cell close by.

11.4.3
Increasing Efficiency with Photonic Structures

The considerations in Section 11.2 lead to the conclusion that photonic structures should increase the performance of fluorescent collector systems. As the detailed balance assumptions point out, the reflectance spectrum of the PBS filter has to be carefully matched to the emission and absorption behavior of the fluorescent collector and to the quantum efficiency of the solar cell in order to achieve highest increases in efficiency. In the following we present the output power increase for both, the side- and the bottom-mounted, geometry.

11.4.3.1 Systems with Side-Mounted III–V Solar Cell

An example, as to how the efficiency of fluorescent concentrator systems can be increased by the use of photonic structures was published in Refs. [23, 25]. The investigated system used a 5 cm × 10 cm, 5-mm-thick luminescent collector of PMMA doped with an organic dye denoted BA241. One GaInP solar cell was coupled to one short edge with silicone. The solar cell had an active area of 5 mm × 49 mm. Hence, the ratio of illuminated luminescent concentrator area and solar-cell area constitutes a geometric concentration ratio of 20×. The solar cell had an efficiency of 16.7% under AM1.5G illumination. White PTFE served as a bottom reflector and also as a reflector at the edges, which were not covered by solar cells. The used photonic structure was produced by the company Mso-Jena by ion-assisted deposition (IAD) and is tuned for high reflection in the emission range of the organic dye (see Figure 11.18). Without this structure, the system had an efficiency of $2.6 \pm 0.1\%$ in reference to the 50 cm² area of the system. The structure increased the efficiency to $3.1 \pm 0.1\%$, which constitutes a relative efficiency increase of around 20%. Figure 11.19 shows the result from a light beam-induced current (LBIC) scan of the system, illustrating how the light collection efficiency is increased over most of the luminescent collector area. With the achieved efficiency of 3.1% and the concentration ratio of 20, the realized luminescent concentrator delivers about 3.7 times more power than the GaInP solar cell delivered on its own.

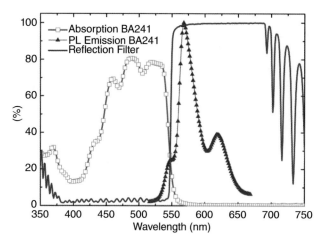

Figure 11.18 Reflection spectrum of the used photonic structure and the absorption and photoluminescence of the luminescent concentrator the filter was designed for. The reflection of the structure very nicely fits the emission peak of the dye in the concentrator.

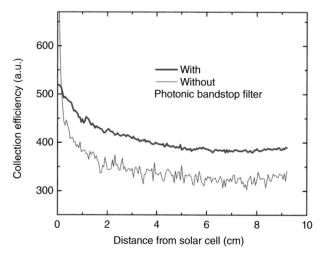

Figure 11.19 Averaged linescans in x-direction from an LBIC scan with and without photonic structure. Close to the solar cell the efficiency is lower with the photonic structure, because it reduces the effectiveness of the bottom reflector for small distances. Over most of the luminescent concentrator, however, collection efficiency is significantly higher with a photonic structure, resulting in a relative efficiency increase of 20%.

The observed efficiency increase of 20% can be already considered as a great success since it shows that photonic structures reduce the escape cone losses significantly. However, the used filter is a multilayer system and therefore costly to produce. Three-dimensional photonic structures are a potential alternative to the presented multilayer systems. A special three-dimensional photonic structure is the opal. The opal has the advantage that it can be produced by a dip-coating process using self-organization of mono-disperse PMMA beads [149]. This is a potentially low-cost process that could be applied on large area concentrators. However, the achieved quality of opaline films is still too low, to achieve optical properties that allow for an increase in efficiency [31].

Another, probably more promising option, is the use of chiral nematic (cholesteric) liquid crystals that act as spectrally selective mirrors. In Ref. [150] such chiral nematic liquid crystals were applied onto luminescent collectors, with a small air gap in between. It was observed that the highest output is achieved using a scattering background and cholesteric mirror with a reflection band significantly redshifted (similar to 150 nm) from the emission peak of the luminescent dye. The use of an air gap results in light bending away from the waveguide surface normal and, consequently, a redshift of the cholesteric mirrors is required. Also, the importance of considering the angular dependence of the spectrally selective mirrors was analyzed in Ref. [151]. Overall, up to 35% more emitted light exits in the luminescent collector edge after application of the cholesteric mirror.

11.4.3.2 Systems with Bottom-Mounted Amorphous Silicon Solar Cell

This section presents a set-up with a fluorescent collector and a PBS filter on top of an amorphous silicon (a-Si) solar cell. Figure 11.20a,b characterizes the components for the experiment. As seen in Figure 11.20b, the fluorescent concentrator absorbs photons with energy above 2.4 eV, while its emission range lies between 2.2 and 2.7 eV. Thus, the fluorescent concentrator absorbs photons for which the quantum efficiency QE < 60% of the solar cell is below 65% for the emitted photons

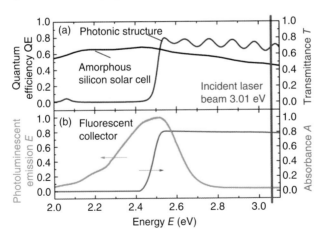

Figure 11.20 (a) Quantum efficiency QE of amorphous silicon (a-Si) solar cell and transmittance TD of photonic band stop (PBS) filter. (b) The fluorescent collector absorbs and emits photons such that the QE of the a-Si cell is more suitable for the emitted photons. The photonic structure reflects most of the emitted photons and transmits 68% of the incoming photons with energy 3.06 eV [153].

(Figure 11.20a also shows that the PBS filter reflects most of the emitted photons from the fluorescent concentrator but transmits 68% of the incoming photons with 3.01 eV.

In this set-up, the a-Si cell of size $14 \times 14\,mm^2$ lies in the center of a $45 \times 45\,mm^2$ area on which LBIC-measurements are performed with a laser beam of 3.01 eV. Summarized over all LBIC values, the solar cell alone achieves the current $I = 8.5\,mA$. Applying a fluorescent concentrator on top of the solar cell increases the current to $I = 9.1\,mA$. Additionally applying the PBS leads to $I = 16.6\,mA$ [152].

Figure 11.21 compares line-scans of the three performed LBIC measurements. Obviously, for the solar cell alone the photovoltaic active area is restricted to the solar cell area, because rays hitting the surrounding area are transmitted. The fluorescent concentrator increases the photon collecting area. The area surrounding the solar cell is covered with white paint which acts almost like a Lambertian reflector. The fluorescent concentrator is optically coupled directly above the solar cell. Thus, part of the rays emitted by the dye with spatially randomized direction is guided to the solar cell via total internal reflection. While the current induced above the solar cell significantly decreases, the larger area compensates this loss. Compared to the fluorescent concentrator alone, the application of the PBS filter not only increases the current collected above the solar cell but also the photovoltaic active area. At the edges of the measured area, the current collected is still above zero. Thus, even points further away from the solar cell than measured contribute to the current. As expected from the Monte-Carlo simulations, the main factor for the success of this photovoltaic system is the scaling. The enlargement of the photovoltaic active area has an absolute value depending on the emission and absorption behavior of the fluorescent concentrator, the transmission, reflection at the PBS, and the back and front side quality of the fluorescent concentrator. The larger the solar cell area the less compensative the enlargement of the current collecting area.

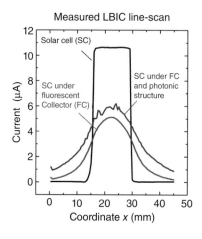

Figure 11.21 Comparison of LBIC line-scans. Above the solar cell the application of the fluorescent concentrator as well as the PBS decreases the current significantly. However, both components enlarge the photovoltaic active area by collecting photons also from the surrounding area [153].

11.5
Conclusion

In this chapter, we have given an overview of the concept of fluorescent concentrators, including the theory, materials for the realization of fluorescent collectors, and the characterization of fluorescent concentrator systems with attached solar cells. We have shown that fundamental aspects, such as efficiency limits, can be investigated using an elegant and detailed balance approach. One major finding, both from theory and experimental results, is that spectrally selectively reflecting structures can increase the efficiency of fluorescent concentrators systems significantly. While best results have been achieved using commercially available multi-layer systems, cheaper processes for the realization of structures with the wanted properties are being developed. A future way to further increase efficiencies might be to embed the luminescent species directly into photonic structures.

We presented the different material systems from which fluorescent collectors can be realized. For all materials, it is important to carefully match luminescent species and matrix material. On the one hand, fluorescent collectors can be formed with organic matrix materials. In these, either organic dyes or luminescent nanocrystalline semiconductor QDs can be embedded. While the organic dyes feature the highest efficiencies, the QDs offer the possibility to extend the used spectral range into the infrared. On the other hand, completely inorganic systems based on rare earths and glasses or glass ceramics allow realizing very stable collectors.

Fluorescent concentrator systems with III–V solar cells, mounted on the edge of fluorescent collectors have shown the highest efficiencies. It has also been demonstrated that the power output of a GaInP can be increased by a factor of 3.7 if it is attached to a fluorescent collector. Nevertheless, also for other types of solar cells out of amorphous or c-Si, which are more relevant for commercial applications, fluorescent collectors can increase the power output by harvesting light from a larger area.

Acknowledgments

The research leading to these results has received funding from the German Federal Ministry of Education and Research in the project "Infravolt" (BMBF, project number 03SF0401B), and from the European Community's Seventh Framework Programme (FP7/2007-2013) in the project NanoSpec under grant agreement no [246200]. This work was also supported by the FhG Internal Programs under Grant No. Attract 692 034. In addition, the authors would like to thank the German Federal Ministry for Education and Research (BMBF) for the financial support within the Centre for Innovation Competence SiLi-nano® (project no. 03Z2HN11). Marie-Christian Wiegand is grateful for the financial support by the German Science Foundation (DFG project PAK 88). Stefan

Fischer gratefully acknowledges the scholarship support from the Deutsche Bundesstiftung Umwelt DBU.

References

1. Shurcliff, W.A. and Jones, R.C. (1949) The trapping of fluorescent light produced within objects of high geometrical symmetry. *J. Opt. Soc. Am.*, **39** (11), 912–916.
2. Birks, J.B. (1964) *The Theory and Practice of Scintillation Counting*, Pergamon Press, London.
3. Weber, W.H. and Lambe, J. (1976) Luminescent greenhouse collector for solar radiation. *Appl. Opt.*, **15** (10), 2299–2300.
4. Goetzberger, A. and Greubel, W. (1977) Solar energy conversion with fluorescent collectors. *Appl. Phys.*, **14**, 123–139.
5. Wittwer, V. et al. (1981) Theory of fluorescent planar concentrators and experimental results. *J. Lumin.*, **24/25**, 873–876.
6. Seybold, G. and Wagenblast, G. (1989) New perylene and violanthrone dyestuffs for fluorescent collectors. *Dyes Pigm.*, **11**, 303–317.
7. van Roosmalen, J.A.M. (2004) Molecular-based concepts in PV towards full spectrum utilization. *Semiconductors*, **38** (8), 970–975.
8. van Sark, W.G.J.H.M. et al. (2008) Luminescent solar concentrators – a review of recent results. *Opt. Express*, **16** (26), 21773–21792.
9. Luque, A. et al. (2005) FULLSPECTRUM: a new PV wave of more efficient use of solar spectrum. *Sol. Energy Mater. Sol. Cells*, **87** (1-4), 467–479.
10. Goldschmidt, J.C. et al. (2006) Advanced fluorescent concentrators. Proceedings of the 21st Photovoltaic Solar Energy Conference, Dresden, Germany.
11. Richards, B.S. and Shalav, A. (2005) The role of polymers in the luminescence conversion of sunlight for enhanced solar cell performance. *Synth. Met.*, **154**, 61–64.
12. Richards, B.S., Shalav, A., and Corkish, R. (2004) A low escape-cone-loss luminescent solar concentrator. Proceedings of the 19th European Photovoltaic Solar Energy Conference, Paris, France.
13. Rau, U., Einsele, F., and Glaeser, G.C. (2005) Efficiency limits of photovoltaic fluorescent collectors. *Appl. Phys. Lett.*, **87** (17), 171101-1–171101-3.
14. Gläser, G.C. and Rau, U. (2006) Collection and conversion properties of photovoltaic fluorescent collectors with photonic band stop filters. *Proc. SPIE* **61970L** 1–11.
15. Goldschmidt, J.C. et al. (2007) Advanced fluorescent concentrator system design. Proceedings of the 22nd European Photovoltaic Solar Energy Conference, Milan, Italy.
16. Danos, L. et al. (2006) Characterisation of fluorescent collectors based on solid, liquid and langmuir blodget (LB) films. Proceedings of the 21st European Photovoltaic Solar Energy Conference, Dresden, Germany.
17. Debije, M.G., Broer, D.J., and Bastiaansen, C.W.M. (2007) Effect of dye alignment on the output of a luminescent solar concentrator. Proceedings of the 22nd European Photovoltaic Solar Energy Conference. Milan, Italy.
18. Slooff, L.H. et al. (2007) The luminescent concentrator: stability issues. Proceedings of the 22nd European Photovoltaic Solar Energy Conference, Milan, Italy.
19. Slooff, L.H. et al. (2008) A luminescent solar concentrator with 7.1% power conversion efficiency. *Phys. Status Solidi RRL*, **2** (6), 257–259.
20. Chatten, A.J. et al. (2003) The quantum dot concentrator: theory and results. Proceedings of the 3rd World Conference on Photovoltaic Energy Conversion, Osaka, Japan.
21. Chatten, A.J. et al. (2001) Novel quantum dot concentrators. Proceedings of

the 17th European Photovoltaic Solar Energy Conference, Munich, Germany.
22. Chatten, A.J. et al. (2003) A new approach to modelling quantum dot concentrators. *Sol. Energy Mater. Sol. Cells*, **75** (3–4), 363–371.
23. Goldschmidt, J.C. et al. (2008) Theoretical and experimental analysis of photonic structures for fluorescent concentrators with increased efficiencies. *Phys. Status Solidi A*, **205** (12), 2811–2821.
24. Peters, M. et al. (2009) The effect of photonic structures on the light guiding efficiency of fluorescent concentrators. *J. Appl. Phys.*, **105**, 014909.
25. Goldschmidt, J.C. et al. (2009) Increasing the efficiency of fluorescent concentrator systems. *Sol. Energy Mater. Sol. Cells*, **93**, 176–182.
26. Goldschmidt, J.C. (2009) Novel solar cell concepts. Dissertation, Faculty of Physics, Universität Konstanz. Verlag Dr. Hut, München.
27. Peters, M. et al. (2010) Spectrally-selective photonic structures for PV applications. *Energies*, **3** (2), 171–193.
28. Peters, I.M. (2009) Photonic concepts for solar cells. Dissertation, Faculty of Mathematics and Physics, Albert-Ludwigs Universität Freiburg.
29. Barnham, K. et al. (2000) Quantum-dot concentrator and thermodynamic model for the global redshift. *Appl. Phys. Lett.*, **76** (9), 1197–1199.
30. Currie, M.J. et al. (2008) High-efficiency organic solar concentrators for photovoltaics. *Science*, **321**, 226–228.
31. Goldschmidt, J.C. (2012) 1.27 – luminescent solar concentrator, in *Comprehensive Renewable Energy* (ed S. Ali), Elsevier, Oxford.
32. Zastrow, A. (1981) Physikalische Analyse der Energieverlustmechanismen im Fluoreszenzkollektor. Dissertation, Faculty of Physics, Albert-Ludwigs-Universität Freiburg.
33. Goetzberger, A. (2009) *Fluorescent Solar Energy Concentrators: Principle and Present State of Development. High-Efficient Low-Cost Photovoltaics: Recent Developments*, Springer-Verlag GmbH.
34. Viehmann, W. and Frost, R.L. (1979) Thin film waveshifter coatings for fluorescent radiation converters. *Nucl. Instrum. Methods*, **167**, 405–415.
35. Bose, R., et al. (2009) Luminescent solar concentrators: cylindrical design. Proceedings of the 24th European Photovoltaic Solar Energy Conference, Hamburg, Germany.
36. Markvart, T. (2006) Detailed balance method for ideal single-stage fluorescent collectors. *J. Appl. Phys.*, **99**, 026101.
37. Keil, G. (1969) Radiance amplification by a fluorescence radiation converter. *J. Appl. Phys.*, **40** (9), 3544–3547.
38. Smestad, G. et al. (1990) The thermodynamic limit of light concentrators. *Sol. Energy Mater.*, **21**, 99–111.
39. Yablonovitch, E. (1987) Inhibited spontaneous emission in solid-state physics and electronics. *Phys. Rev. Lett.*, **58** (20), 2059–2062.
40. Yablonovitch, E. (1993) Photonic band-gap structures. *J. Opt. Soc. Am. B: Opt. Phys.*, **10** (2), 283–295.
41. Yablonovitch, E. (1998) Engineered omnidirectional external-reflectivity spectra from one-dimensional layered interference filters. *Opt. Lett.*, **23** (21), 1648–1649.
42. Southwell, W.H. (1989) Using apodization functions to reduce sidelobes in rugate filters. *Appl. Opt.*, **28** (23), 5091–5094.
43. Martinu, L. and Poitras, D. (2000) Plasma deposition of optical films and coatings: a review. *J. Vac. Sci. Technol., A: Vac. Surf. Films*, **18**, 2619 1–2619 27.
44. Chigrin, D.N. and Sotomayor Torres, C.M. (2001) Periodic thin-film interference filters as one-dimensional photonic crystals. *Opt. Spectrosc.*, **91** (3), 484–489.
45. Wittwer, V., Stahl, W., and Goetzberger, A. (1984) Fluorescent planar concentrators. *Sol. Energy Mater.*, **11**, 187–197.
46. Prönneke, L. and Rau, U. (2008) Geometry effects on photon collection in photovoltaic fluorescent collectors. Proceedings of SPIE, Photonics for Solar Energy Systems II, Strasbourg, France.

47. Schöfthaler, M., Rau, U., and Werner, J.H. (1994) Direct observation of a scaling effect on effective minority carrier lifetimes. *J. Appl. Phys.*, **76** (7), 4168–4172.
48. Gläser, G.C. (2007) Fluoreszenzkollektoren für die Photovoltaik. Dissertation, Faculty of Physics, Universität Stuttgart. Shaker.
49. Prönneke, L., Gläser, G.C., and Rau, U. (2012) Simulations of geometry effects and loss mechanisms affecting the photon collection in photovoltaic fluorescent collectors. *EPJ Photovoltaics*, **3**, 30101-1–30101-11.
50. Markvart, T. (2006) Comment on efficiency limits of photovoltaic fluorescent collectors. *Appl. Phys. Lett.*, **87**, 171101; *Appl. Phys. Lett.*, (2005), **88** (17).
51. Rau, U., Einsele, F., and Glaeser, G.C. (2006) Response to "Comment on 'Efficiency limits of photovoltaic fluorescent collectors' [Appl. Phys. Lett., 87, 171101 (2005)]". *Appl. Phys. Lett.*, **88** (17), 176102.
52. Goldschmidt, J.C. et al. (2010) Increasing fluorescent concentrator light collection efficiency by restricting the angular emission characteristics of the incorporated luminescent material – the "nano-fluko" concept. *Proc. SPIE* (Photonics for Solar Energy Systems III, Brussels, Belgium), **7725**, 77250S1–11.
53. Bykov, V.P. (1972) Spontaneous emission in a periodic structure. *Sov. Phys. JETP*, **35**, 269.
54. Vos, W.L., Koenderink, A.F., and Nikolaev, I.S. (2009) Orientation-dependent spontaneous emission rates of a two-level quantum emitter in any nanophotonic environment. *Phys. Rev. A*, **80**, 053802-1–053802-7.
55. Wang, R. et al. (2003) Local density of states in three-dimensional photonic crystals: calculation and enhancement effects. *Phys. Rev. B*, **67**, 155114-1–155114-7.
56. Busch, K. and John, S. (1998) Photonic band gap formation in certain self-organizing systems. *Phys. Rev. B*, **58** (3), 3896–3908.
57. Vats, N., John, S., and Busch, K. (2002) Theory of fluorescence in photonic crystals. *Phys. Rev. A*, **65**, 043808-1–043808-13.
58. Barth, M., Gruber, A., and Cichos, F. (2005) Spectral and angular redistribution of photoluminescence near a photonic stop band. *Phys. Rev. B*, **72**, 085129-1–085129-10.
59. Bovero, E. and Van Veggel, F.C.J.M. (2008) Wavelength redistribution and color purification action of a photonic crystal. *J. Am. Chem. Soc.*, **130** (46), 15374–15380.
60. Frezza, L. et al. (2011) Directional enhancement of spontaneous emission in polymer flexible microcavities. *J. Phys. Chem. C*, **115** (40), 19939–19946.
61. Yee, K.S. (1966) Numerical solution of initial boundary value problems involving Maxwell's equations in isotropic media. *IEEE Trans. Antennas Propag.*, **14** (3), 302–307.
62. Oskooi, A.F. et al. (2010) MEEP: a flexible free-software package for electromagnetic simulations by the FDTD method. *Comput. Phys. Commun.*, **181** (3), 687–702.
63. Taflove, A. and Hagness, S.C. (2005) *Computational Electrodynamics: The Finite-Difference Time-Domain Method*, Artech House.
64. Gutmann, J. et al. (2012) Electromagnetic simulations of a photonic luminescent solar concentrator. *Opt. Express*, **20** (S2), A157–A167.
65. Tsakalakos, L. (2010) *Nanotechnology for Photovoltaics*, CRC Press, Boca Raton, FL.
66. Kaino, T. (1985) Influence of water absorption on plastic optical fibers. *Appl. Opt.*, **24** (23), 4192–4195.
67. Hornak, L. (1992) *Polymers for Lightwave and Integrated Optics: Technology and Applications*, Optical Engineering Series, vol. **32**, Marcel Dekker, New York.
68. Ma, H., Yen, A.K.-J., and Dalton, L.R. (2002) Polymer-based optical waveguides: materials, processing, and devices. *Adv. Mater.*, **14** (19), 1339–1365.
69. Polishuk, P. (2006) Plastic optical fibers branch out. *IEEE Commun. Mag.*, **44** (9), 140–148.

70. Kastelijn, M.J., Bastiaansen, C.W.M., and Debije, M.G. (2009) Influence of waveguide material on light emission in luminescent solar concentrators. *Opt. Mater.*, **31**, 1720–1722.
71. Earp, A.A. et al. (2004) Maximising the light output of a luminescent solar concentrator. *Sol. Energy*, **76** (6), 655–667.
72. Wilson, L.R. et al. (2014) Characterization and reduction of reabsorption losses in luminescent solar concentrators. *Appl. Opt.*, **49** (9), 1651–1661.
73. Gallagher, S.J. et al. (2007) Quantum dot solar concentrator: device optimisation using spectroscopic techniques. *Sol. Energy*, **81**, 540–547.
74. Büchtemann, A. et al. (2005) Spectroscopic measurements on matrix materials for concentrator plates. Proceeding of the 16th European Symposium Polymer Spectrosc, Kerkrade, The Netherlands.
75. Mansour, A.F. (2004) Photostability and optical parameters of copolymer styrene/MMA as a matrix for the dyes used in fluorescent solar collectors. *Polym. Test.*, **23**, 247–252.
76. Bose, R. et al. (2007) Novel configurations of Luminescent Solar Concentrators. Proceedings of the 22nd European Photovoltaic Solar Energy Conference, Milan, Italy.
77. Dienel, T. et al. (2010) Spectral-based analysis of thin film luminescent solar concentrators. *Sol. Energy*, **84** (8), 1366–1369.
78. Giovanella, U. et al. (2010) Core-type polyfluorene-based copolymers for low-cost light-emitting technologies. *Org. Electron.*, **11** (12), 2012–2018.
79. Mulder, C.L. et al. (2009) Luminescent solar concentrators employing phycobilisomes. *Adv. Mater.*, **21** (31), 3181–3185.
80. Gonzalez, M. et al. (2010) Bio-derived luminescent solar concentrators using phycobilisomes. Proceedings of the 25th European Photovoltaic Solar Energy Conference and Exhibition/5th World Conference on Photovoltaic Energy Conversion, Valencia, Spain.
81. Rowan, B.C., Wilson, L.R., and Richards, B.S. (2008) Advanced material concepts for luminescent solar concentrators. *IEEE J. Sel. Top. Quantum Electron.*, **14** (5), 1312–1322.
82. Reisfeld, R. (2002) Fluorescent dyes in sol–gel glasses. *J. Fluoresc.*, **12** (3-4), 317–325.
83. Zastrow, A. et al. (1980) On the conversion of solar radiation with fluorescent planar concentrators. Proceedings of the 3rd European Photovoltaic Solar Energy Conference.
84. Sah, R.E. (1981) Stokes shift of fluorescent dyes in the doped polymer matrix. *J. Lumin.*, **24/25**, 869–872.
85. Mugnier, J. et al. (1987) Performances of fluorescent solar concentrators doped with a new dye (benzoxazinone derivative). *Sol. Energy Mater.*, **15**, 65–75.
86. Reisfeld, R., Shamrakov, D., and Jorgensen, C. (1994) Photostable solar concentrators based on fluorescent glass films. *Sol. Energy Mater. Sol. Cells*, **33**, 417–427.
87. Hammam, M. et al. (2007) Performance evaluation of thin-film solar concentrators for greenhouse applications. *Desalination*, **209**, 244–250.
88. Colby, K.A. et al. (2010) Electronic energy migration on different time scales: concentration dependence of the time-resolved anisotropy and fluorescence quenching of lumogen red in poly(methyl methacrylate). *J. Phys. Chem. A*, **114** (10), 3471–3482.
89. Debije, M.G. et al. (2011) Promising fluorescent dye for solar energy conversion based on a perylene perinone. *Appl. Opt.*, **50** (2), 163–169.
90. Calzaferri, G., Li, H.R., and Bruhwiler, D. (2008) Dye-modified nanochannel materials for photoelectronic and optical devices. *Chem. Eur. J.*, **14** (25), 7442–7449.
91. Bozdemir, O.A. et al. (2011) Towards unimolecular luminescent solar concentrators: bodipy-based dendritic energy-transfer cascade with panchromatic absorption and monochromatized emission. *Angew. Chem. Int. Ed.*, **50** (46), 10907–10912.
92. Wu, W.X. et al. (2010) Hybrid solar concentrator with zero self-absorption loss. *Sol. Energy*, **84** (12), 2140–2145.

93. Wilson, H.R. (1987) Fluorescent dyes interacting with small silver particles; a system extending the spectral range of fluorescent solar concentrators. *Sol. Energy Mater.*, **16**, 223–234.
94. Hinsch, A., Zastrow, A., and Wittwer, V. (1990) Sol–gel glasses: a new material for solar fluorescent planar concentrators? *Sol. Energy Mater.*, **21**, 151–164.
95. Ishchenko, A.A. (2008) Photonics and molecular design of dye-doped polymers for modern light-sensitive materials. *Pure Appl. Chem.*, **80** (7), 1525–1538.
96. Gallagher, S.J., Norton, B., and Eames, P.C. (2007) Quantum dot solar concentrators: electrical conversion efficiencies and comparative concentrating factors of fabricated devices. *Sol. Energy*, **81**, 813–821.
97. Sholin, V., Olson, J.D., and Carter, S.A. (2007) Semiconducting polymers and quantum dots in luminescent solar concentrators for solar energy harvesting. *J. Appl. Phys.*, **101** (12), 123114.
98. Hyldahl, M.G., Bailey, S.T., and Wittmershaus, B.P. (2009) Photostability and performance of CdSe/ZnS quantum dots in luminescent solar concentrators. *Sol. Energy*, **83** (4), 566–573.
99. Bomm, J. *et al.* (2011) Fabrication and full characterization of state-of-the-art quantum dot luminescent solar concentrators. *Sol. Energy Mater. Sol. Cells*, **95** (8), 2087–2094.
100. van Sark, W.G.J.H.M. (2012) Luminescent solar concentrators – A low cost photovoltaics alternative. *Renewable Energy*, **49**, 207–210
101. Carbone, L. *et al.* (2007) Synthesis and micrometer-scale assembly of colloidal CdSe/CdS nanorods prepared by a seeded growth approach. *Nano Lett.*, **7** (10), 2942–2950.
102. Talapin, D.V. *et al.* (2007) Seeded growth of highly luminescent CdSe/CdS nanoheterostructures with rod and tetrapod morphologies. *Nano Lett.*, **7** (10), 2951–2959.
103. Bomm, J. *et al.* (2010) Fabrication and spectroscopic studies on highly luminescent CdSe/CdS nanorod polymer composites. *Beilstein J. Nanotechnol.*, **1**, 94–100.
104. Xie, R. *et al.* (2005) Synthesis and characterization of highly luminescent CdSe—core CdS/Zn0.5Cd0.5S/ZnS multishell nanocrystals. *J. Am. Chem. Soc.*, **127** (20) pp 7480–7488.
105. Koole, R. *et al.* (2008) On the incorporation mechanism of hydrophobic quantum dots in silica spheres by a reverse microemulsion method. *Chem. Mater.*, **20** (7), 2503–2512.
106. Schüler, A. *et al.* (2006) Quantum dot containing nanocomposite thin films for photoluminescent solar concentrators. *Sol. Energy*, **81**, 1159–1165.
107. Reda, S.M. (2008) Synthesis and optical properties of CdS quantum dots embedded in silica matrix thin films and their applications as luminescent solar concentrators. *Acta Mater.*, **56**, 259–264.
108. Wang, C.H., *et al.* (2010) Efficiency improvement by near-infrared quantum dots for Luminescent Solar Concentrators. Proceedings of SPIE, Next Generation (Nano) Photonic and Cell Technologies for Solar Energy Conversion, Bellingham, Spie-Int Soc Optical Engineering.
109. Schaller, R.D. and Klimov, V.I. (2004) High efficiency carrier multiplication in pbse nanocrystals: implications for solar energy conversion. *Phys. Rev. Lett.*, **92** (18), 186601-1–186601-4.
110. Kennedy, M. *et al.* (2009) Improving the optical efficiency and concentration of a single-plate quantum dot solar concentrator using near infra-red emitting quantum dots. *Sol. Energy*, **83** (7), 978–981.
111. van Sark, W.G.J.H.M., de Mello Donega, C., and Schropp, R.E.I. (2010) Towards quantum dot solar concentrators using thin film solar cells. 25th European Photovoltaic Solar Energy Conference, Munich, Germany.
112. Alivisatos, A.P. (1998) Electrical studies of semiconductor-nanocrystal colloids. *MRS Bull.*, **23** (2), 18–23.
113. Du, H. *et al.* (2002) Optical properties of colloidal PbSe nanocrystals. *Nano Lett.*, **2** (11), 1321–1324.

114. Debije, M.G. and Verbunt, P.P.C. (2012) Thirty years of luminescent solar concentrator research: solar energy for the built environment. *Adv. Energy Mater.*, **2** (1), 12–35.
115. Erogbogbo, F. et al. (2008) Biocompatible luminescent silicon quantum dots for imaging of cancer cells. *ACS Nano*, **2** (5), 873–878.
116. Intartaglia, R. et al. (2012) Luminescent silicon nanoparticles prepared by ultra short pulsed laser ablation in liquid for imaging applications. *Opt. Mater. Express*, **2** (5), 510–518.
117. Nagarajan, R. (2008) Nanoparticles: building blocks for nanotechnology, in *Nanoparticles: Synthesis, Stabilization, Passivation, and Functionalization* (eds R. Nagarajan and T. Alan Hatton), American Chemical Society.
118. Inman, R.H. et al. (2011) Cylindrical luminescent solar concentrators with near-infrared quantum dots. *Opt. Express*, **19** (24), 24308–24313.
119. Schüler, A. et al. (2007) Quantum dot containing nanocomposite thin films for photoluminescent solar concentrators. *Sol. Energy*, **81**, 1159–1165.
120. Juodkazis, S. et al. (1998) Optical properties of CdS nanocrystallites embedded in $(SiO.2TiO.8)O_2$ sol–gel waveguide. *Opt. Commun.*, **148** (4–6), 242–248.
121. Pang, L. et al. (2005) PMMA quantum dots composites fabricated via use of pre-polymerization. *Opt. Express*, **13** (1), 44–49.
122. Lee, J. et al. (2000) Full color emission from II-VI semiconductor quantum dot-polymer composites. *Adv. Mater.*, **12** (15), 1102–1105.
123. Peres, M. et al. (2005) A green-emitting CdSe/poly(butyl acrylate) nanocomposite. *Nanotechnology*, **16** (9), 1969.
124. Zhang, H. et al. (2003) From water-soluble CdTe nanocrystals to fluorescent nanocrystal-polymer transparent composites using polymerizable surfactants. *Adv. Mater.*, **15** (10), 777–780.
125. Alivisatos, A.P. (1998) Electrical Studies of Semiconductor-Nanocrystal Colloids. *MRS Bull.- Mater. Res. Soc.*, **2** (23), 18–23.
126. Chandra, S. et al. (2012) Enhanced quantum dot emission for luminescent solar concentrators using plasmonic interaction. *Sol. Energy Mater. Sol. Cells*, **98**, 385–390.
127. Farrell, D.J. et al. (2006) Fabrication, characterisation and modelling of quantum dot solar concentrator stacks. Conference Record of the 2006 IEEE 4th World Conference on Photovoltaic Energy Conversion (IEEE Cat. No. 06CH37747), pp. 1–4.
128. Fisher, M. et al. (2011) Luminescent solar concentrators utilizing aligned CdSe/CdS nanorods. Proceedings of the 37th IEEE Photovoltaic Specialists Conference (PVSC) Seattle.
129. Kang, I. and Wise, F.W. (1997) Electronic structure and optical properties of PbS and PbSe quantum dots. *J. Opt. Soc. Am. B*, **14** (7), 1632–1646.
130. Micic, O.I. et al. (1997) Size-dependent spectroscopy of InP quantum dots. *J. Phys. Chem. B*, **101** (25), 4904–4912.
131. Sargent, E. (2005) Infrared quantum dots. *Adv. Mater.*, **17** (5), 515–522.
132. Shcherbatyuk, G.V. et al. (2010) Viability of using near infrared PbS quantum dots as active materials in luminescent solar concentrators. *Appl. Phys. Lett.*, **96** (19), 191901-1–191901-3.
133. Wang, Y. et al. (1987) PbS in polymers. From molecules to bulk solids. *J. Chem. Phys.*, **87**, 7315–7322.
134. Borrelli, N.F. and Smith, D.W. (1994) Quantum confinement of PbS microcystals in glass. *J. Non-Cryst. Solids*, **180** (1), 25–31.
135. Gao, Y. et al. (2008) Encapsulating of single quantum dots into polymer particles. *Colloid. Polym. Sci.*, **286** (11), 1329–1334.
136. Allen, P.M. and Bawendi, M.G. (2008) Ternary I-III-VI quantum dots luminescent in the red to near-infrared. *J. Am. Chem. Soc.*, **130** (29), 9240–9241.
137. Rath, T. et al. (2011) A direct route towards polymer/copper indium sulfide nanocomposite solar cells. *Adv. Energy Mater.*, **1** (6), 1046–1050.
138. Pham, T.A. (2012) Synthese und Charakterisierung von NIR-emittierenden Quantenpunkten. Master thesis, University of Potsdam.

139. Balaji, S., Misra, D., and Debnath, R. (2011) Enhanced green emission from Er2WO6 nanocrystals embedded composite tellurite glasses. *J. Fluoresc.*, **21** (3), 1053–1060.
140. Reisfeld, R. and Kalisky, Y. (1981) Nd^{3+} and Yb^{3+} germinate and tellurite glasses for fluorescent solar energy collectors. *Chem. Phys. Lett.*, **80** (1), 178–183.
141. Reisfeld, R. (1984) Fluorescence and nonradiative relaxations of rare earths in amorphous media on high surface area supports: a review. *J. Electrochem. Soc.: Solid-State Sci. Technol.*, **131** (6), 1360–1364.
142. Andrews, L.J., McCollum, B.C., and Lempicki, A. (1981) Luminescent solar collectors based on fluorescent glasses. *J. Lumin.*, **24/25**, 877–880.
143. Neuroth, N. and Haspel, R. (1987) Glasses for luminescent solar concentrators. *Sol. Energy Mater.*, **16**, 235–242.
144. Dyrba, M., Wiegand, M.-C., Ahrens, B., and Schweizer, S. (2012) Neodymium-doped barium borate glasses as fluorescent concentrators for the infrared spectral range. *Photonics for Solar Energy Systems IV*, (eds R.B. Wehrspohn and A. Gombert), Proceedings of SPIE, vol. 8438, SPIE, p. 84380Z. doi:10.1117/12.921771
145. Dieke, G.H. and Crosswhite, H.M. (1963) The spectra of the doubly and triply ionized rare earths. *Appl. Opt.*, **2** (7), 675–686.
146. Dejneka, M.J. (1998) Transparent oxyfluoride glass ceramics. *MRS Bull.*, **23** (11), 57–62.
147. Ahrens, B. *et al.* (2008) Structural and optical investigations of Nd-doped fluorozirconate-based glass ceramics for enhanced upconverted fluorescence. *Appl. Phys. Lett.*, **92** (6), 61905.
148. Farrell, D.J. and Yoshida, M. (2012) Operating regimes for second generation luminescent solar concentrators. *Prog. Photovoltaics Res. Appl.*, **20** (1), 93–99.
149. Jiang, P. *et al.* (1999) Single-crystal colloidal multilayers of controlled thickness. *Chem. Mater.*, **11** (8), 2132–2140.
150. Debije, M.G. *et al.* (2010) Effect on the output of a luminescent solar concentrator on application of organic wavelength-selective mirrors. *Appl. Opt.*, **49** (4), 745–751.
151. de Boer, D.K.G. (2010) Optimizing wavelength-selective filters for luminescent solar concentrators. *Proc. SPIE* (Photonics for Solar Energy Systems III, Bellingham), p. 77250Q.
152. Prönneke, L. *et al.* (2009) Measurement and simulation of enhanced photovoltaic system with fluorescent collectors. Proceedings of the 24th European Photovoltaic Solar Energy Conference, Hamburg, Germany.
153. Prönneke, L. *et al.* (2012) Fluorescent materials for silicon solar cells. Diss. Universität Stuttgart, 2012.

12
Light Management in Solar Modules

Gerhard Seifert, Isolde Schwedler, Jens Schneider, and Ralf B. Wehrspohn

12.1
Introduction

From the beginning of the twenty-first century, production and installation of solar cells has seen a continuous, strong growth from a specialized niche market to currently about 35 GW peak power being installed worldwide in the year 2013. Meanwhile in Germany photovoltaic power generation provides more than 5% of the annual electrical energy production, and the accompanying huge price drop brought solar electricity costs well below grid parity. Still further cost reduction is necessary and possible ensuring full competitiveness without any subsidies. The main drivers of cost reduction of crystalline silicon-based photovoltaics in the past years were decrease of prices for silicon feedstock, optimization of the cell process, and general scaling effects. While the solar cell price dropped strongly, the relative share of processing the cells into modules increased to 50% of the modules price. Therefore, module material cost and module performance have become one of the biggest levers in reducing modules costs and thus overall photovoltaic levelized cost of electricity (LCOE). An important aspect of this necessary optimization is the light management in the solar module. In the latest update of the International Technology Roadmap for Photovoltaic (ITRPV, 4th Edition, March 2013), the development of the so-called cell-to-module efficiency is expected to be increased from 100 to 104% by 2023. This ambitious goal means that in addition to the incoming sunlight that has to be admitted to enter the actual solar cell without loss, light being incident on inactive areas such as contact ribbons or gaps between the individual cells also has to be redirected to and trapped into the optically active cell areas by proper light management concepts.

It is the goal of this chapter to provide an overview of the physical challenges and concepts for light management in solar modules, as well as established solutions and current and novel approaches on the way to realize the ambitious goal of the ITRPV roadmap. Focusing on the typical construction of the currently market-dominating crystalline silicon solar modules, we start with a description of the physical phenomena that determine the losses of a light beam on its path through

Photon Management in Solar Cells, First Edition. Edited by Ralf B. Wehrspohn, Uwe Rau, and Andreas Gombert.
© 2015 Wiley-VCH Verlag GmbH & Co. KGaA. Published 2015 by Wiley-VCH Verlag GmbH & Co. KGaA.

the several layers of a module. Next, we introduce the physics of anti-reflection and light trapping/light redirection concepts, and report on experimental and simulation techniques to assess the achievable progress under consideration of the real field situation, expressed in parameters like spectral distribution and varying incidence angle of sunlight, or diffuse lighting. Based on these fundamentals, the second part is devoted to the technological solutions being (i) already implemented, (ii) on the threshold to industrial use, and (iii) novel concepts currently under basic investigation.

12.2
Fundamentals of Light Management in Solar Modules

The vast majority of solar modules on the market and, in particular, those being already installed on roofs or in the field are composed of photovoltaic cells based on 6-in., (nearly) quadratic crystalline silicon wafers. The standard setup of these modules is a rectangular array of typically 60 cells, which are electrically interconnected in series by two or three parallel metal ribbons per cell row. This cell arrangement is laminated by transparent polymer interlayers (mostly EVA: ethylene-vinyl-acetate) to a glass sheet on the sunnyside and another, typically white polymer backsheet. The basic optical phenomena which have to be addressed to use as much of the incoming sunlight as possible for electrical power generation by the module will be discussed in the following text for this standard configuration, as sketched in Figure 12.1. The physical concepts described here can, however, easily be transferred in an analogous way to other module constructions.

12.2.1
Basic Physical Concepts of Light Management in Solar Modules

12.2.1.1 Optical Losses due to Reflection and Absorption

The most basic physics problems that impact the percentage of light being available for generating electron–hole pairs in the photovoltaic diode are the behavior of an electromagnetic wave at an interface of two materials and its absorption along the light path through the layers. The reflectivity r at a plane interface for P or S polarization of the incident light is defined by Fresnel's equations as a function of complex refractive index $\hat{n}_j = n_j + ik_j$ of layer j and the angles of incidence (θ_j) and refraction (θ_{j+1}):

$$r_P = \frac{\hat{n}_j \cos\theta_j - \hat{n}_{j+1} \cos\theta_{j+1}}{\hat{n}_{j+1} \cos\theta_j + \hat{n}_j \cos\theta_{j+1}}; \quad r_S = \frac{\hat{n}_{j+1} \cos\theta_j - \hat{n}_j \cos\theta_{j+1}}{\hat{n}_{j+1} \cos\theta_j + \hat{n}_j \cos\theta_{j+1}}.$$

These formulas describe in general the amplitude and phase changes of the electric field of incident light upon reflection; angles of incidence and refraction are related by Snell's law. In the transparency range of a material, as mostly discussed

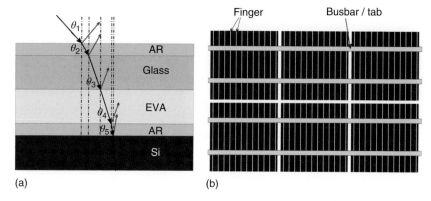

Figure 12.1 (a) Schematic cross section of a typical crystalline silicon-based solar module with incident, refracted, and reflected beams at every interface and (b) typical configuration of c-Si solar cells in a module, in top view, demonstrating the optical loss areas (gray contact fingers and busbars/tabs and white cell interspaces).

here, the absorption index k_j is orders of magnitude smaller than n_j, and can thus be neglected for calculating the reflectivity at an interface. Intensity reflection factors are found by $R = r^2$, and for a single interface the intensity transmission factor follows from energy conservation: $T = 1 - R$. For thin layers (thickness in the order of the light wavelengths), interference effects between multiple reflexes have to be considered in an electric field description to determine the total intensity reflection R. For convenience, this is nowadays mostly calculated numerically, usually in the framework of the transfer matrix formalism [1], which is able to handle an arbitrary number of thin layers in a correct way.

For thicker layers, it is normally sufficient to describe in geometrical optics the refraction angle (by Snell's law) and transmission factor T upon entry of a light ray into the medium, and its absorption losses along its path through the medium (ray tracing approach). Instead of the absorption index k, the most common way to describe absorption is using the absorption constant $\alpha = 4\pi k\lambda/n$, which determines the exponential decay of light intensity I with thickness d of a traversed layer, usually denoted as Beer's law:

$$T = I(d)/I(0) = \exp(-\alpha \cdot d)$$

This equation describes the intensity transmission T of light after traversing a layer of thickness d. With these fundamental ingredients, the part of incident sunlight finally entering the photovoltaic active material (silicon) can be calculated, as long as light scattering is negligible, exactly by summing up all reflected and transmitted beams as indicated in Figure 12.1. Within the silicon layer, the light should in the ideal case be completely absorbed. While a full theoretical optimization of the total transmission under real conditions (e.g., large incidence angles, consideration of polarization effects and diffuse light) may develop into a very complex and time-consuming task, already the simplest case of direct sunlight at vertical incidence provides important insight into the basic requirements for light management in the modules. For $\theta_j = 0°$, the Fresnel equations simplify to only one yielding an intensity reflection of

$$R_{\theta=0°} = (n_{j+1} - n_j)^2/(n_{j+1} + n_j)^2.$$

Owing to the quadratic dependence of R on the refractive index difference between the two layers, the large reflectivity at a silicon/air interface ($n(\text{air}) \approx 1$); $n(\text{Si}) = 3.8$ at $\lambda = 680$ nm) of $R \approx 35\%$ could be reduced considerably by distributing the total Δn over several smaller steps. Even if materials with the optimal refractive indexes were to be available, the total reflective losses would still be above 10%. In reality, there is very good index matching for the materials glass ($n \approx 1.5$) and EVA ($n \approx 1.48$) leading to a negligible reflection of $R < 10^{-4}$ at the glass/EVA interface. For the other interfaces (glass/air and EVA/Si), anti-reflection (AR) coatings are thus the best way to reduce R. By utilizing interference effects, the well-known quarter-wave layers (optical thickness $n_2 \cdot d = \lambda/4$) can decrease reflection to zero for the specified wavelength when the refractive index of the layer is intermediate to the adjacent ones: $n_j = \sqrt{n_{j-1} \cdot n_{j+1}}$. Though the functionality of such AR layers is strongly wavelength dependent, a proper spectral positioning of the design wavelength within the useable spectral range can, for example, reduce the integral reflection loss at the glass/air interface from 4% to below 1%. A good means to achieve the required very low refractive index in this case is to prepare a thin layer of nanoporous SiO_2 on top of the glass, where the value of n can be designed by the volume fraction of the pores [2]. The value of n of such a layer is in general different from a mere volume average, but can be calculated by an effective medium model, for example, the one of Maxwell Garnett being insensitive to shape and spatial distribution of the nanopores [3].

In principle, multi-layer AR coatings could suppress reflection almost perfectly to zero in the whole usable spectral range; but the added costs for the solar module would be much higher than the achievable relative efficiency gain of 1%. Thus, they have never been considered seriously for use in photovoltaic energy conversion. Even a single layer AR coating has only become standard in the photovoltaic (PV) production in 2012 [4].

At the EVA/Si interface the situation is even more special, because here the AR coating at the same time has to fulfill the requirement to act as an electrical passivation layer. This limits the choice of materials, and thus refractive indices. Furthermore, the relatively low absorption coefficient of Si at photon energies directly above its band gap allows some percent of the incident photons even after reflection at the rear side of the cell to leave the module again through the anti-reflective Si/EVA interface. Therefore, most of the currently produced Si solar cells have textured surfaces enabling light trapping effects, the principle of which will be detailed in the next section.

12.2.1.2 Optical Description of Textured Interfaces or Surfaces

The previous section described the optical problems and loss minimization concepts related to plane interfaces. This section implies that the average surface or interface roughness is much smaller than the involved optical wavelengths.

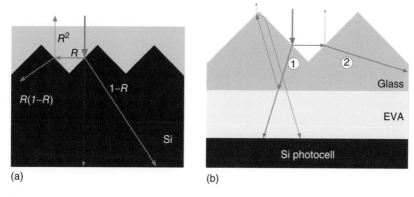

Figure 12.2 Principle of (a) light trapping at textured surface of c-Si solar cell and (b) light trapping at textured glass surface (module front sheet).

As soon as the typical lengths of an interface (surface) relief approach or exceed the wavelengths of the incident light, new physical phenomena arise, which provide additional options for light management. The physical description which is best suited to characterize the optical functionality of a surface texture depends on its length scale and regularity.

We start at the macroscopic scale, where a discussion in terms of geometrical ray optics is valid and wave optical effects (in particular, diffraction) can be neglected. This is correct for characteristic structure dimensions clearly above the involved light wavelengths; usually, $>10\lambda$ is used as limiting condition. The typical example for this case is the pyramidal texture of the order of 10 μm height prepared by etching, which is part of the current standard processing of crystalline Si solar cells with the goal of (geometrical) *light trapping* [5]. The sketch in Figure 12.2a shows what is meant by this notion: an incoming beam at vertical incidence is refracted into the silicon such that its effective path length in the absorber material is increased. This is particularly useful in case of low absorption. The second important effect is that the reflected beam hits a second interface at the neighbor pyramid, giving the incident light a second chance to enter the active material; this reduces the effective reflectivity from R to R^2, for example, from 10 to 1% reflection loss. Overall, the texture increases both the amount of light entering the active layer and the effective optical path length (absorptivity) in the material.

While the surface texture of the cell primarily has the function of light trapping within the active layer, it also creates some challenges for the light management on module level: first, the lamination process has to guarantee a perfect mold of the EVA layer into the Si cell texture; otherwise the residual air gaps could dramatically increase reflection losses [6]. Second, when the texture conformity is maintained, the optimal refractive index of the intermediate AR coating is much higher ($n_{AR} \approx 2.37$) than against air ($n_{AR} \approx 1.95$).

Of course, also the outer surface of the module (usually the front glass) can be textured to provide a light trapping effect within the module (Figure 12.2b).

Here the primary goal is not to obtain an increase of the optical path length (which could in fact even be counterproductive, if glass or encapsulation foil have non-negligible absorption), but find an alternative to a standard $\lambda/4$ AR coating. In the mentioned limit of geometrical optics, the effects of such structures can be analyzed quantitatively by help of appropriate *ray tracing* software [7]. In this way, it has been demonstrated that pyramids [7] or sinusoidal gratings [8] as module surface texture can provide superior reflex reduction at higher incidence angles.

For smaller, but still well-defined patterns with dimensions approaching the wavelength, interference effects have to be accounted for. This can be done analytically in simplified, well-established descriptions, for example, the Fraunhofer diffraction, which can predict very well the angular distribution of light in the far field after passing a slit or a regular grating structure. The optical effect of more complex regular diffractive structures and, in particular, a quantitative solution of the resulting angular distribution of intensities has to be done on the fundamental level of the electric field (considering amplitude and phase) according to Maxwell's equations. The respective diffraction integrals can in some special cases still be solved analytically, but are nowadays mostly treated by numerical calculations, for instance using Finite Element methods. The latter is actually applicable for any kind and size of structure, when simplified descriptions fail.

The case of irregular or statistical patterns is often mainly characterized by its effect on the light, that is, which scattering distribution it causes. The most prominent case is a rough surface intended to act as a Lambertian scatterer, which is an idealized model for diffuse light scattering. Only recently, a quantitative measure how "Lambertian" a nano-textured surface is, has been published [9]. Typically, the rear sides of crystalline Si solar cells are currently produced with a statistical texture to enhance light trapping within the cell. The maximum gain factor achievable for extending the average light path length within the "trap" is $4n^2$ (n being the material's refractive index) for an ideal Lambertian scatterer in the weak absorption limit [10]. For wafer-based silicon solar cells, the optimum enhancement factor of about 50 is by far not necessary, as only light in a quite narrow wavelength range close to the Si band gap has penetration lengths larger than the current minimum wafer thickness of $\approx 150\,\mu m$. For thin-film photovoltaics, an optimized light trapping is apparently much more relevant. Several recent studies predict that very specialized nanostructures can beat the Yablonovitch limit by providing a light trapping factor larger than $4n^2$ [11–14]. Experimentally, however, the principal optical advantage is often compensated by decreased electrical performance [15].

In general, for nanostructures it has to be decided from case to case which physical approach provides the simplest albeit convenient description for its optical properties. For instance, in some cases effective medium approaches turn out to be applicable even for strongly ordered nanostructures, as has been shown, for example, for an AR layer on GaAs, which consisted of a very regular nanograting [16].

12.2.1.3 Spectral Effects and Solar Concentration

With the fundamental physical ingredients described briefly in the previous sections, the optical behavior of a solar module can be analyzed quantitatively, provided that the required material parameters and interface structures are known. However, also the properties of the incident light are of central importance; in particular, the spectral distribution of light intensity (power per area) has to be known and considered in any optimization approach. The goal of converting solar radiation with optimal efficiency into electrical power can be addressed by three main strategies for light management in a PV module:

1) Admit all photons principally available for energy conversion (i.e., having energies above the band gap of the absorber material) to enter the active material (and trap them there).
2) Avoid all deleterious effects by photons which are not available for energy conversion (e.g., efficiency decrease by cell heating via infrared radiation).
3) Manipulate the spectral characteristics (by up- and down-conversion) or the average intensity (by concentrating optics) such that the active material can convert power with higher efficiency than when using normal sunlight.

The basis for all three strategies is, of course, the solar spectrum after passing our Earth's atmosphere, normalized to the air mass (AM) at vertical incidence. Larger incidence angles cause a longer pathway through the atmosphere, changing the spectrum and decreasing the total incident intensity. For instance, a 1.5-fold longer path is denoted by AM1.5 (Figure 12.3). This spectrum could be the basis to select materials and design AR/light trapping functionality of a module. In reality however, any assessment has to account for the absorption and conversion characteristics of the cell: only photons with energy $h\nu$ above the material's band gap E_g can create electron–hole pairs at all; even above E_g the probability for that process

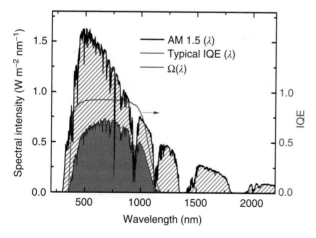

Figure 12.3 Solar spectral intensity AM1.5 (black) and solar response function $\Omega(\lambda)$ (red/orange) for a typical c-Si solar module with internal quantum efficiency IQE (blue).

is not constant, but strongly wavelength dependent, expressed by the internal quantum efficiency factor IQE(λ); in addition, with increasing $h\nu > E_g$ an increasing part of the photon energy does not contribute to the electrical power, but only heats the cell. Therefore, the relevant quantity to be used for the optical design of a module is a weighted intensity, sometimes denoted as the solar response function $\Omega(\lambda)$ (e.g., [6]):

$$\Omega(\lambda) = I_{AM1.5}(\lambda) \cdot \text{IQE}(\lambda) \cdot \frac{\lambda}{\lambda_{Gap}}$$

With this weighting by IQE of the cell and heating losses (last factor), the relevant contribution of a wavelength interval to the total conversion efficiency can be seen.

While the technological part of this article will focus on the strategy (1), the solar response function is also the benchmark for the other two. Spectral up-conversion of infrared to visible light increases the total conversion efficiency, spectral down-conversion of photons increases the photon wavelength from 400 to 800 nm. More generally spoken, the spectral conversion characteristics of the photovoltaic diode are the scale for which light management in the module has to be optimized.

The most important points within strategy (b) are material deterioration by UV photons and temperature rise by absorption of IR photons leading to reduced efficiency in the case of crystalline silicon technology (where power conversion efficiency decreases by $\approx 0.5\%/K$ [5]). While the first point is a well-established criterion for the choice and design of encapsulation materials to guarantee durability of the produced modules, the second one is a known problem [17, 18], which has up to now not normally been included in module optimization strategies. This may be due to the fact that the standard characterization of solar modules is to measure I–V-characteristics upon a short flash of light similar to solar radiation under vertical incidence. Under these circumstances no considerable temperature increase will occur.

Finally, the more future-oriented strategy (c) is currently not relevant for the c-Si technology. First, sunlight concentrated by lenses or other simple optical systems to much higher intensity (e.g., 100 or 1000 "suns"), can principally provide higher efficiency by using much less cell material [5]. Owing to the recent price drop, this option is however far from being economically attractive for silicon. There is a first commercial system on the market using triple-junction cells based on III–V semiconductors with efficiency up to more than 44% [19].

Materials for up-/down-conversion, typically rare-earth ions in crystalline or glassy hosts, are still a subject of basic research. This approach for photon management in solar modules is quite attractive because, in principle, it could be sufficient to replace standard components by, for example, properly doped cover glass or encapsulation foils. Currently however, although there are first promising up-conversion results under intense monochromatic illumination [20, 21], the efficiency of the frequency conversion processes under normal solar illumination is far too low for any reasonable application. Therefore, research in that field is still mainly focused on understanding the fundamental processes determining the conversion efficiency (e.g., [22–24]).

12.2.1.4 Effects of Incidence Angle Variation and Diffuse Light

The final generalization which has to be done for the design and assessment of any light management in a solar module is the real field situation it will be exposed to. In the European countries which were the largest PV market in the past years, almost 100% of the solar modules are installed in a fixed position. This means that, no matter if experimentally or by numerical simulation, the determination of module energy yield over the course of a year has to account for the daily and seasonal variation of the incidence angle of sunlight, as well as for the large relative contribution of diffuse light.

Such a comprehensive, realistic treatment of the "real" module efficiency is not a principal problem, since it can be handled numerically well with all the theoretical ingredients listed in the previous sections. It can, though, become a very time-consuming task when, for example, a ray tracing approach has to be solved as a function of incidence angle and wavelength for realistic illumination conditions and correct averaging over the whole typical year. Nonetheless, such studies have been conducted successfully [25], showing for instance that without AR coating, c-Si modules in fixed orientation exhibit more than 3% additional reflection loss averaged over the day as compared to standard test conditions (STCs). Correspondingly, a module with structured glass surface was shown to reduce these losses most effectively [8]. A very recent study demonstrated that even the use of the standard AM spectra in simulations can cause deviations up to 20% compared with real outdoor performance (for a-Si), because a significant part of the light reaching the solar modules over the year is spectrally modified by the spectral albedo of its surroundings (e.g., grass and clouds) [26].

12.2.2
Assessment of the Optical Performance of Solar Modules

12.2.2.1 Experimental Techniques

For researchers and engineers who want to design and assess novel approaches for improved light management in solar modules, there is an obvious demand for precise and reliable experimental methods to determine the optical and electrical performance of the new module concept. The required techniques and equipment can be divided into three groups:

- The first and most straightforward approach is to qualify the individual components and the completely mounted module by optical spectroscopy yielding spectra of transmission and reflection factors and, if necessary, scattering intensity.
- The second step toward the relevant information for the potential customers is to assess the efficiency gain by a light management concept by comparing the electrical output under simulated solar radiation in the lab.

- The most reliable, but also most expensive and time-consuming approach is to install the interesting modules in the field together with monitoring equipment for the solar radiation behavior and resulting electrical power generation.

The first point can for the simplest case – direct sunlight and flat surfaces – be handled using standard spectrophotometers with collimated beams, monitoring the intensities of transmitted beam and specular reflection with small-area detectors. If additionally the amount of scattered light is no longer negligible, for example, when light trapping structures are applied, usually an integrating (Ulbricht) sphere has to be used for the detection in order to register quantitatively also the scattered light over the whole dihedral angle [27]. The technical problems to be solved for such measurements are (i) the spatial separation of specular and diffuse reflection and (ii) the size of the Ulbricht sphere. A quantitatively correct detection of the scattered light works well only when less than 5% of the Ulbricht sphere (inner) surface are used for "ports" (light entrance, sample, detector). Therefore, it is normally not possible to measure whole modules with ≈1.6 m height, but only relatively small areas of them at a time [28]; reliable mapping of modules with the help of automated translation has been demonstrated [29].

As already mentioned above in Section 12.1.2.3, an apparent optical improvement does not automatically mean an improvement also in terms of electrical power generation efficiency of a module. Only when the additional photons entering the solar cell can be converted into additional current, there will be an increase of the overall module efficiency. Experimentally, this information is usually obtained by measuring the photovoltaic conversion efficiency under standardized illumination comparable to natural sunlight. At present, such solar simulators are mostly based on Xe arc lamps or metal-halide arc lamps, which can be operated continuously or flashed. The emission spectra of these light sources are fairly similar to sunlight or can be reasonably adjusted to it by optical filters (the standards for solar simulators are defined in IEC 60904-9). However, the difference to natural sunlight remains considerable, and, in particular, upon flashed operation the unavoidably transient character of the light always restricts the comparability of an efficiency measured with solar simulator to the real one obtained in the field. In the special case of concentrating photovoltaics, it has been reported recently that also the knowledge (and simulation) of the solar divergence is important for correct testing of the concerntrated photovoltaics (CPV) modules [30].

Following the demand for better solar simulators, light emitting diode (LED)-based solutions promise far better spectral match and stability, more flexibility in the test conditions, as well as drastically reduced energy consumption for conversion efficiency testing.Small-size LED solar simulators (for standard 6 in. cells) are commercially available, and larger devices feasible for standard module sizes are under development.

Another problem comes into the game when inactive areas (in particular, gaps between individual cells or contact strings) are addressed by light management techniques. The redirected (scattered) light cannot quantitatively be determined in specular reflection, and even using an integrating sphere gives no proof that the scattered amount of light is really absorbed in an active solar cell area. In those cases, spatially resolved technique such as laser (or more general, light) beam induced current (LBIC) is a good alternative to measure the relative amount of redirected light via the electrical response of the photovoltaic converter.

The ultimate test for the beneficial effects of light management concepts in a module is to assess the module efficiency in an outdoor location over a long time period of, at least, 1 year. The required experimental equipment ideally comprises devices to monitor in detail generated power, direct and diffuse illumination intensity, and spectral characteristics, and further weather conditions like temperature, wind velocity, and precipitation. Once these data are available, they can be used as references to check the validity of any numerical modeling approaches (see below), which apparently enable a much faster analysis and assessment of approaches to improve the light management in a module.

12.2.2.2 Simulation Approaches and Studies

While a precise measurement of the optical properties and resulting power conversion efficiency is definitely the ultimate proof of the success of any light management in solar modules, there is an apparent need for numerical simulations to evaluate the prospect of novel light management concepts prior to their technical realization. Similarly to the classification of experimental methods, numerical approaches can (i) be restricted to the optical properties (with the goal to predict total transmittance) under directed illumination, (ii) additionally include the electrical response of the system to an AM$x.y$ spectrum under directed illumination, or (iii) try to account for the full averaging over the expected illumination conditions within a longer time period (ideally, 1 year) in the outdoor situation.

Instead of a detailed description of different simulation approaches, which would go far beyond the scope of this chapter, only few examples of published studies are mentioned and cited here. A very recent work [31] combining experiments and numerical simulation addressed the above mentioned stage (ii), aiming at the optimal combination and/or interplay of the optical improvements achievable in a solar module by using thinner glass (TG), AR coating, encapsulation with enhanced UV-transparency [32], and "light harvesting strings" (LHSs). The authors could show that a numerical prediction of the gain in power conversion based on the experimental assessment of the optical properties of each module component agrees well with the efficiency measured with usual solar simulators on the final mounted modules.

Several different investigations addressing stage (iii) have been published previously. For instance, an approach based on flat surfaces and Fresnel equations including the daily incidence angle variation and diffuse radiation [25] yielded about 3% yearly reflection losses (for front glass without AR coating) for tilt angle corresponding the latitude, in good agreement with measurements. A similar method for numerical treatment of optical module losses was described in more detail in Ref. [33]. Extension to textured surfaces have also been made, showing that a proper texture of the front glass can improve the daily energy output by capturing more light at high incidence angles (i.e., in the morning and evening) [8]. Other work included also the temperature effects, that is, thermal losses in the analysis [17, 18], and it was found that the seemingly optimal performance at almost vertical incidence around noon is often counteracted or even overcompensated by the negative temperature coefficient of solar power conversion in silicon. Therefore, strategies to lower average module temperatures are highly desirable; from the viewpoint of light management, this could be realized by surface coatings with high reflectivity for near-infrared wavelengths (photon energies below the Si band gap).

12.3
Technological Solutions for Minimized Optical Losses in Solar Modules

In order to maximize the photoelectrical conversion efficiency of a solar module per illuminated area light management has to maximize the fraction of the incident photon flux absorbed in the high photo response part of the solar cells. As detailed in the previous section, approaches to overcome this challenge have to address minimization of absorption, scattering, and reflection losses of light on its way through the cover layers of solar modules, as well as redirection of light incident on inactive areas. In this chapter an overview of the state of the art of technological solutions in this field is given, ranging from material selection to surface nanostructuring techniques. The various optimization approaches will be ordered mainly along the light path through the module components, from entering the module to absorption in the c-Si solar cell.

12.3.1
Minimization of Optical Losses in Front Glass Sheets

Mineral glass, in particular, the so-called extraclear or extrawhite low-iron soda-lime-silica glass, is the current standard front protection material for PV modules because of its superior physical and chemical protection characteristics at acceptable costs. Only for special modules, for example, for light-weight applications and certain consumer-market products, polymers having high transmittance are being used alternatively. The reason is that it is a challenge to upgrade polymer materials to the high PV standards especially with respect to long-term outdoor durability and transmittance characteristics. Only a few, patented high-tech modifications based on glass-like transparent Polymethyl methacrylate (PMMA) thermoplastics or Ethylene tetrafluoroethylene (ETFE) fluoropolymers meet these requirements. In this chapter we restrict the discussion to the market-dominating mineral glasses.

During the past two decades of industrial PV module production the low-iron soda-lime-silica glass with typical transmittance values above 90% has been the standard front sheet material. This glass contains at least the following basic compounds given in mass percent: 60–75% of silica (SiO_2 stoichiometry within a tetrahedral bonded network) and as network modifier oxides at least 5–20% of soda

(Na$_2$O) and 5–15% of lime (CaO). A lot of other components are often added to modify the tetrahedral bonded silica glass network for dedicated applications. With respect to the desired broadband high transmittance of the front sheet, the network modifiers are mainly relevant for the UV absorption edge, shifting it to longer wavelengths compared to pure silica glass where the UV cut-off is located at about 280 nm. In the visible and near-infrared part of the spectrum relevant for power generation in c-Si solar cells (300–1100 nm wavelength), absorption due to alkaline and alkaline-earth network modifiers is mostly negligible. For PV front glass, the UV cut-off can be optimized to wavelengths somewhere below 350 nm depending on the best compromise between UV protection (adapted to the encapsulation material) and power generation efficiency. The only other PV-relevant absorption loss can arise from transition metal ion impurities, normally iron oxides; all other glass coloring agents can be avoided. The ferric oxidic state Fe^{3+} absorbs in the UV (240–340 nm [34]) whereas the ferrous ions Fe^{2+} absorb in the infrared. Ferric ions are preferred for PV glass compositions because the main absorption edge is nearly equivalent with the silica glass intrinsic UV cut-off. Some float glass manufacturers increase the ferric content thus minimizing the redox ratio of FeO:Fe$_2$O$_3$ by adding oxides of Ce, Sb, As, W, and others [35]. Over all, a maximum transmittance of about 92% (owing to the 4% reflection on both air–glass surfaces) throughout the PV-relevant spectrum can be achieved (see, e.g., Ref. [36]) by keeping the iron content below 0.02 wt% (<200 ppm). This "extraclear" glass quality has only residual absorption losses of about 0.1% per millimeter of glass thickness within the PV relevant spectrum [37]. Besides glass composition-based absorptive losses, glass defects like bubbles or variations in surface geometry, due to imperfect production or glass aging can lead to a distortion of the photon flux mainly by Rayleigh and Mie scattering effects. Therefore, a regular quality control of the glass sheet itself and any surface modification to it (see below) are essential.

Possible techniques for surface modification (e.g., to decrease its reflectivity) depend on the glass manufacturing process. Two types are currently industrially relevant for PV front sheets, the roll glass, and the float glass process. Most of the current PV modules are manufactured with polymer backsheets and roll glass as front sheet. For roll glass, the surface microstructuring is more comfortable because the final glass rollers can easily emboss their individual surface structures to the hot viscous glass sheets. Surface texturing for float glass requires different techniques like post-treatment, for example, by etching. For c-Si glass-glass modules, and also for the majority of thin film modules the outstanding planarity of float glass is preferred or even required.

Most c-Si PV modules are nowadays equipped with tempered front glasses to guarantee durability of the front glass at minimum weight. This thermal hardening, which is done at temperatures up to 700 °C, can often be combined with surface modification processes, in particular, for the essential sintering of the recently market-wide introduced, sol–gel based, PV glass AR coatings. Another option for glass hardening is chemical modification of the surface glass composition with the goal to increase the internal compression stress gradient to a maximum at the air–glass side. Owing to much lower cost, thermal toughening is the prevalent and preferred technology for solar modules front glass.

12.3.2
Anti-reflection (AR) Technologies for PV Module Front Surface

Using the above described glasses with negligible absorptive and scattering losses, as well as the usual, refractive index-matched encapsulation sheets, the main issue for light management at the outside of a PV module is the air–glass reflection loss of typically 4% (relevant for normal incidence). To minimize these losses, a broad range of surface modifications from simple single-layer, quarter-wavelength AR layer to sophisticated light-trapping microstructures can be applied. Quarter-wavelength AR layers are optimized for a design wavelength (typically 600 nm) at the maximum of the solar response function of the individual cell, in order to maximize the electrical efficiency at standard module testing conditions (STC: direct incidence of 1000 W m^{-2} of AM1.5 spectrum at 25 °C). The potential to increase the effective photon flux to the photoactive cells is even more than 4%, because the glass/air interface reflectivity increases strongly at higher incidence angles (in particular, above $\theta = 50°$), important in the typical real outdoor situation of installed PV modules. In the following text, an overview of the existing variety of industrial standard and lab-scale process technologies for anti-reflective surface modifications will be provided.

12.3.2.1 Nano-scale Technologies for Anti-reflective Treatment of PV Front Glass
AR treatment of glass can be traced back to the early nineteenth century: Searching for the improvement of optical lenses and instruments, J. Fraunhofer discovered in 1817 a reduced reflectance for some glass surfaces after he had immersed them for 24 h in sulfuric or nitric acid [38]. During the last century this wet etching process has been developed further by using acidic fluorine derivates and later also vapor phase or plasma etching. For PV the AR glass etching, despite its high durability, could not compete with the large-scale developments of sol–gel coatings and physical vapour deposition (PVD) technologies adapted to PV within the last two decades. Detailed overviews of several AR technologies are provided, for example, by Cathro *et al.* [39], Raut *et al.* [40], and Askar *et al.* [41].

For the broadband AR coating of glass in the wavelength range (300–1100 nm) relevant for c-Si based solar modules, there are mainly three physical concepts:

- single AR layer of quarter-wave thickness with homogeneous refractive index (nanostructure much smaller than wavelength),
- gradient single-layer AR coatings
- multi-layer AR coatings.

For the first option, the layer material should ideally have a refractive index of $n \approx 1.23$ and a thickness of about 125 nm (with respect to the standard design wavelength 600 nm). Since durable natural materials with such low refractive index do not exist, the best way to approach the theoretical optimum is to prepare a nanoporous layer – when the pores are small enough, the layer behaves like an effective medium having a constant refractive index intermediate between that of the massive material and the enclosed air. In principle, typical AR coating materials like MgF_2 can be prepared also as a porous layer [42], but the high resistance against outdoor conditions required for PV application is best achieved by porous silica. Since the high porosity of about 50% required for the optimum $n \approx 1.23$ is a challenge regarding long-term stability, current industrially produced AR porous silica layers usually feature a slightly higher refractive index as compromise. Still, these layers reduce the total average reflectivity of the module surface from 4% to below 1% at least in the region around the AR design wavelength.

The second concept, single AR layers showing an index gradient, is principally very attractive, because it provides a means to suppress reflection completely when a well-defined index profile varying from n_{air} to n_{glass} can be provided [40]. Approximations to this idealized case can be produced in the form of porous layers with depth-dependent gradient in porosity [40], but due to both complexity of the production process and questionable long-term stability such layers are currently not considered for application in industrial PV production.

The same statement holds for multi-layer AR coatings, which have the principal potential to optimize the suppression of surface reflection further, but are complicated and thus too expensive processes in relation to the additional reduction of optical losses they provide. Multi-layer coatings might become an issue for coloration of solar modules in view of special design of esthetic reasons; but it might be more convenient, however, to realize these requirements on the solar cell level [43].

There are various established technologies for production of AR coatings. The favored production process for porous silica AR layers is currently the sol–gel coating for roll glass and the magnetron sputtering for float glass. Both AR technologies provide stable high-throughput processes with good control over pore structure and size. After the film preparation a typical post-treatment sintering (at 500–700 °C) is applied to stabilize the tetrahedral Si–O bonds within the AR layer and also for better connection to the front glass interface. Industrially, this annealing process can nicely be combined with the typical thermal hardening of the PV front glass, keeping cycle times low, for example, at less than 1 min at about 650 °C. In general, however, almost any other established deposition or etching process for preparing homogeneous or porous thin films has been demonstrated, at least in principle, to be applicable. These technologies comprise physical and chemical vapor deposition techniques as well as wet chemistry processes. Details about the wet chemistry AR coating can be found in several publications of colloidal Science and are in general applicable to PV glasses [45, 44, 2]. Sol–gel AR coatings are currently prepared by roller coating, less frequently by spin coating and spraying. Details of sputtering and PVD coatings can be found, for instance, in Ref. [46].

If the PV front glass would be replaced by polymers, some imprinting and (e.g., laser) lithography and self-assembling Langmuir–Blodgett layers as they are already known for polymer lenses could be adapted for future PV applications.

Especially the moth eye and honeycomb structures are a continuous topic of AR research [40].

12.3.2.2 Micro-scale Structures for Light-Trapping on PV Front Glasses

As already mentioned above, an alternative or additional microstructure applied to the air/glass interface can provide further increase of PV efficiency. A lot of industrial PV AR microstructures have been designed for the patterned roll glass dominating the PV front glass market. The light-trapping microstructures are mostly featuring rounded and channel-, wave-, rippled, or dot- (or hill-) like shapes; the variety is further extended by different degree of order, ranging from completely stochastically distributed to perfectly regular patterns of interconnected pyramidal (especially inversed pyramids), V-shaped, or mainly concave shapes toward the air surface. Some of these micro-patterns are geometrically and chemically optimized to enhance self-cleaning. The characteristic microstructure sizes are typically above 10 μm to enable multiple internal reflections targeting and trapping photons within the front glass toward the cell.

The electrical gain of micro-structured front glasses compared to flat glass is about 1% in maximum power point performance at STC. The real outdoor performance, however, can be improved significantly further, because the advantages of the microstructure are particularly important for operation at diffuse and grazing light conditions. Quantifying these advantages is an often highly complex challenge, which forms a current field of intensive research including simulations and outdoor performance validations of different AR-modified front glass modules and their reference glasses (some recent work has already been cited in Section 12.2). Technologically, for the current standard glass-foil modules in the market, micro-structures are mainly produced on roll glass directly by the rollers of the last roll glass step, although there are other micro-scale AR surface structuring options for roll glass and for float glass.

12.3.3
Material Selection and Optimization for Encapsulation Film

Important encapsulation materials for conventional c-Si PV modules are the duroplastic cross-linked EVA especially used for modules with polymer backsheets whereas the thermoplastic polyvinyl butyral (PVB) is the typical encapsulation material for glass-glass modules mainly for building integrated PV applications (BIPV). Still less frequently there are thermoplastic silicones or silicone elastomers (TPSE) or derivates of polyethylene (PE) and polyurethane (PU). All of them are comparable, with respect to the main criterion, for their optical properties, namely to provide a refractive index matching that of the PV module front glass. Typically, the refractive index of PV encapsulants is 1.48–1.50, and therefore also the transmittance of the free-standing films is in the same range as that of the front glass (90–92%). The polymer foil, film, and pottant encapsulants are typically modified to provide high long-term stability and chemical durability against yellowing and browning effects mainly caused by UV radiation. For the reactive EVA, it is

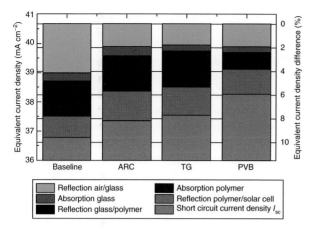

Figure 12.4 Summary of calculated values of the optical loses and estimated current density for the module series baseline, anti-reflective coating (ARC), thin glass (TG), and polyvinyl butyral (PVB) at AM1.5 sun radiation [31].

often necessary to add UV blockers resulting in a UV cut-off already at 380 nm. Some PVB or silicones are more stable to UV and can be used without such UV-absorbers. The gain in electrical conversion efficiency depends on the particular module setup. Experimental investigations and optical simulations are used for individual comparison. Figure 12.4 shows simulation results for solar module short circuit current densities based on measurements of individual solar module components from Schneider *et al.* [31]. Without optical losses the solar cell would generate a short circuit current of 40.6 mA cm^{-2}. The magnitude of the different optical loss mechanism has been calculated for different setups. In the baseline process under investigation in this publication, the module has a short circuit current of 36.8 mA cm^{-2}. This corresponds to a relative short circuit current loss of over 9%. The largest part of optical losses comes from the reflection at the air/glass interface. Introducing an anti-reflection coating (ARC) these losses are reduced strongly and the shirt circuit current is increased. With TG the absorption in the glass is reduced slightly. High UV transmission PVB additionally reduces the absorption in the polymer encapsulant and the resulting total shirt circuit current losses are reduced to below 6%.

From the optical point of view, index matching with the front glass and broadband transparency are the main optimizing criteria for the encapsulation materials. The interface between encapsulation and c-Si cell is determined by the textured surface and the silicon nitride AR layer of the cell. Encapsulant materials could be optimized to better fit the refractive index of the silicon nitride, but resulting additional reflection at the glass encapsulant interface and potentially reduced transmission properties would need to be taken into account.

Still a topic of basic research is the inclusion of luminescent particles in the encapsulation for the wavelength conversion of photons, which potentially increase the photoelectrical efficiency, but the cost efficiency and long-term durability still have to be improved.

12.3.4
Minimization of Losses due to Metallization and Contact Tabs

Solar cell metallization fingers, busbars, and contacting tabs (compare Figure 12.1b) are required for extracting the electrical power. But, since metals are strongly absorptive for visible and near infrared (NIR) light, the solar cell area under these contacts is shaded and does not contribute to charge carrier generation. Therefore, in agreement with the goal to reduce material costs, it is desirable to minimize the width of the metallization structures. On the other hand, the electrical functionality is best when the series resistance losses can be reduced as far as possible, which in turn requires metallization with larger cross section. Optimization thus leads to a high aspect ratio for metallization with small width and large height. Also for the cell connection tabs, a high aspect ratio is desirable to achieve high opto-electrical efficiency. On the downside, however, high aspect ratio tabs are considerably stiffer, which can induce cell breakage during soldering, lamination, and module operation.

The minimum width of the grid fingers depends on the specific metallization design, in particular, the number of busbars: if the number of busbars is increased without increasing the total area covered by them (i.e., making the individual busbar proportionally narrower), the grid finger length between the busbars is reduced resulting in less series resistance. This allows further reducing the grid finger width and thus the shading losses. In recent years the number of busbars was increased from two to three busbars by most solar cell producers. New generation tabber stringer equipment offers up to five busbars technology (Meyer Burger, Technological developments in module production Article PV Production Annual 2013). For the three busbar design, the typical solar cell has about 86 grid fingers (spacing of 1.8 mm) with a grid finger width of approximately 0.1 mm. Thus, a total of 8.6 mm out of 156 mm cell length are shaded, which corresponds to about 5.5% of the total cell area. The three busbars (and tabs on top) with a width of 2 mm each, therefore cover 6 mm out of 156 mm of the cell width, corresponding to another 3.8% of the cell being shaded. When, keeping the latter percentage constant, four or five busbars are used, the collection area and thus current conducted in each grid finger is reduced to 3/4 or 3/5, respectively. Since ohmic series resistance losses are determined by the series resistance multiplied with the square of the current, the electrical conductivity, and thus width of each grid finger can be reduced to 3/4 squared or 3/5 squared respectively. With an initial grid finger width of 100 µm, the grid fingers for a four busbar cell can be reduced to 56.25 µm width and for a five busbar cell to 36 µm without adding ohmic losses. This corresponds to a reduction of shading percentage (due to the grid fingers) to 3.1 and 2.0%, respectively.

Three different approaches for the geometry of contacting tabs are shown in Figure 12.5, standard tabs (left), texture tabs (middle), and wire contacts (right). Standard contacts reflect normal incident light straight back away from the module and thus shade the solar cell effectively over the entire area.

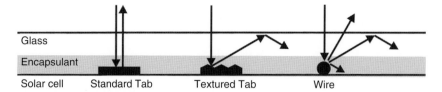

Figure 12.5 Schematics of reflection path with different tab technologies. Standard tabs reflect normal incident light straight back. Textured tabs reflect normal incident light sideways at low angles for total internal reflection at the glass surface. Contacting wires reflect light back through the glass, sideways for internal reflection at the glass surface or sideways onto the cell depending on the specific position of the incident light on the round surface of the wire.

With textured tabs (middle), normal incident light is scattered sideways and reflected at the inside of the front glass and back onto the solar cell. The texture tabs are available with zinc (low cost) and silver (high reflection) coatings on the copper tab. Company Schlenk has branded their tabs as LHSs, whereas Heraeus offers Light Capturing ribbons (LCRs).

LHS lead to an increase of the short circuit current and power by 2.5–3% [31]. LBIC measurements at several wavelengths have been performed. In case of a standard string, the current generated in the string region is <5% of the current in the cell region. In contrast, the current generated by a local light spot on a LHS is strongly increased due to the optimized reflection of the incident light into the module. Here, we found that the current density generated by 960 nm light incident on the string is about 75% of the current obtained in the cell region which would imply an overall increase in current density of the entire cell by 3%. Thus, the LBIC analysis is consistent with the directly measured short circuit current values. It has to be noted that these measurements are taken with normal incident light which is the ideal case for textured tabs. In case of lower light incident angles also standard tabs will contribute to current generation. In order to verify textured tab's benefit, energy yield measurements are recommended.

The third approach is to replace planar tabs by round wires (see right side in Figure 12.5). Three regions on the wire can be distinguished [47]:

1) At angles above 45° between wire surface normal and light at normal incidence on the module, the light is scattered toward the solar cells reducing the effective shaded area to sin 45°, that is, 70.7%.
2) At angles below 45° and above half the angle of total reflectance at the glass/air interface (i.e., 20.4° for an refractive index of the glass of 1.5) normal incident light on the module is scattered backwards and internally reflected at the glass reducing the effective shaded area to 35.7% in this case.
3) Light with an angle between surface normal of the wire and the light with normal incidents on the module below 20.4° is partially reflected and transmitted at the glass surface depending on the angle.

Braun et al. [47] have shown that a wire contacting scheme with 14 wires with a diameter of 250 μm leads to a shaded area of 2.24% (geometrically), but the effective shading area is reduced to 0.8% due to the light redirection effects described above.

Another, principally different approach to reduce shading losses of front side metallization is to place most or all of the metallization on the rear side of the cell. In Metal Wrap Through (MWT) and Emitter Wrap Through (EWT) solar cells holes are typically drilled with lasers into the silicon wafers. Front side metallization and emitter are "wrapped through" the resulting so-called vias. Charge carriers are collected at the sunny side of the solar cells and transported through the vias. Interconnection of cells is realized on the rear side of the cells only.

Interdigitated Back Contact solar cells developed by Richard Swanson since 1984 [48] to date at Sunpower Corp do not use front side metallization at all, but instead place both electrodes as well as the p–n junction on the rear side of the solar cell. The entire sunny side of the solar cell is covered with a highly optimized anti-reflective coating which must also provide excellent electronic passivation to avoid front surface recombination of minority charge carriers generated close to the sunny side of the solar cell far away from the p–n junction. This technology offers many challenges related to material and production cost of the solar cells, but it has offered for the past decades the highest commercial efficiencies at both solar cell (24.2%) and module level (20.4%) especially driven through the highest short circuit current densities partially obtained due to elimination of shading losses [49].

12.3.5
Optical Optimization of Cell Front and Back Interface

The last interface before the incident photons finally enter the solar cell itself, has usually a textured surface for the purpose of light trapping, and an anti-reflection coating for reduction of reflection losses at the encapsulant/solar cell interface. Typically, the anti-reflection coating is a silicon nitride layer with a refractive index between 1.9 and 2.0 at a light wavelength of 600 nm. Besides its anti-reflection properties, the silicon nitride also serves as front surface passivation at the interface between silicon nitride and silicon, and as a hydrogen source for passivation of bulk defects in the silicon. Furthermore, the front side metallization is fired through the silicon nitride layer to form an ohmic contact to the silicon. Thus, the silicon nitride layer has to fulfill many boundary conditions. It is important that the optical properties of the nitride should not be optimized on cell level with an interface to air, but rather on module layer with an interface to the encapsulant.

In 2013, typical crystalline silicon solar cells have been 180 μm thick [4]. Light with a wavelength of 1006 nm has an absorption depth of 180 μm [50]. Thus, light with energies both below and above the silicon band gap reaches the back contact of standard solar cells. From an electrical point of view, the back contact should mostly form a low surface recombination interface and provide excellent electrical conductivity; however, in terms of light management, it should additionally reflect the light reaching the back contact back into the solar cell with as little absorption as possible. Light with sufficient energy can contribute to charge carrier generation on a second path through the silicon while low energy infrared light can leave the solar cell at the front surface without heating the solar cell. In thinner solar cells, more light reaches the rear side and thus they require lower surface recombination and higher reflection from the rear side of the cell. Bifacial solar cells with grid-like metallization on both sides and thus without a back reflector require sufficient thickness for absorbing the light with a single path.

12.3.6
Redirection of Light from Cell Interspaces

The spaces between solar cells are non-active areas and do not contribute to photocurrent generation directly. Thus, spaces should be minimized. Solar cell space within a string and between strings is typically 2–5 mm. Often equal spaces in both directions are chosen for esthetical reasons. The space is mostly limited by mechanical stress induced by the tabs from one cell front to the adjacent cell back contact. Steps in the tabs produced by crimping are used as a measure to relieve the stress and move cells closer together. But it has been shown that such crimping steps themselves can lead to tab breakage and performance loss of modules [51]. The minimum spacing between strings is mostly limited by positioning accuracy of the strings and foil shrinkage during or after lamination, which can potentially result in shorts when cells in neighboring strings are in contact. Light between solar cells is not lost entirely though, and as a matter of fact the recent expectation for solar module power to exceed the sum of the solar cells power within the module is mostly driven by extra current generation from light scattering from the white back sheet between solar cells. The ITRPV 2014 expects up to 104% of power from modules as compared to cells. LBIC measurements show that the back sheet area next to a solar cell has up to 20% short circuit current generation of the actual cell. With two adjacent cells this number is doubled to 40%. Assuming a solar cell efficiency of 18% the gaps between the cells have an efficiency of above 7%. A module with 5 mm gaps between cells in both directions would add about 0.1 m^2 additional space at 7% efficiency, which corresponds to about 7 W or up to 3% of additional power.

Glass–glass modules offer the opportunity for partially transparent modules which still illuminate a space underneath which is used in building integrated application especially for covering large spaces like, for example, the main train station in Berlin, Germany. Here, the light between cells is not available for current generation in the solar cells. In order to allow glass–glass modules to make use of the light from cell interspaces both dispersion of white pigments in the rear side encapsulant as well as direct application of the pigments on the rear side glass are under investigation.

12.4
Outlook

In this chapter, we elucidated the opportunities for improving light management in solar modules from a physical perspective. We also described the current technology trends on how to realize the opportunities. In the coming years advanced light management will lead to optical gains exceeding optical losses in solar modules and thus enabling solar module powers to exceed the sum of the power of the solar cells used in them. A key challenge in light management is the cost effectiveness, that is, that additional costs are compensated by the additional revenues. With the gradual reduction of losses however, the potential of further improvements is becoming smaller and smaller, which puts additional pressure on advanced novel concepts. As a result the PV industry has been looking for light management solutions which are at the same time better and cheaper. But the optical improvements do not only improve the power and energy yield of the modules, they must also meet module durability goals. Solar modules and their components must last for 25 and more years in various climates. Thus, for example, optical coatings for solar glass in desert applications must also provide protection from abrasion and at the same time have self-cleaning properties. Many interesting challenges with multiple boundary conditions must be met, and advanced light management solutions will be required for making the most from solar power.

References

1. Born, M. and Wolf, E. (1964) *Principles of Optics: Electromagnetic Theory of Propagation, Interference and Diffraction of Light*, Pergamon Press, Oxford.
2. Ballif, C., Dicker, J., Borchert, D., and Hofmann, T. (2004) Solar glass with industrial porous SiO_2 antireflection coating: measurements of photovoltaic module properties improvement and modelling of yearly energy yield gain. *Sol. Energy Mater. Sol. Cells*, **82**, 331–344.
3. Braun, M.M. and Pilon, L. (2006) Effective optical properties of non-absorbing nanoporous thin films. *Thin Solid Films*, **496**, 505–514.
4. SEMI PV Group (2013) *International Technology Roadmap for Photovoltaic*, 4th edn, http://www.itrpv.net/Reports/Downloads/2013 (accessed 20 December 2013).
5. Green, M.A. (1995) *Silicon Solar Cells: Advanced Principles and Practice*, Centre for Photovoltaic Devices and Systems, University of New South Wales, Sydney, pp. 92–116.
6. Pfeifer, M. (2008) Optimierung von light-trapping-Strukturen zur Effizienzsteigerung multikristalliner Siliziumsolarzellen. Diploma thesis. University of Freiburg, http://www.ise.fraunhofer.de/de/geschaeftsfelder-und-marktbereiche/angewandte-optik-und-funktionale-oberflaechen/marktbereiche/mikrostrukturierte-oberflaechen/dokumente-ordner-nicht-sichtbar-nur-fuer-links/diplomarbeit-marcel-pfeifer (accessed 5 November 2014).
7. Sánchez-Illescas, P.J., Carpena, P., Bernaola-Galván, P., Sidrach-de-Cardona, M., Coronado, A.V., and Álvarez, J.L. (2008) An analysis of geometrical shapes for PV module glass encapsulation. *Sol. Energy Mater. Sol. Cells*, **92**, 323–331.
8. Yang, H. and Wang, H. (2011) Performance analysis of crystalline silicon solar modules with the same peak power and the different structure. *Clean Technol. Environ. Policy*, **13**, 527–533.
9. Battaglia, C., Boccard, M., Haug, F.-J., and Ballif, C. (2012) Light trapping in solar cells: when does a Lambertian scatterer scatter Lambertianly? *J. Appl. Phys.*, **112**, 094504.
10. Yablonovitch, E. (1982) Statistical ray optics. *J. Opt. Soc. Am.*, **72**, 899–907.
11. Yu, Z., Raman, A., and Fan, S. (2010) Fundamental limit of light trapping in grating structures. *Opt. Express*, **18**, A366–A380.
12. Yu, Z., Raman, A., and Fan, S. (2010) Fundamental limit of nanophotonic light

trapping in solar cells. *Proc. Natl. Acad. Sci. U.S.A.*, **107**, 17491–17496.

13. Callahan, D.M., Munday, J.N., and Atwater, H. (2012) Solar cell light trapping beyond the ray optic limit. *Nano Lett.*, **12**, 214–218.

14. Wang, C., Yu, S., Chen, W., and Sun, C. (2013) Highly efficient light-trapping structure design inspired by natural evolution. *Sci. Rep.*, **3**, 1025. doi: 10.1038/srep01025.

15. Battaglia, C., Hsu, C.-M., Söderström, K., Escarré, J., Haug, F.-J., Charrière, M., Boccard, M., Despeisse, M., Alexander, D.T.L., Cantoni, M., Cui, Y., and Ballif, C. (2012) Light trapping in solar cells: can periodic beat random? *ACS Nano*, **6**, 2790–2797.

16. Smith, R.E., Warren, M.E., Wendt, J.R., and Vawter, G.A. (1996) Polarization-sensitive subwavelength antireflection surfaces on a semiconductor for 975 nm. *Opt. Lett.*, **21**, 1201–1203.

17. Krauter, S. and Hanitsch, R. (1996) Actual optical and thermal performance of PV-modules. *Sol. Energy Mater. Sol. Cells*, **41** (42), 557–574.

18. Lu, Z.H. and Yao, Q. (2007) Energy analysis of silicon solar cell modules based on an optical model for arbitrary layers. *Sol. Energy*, **81**, 636–647.

19. Soitec (2013) *http://www.soitec.com/en/news/press-releases/world-record-solar-cell-1373/* (accessed 15 January 2014).

20. Fischer, S., Goldschmidt, J.C., Löper, P., Bauer, G.H., Brüggemann, R., Krämer, K., Biner, D., Hermle, M., and Glunz, S.W. (2010) Enhancement of silicon solar cell efficiency by upconversion: optical and electrical characterization. *J. Appl. Phys.*, **108**, 044912.

21. Martín-Rodríguez, R., Fischer, S., Ivaturi, A., Froehlich, B., Krämer, K.W., Goldschmidt, J.C., Richards, B.S., and Meijerink, A. (2013) Highly efficient IR to NIR upconversion in $Gd_2O_2S:Er^{3+}$ for photovoltaic applications. *Chem. Mater.*, **25**, 1912–1921.

22. Schweizer, S., Henke, B., Ahrens, B., Paßlick, C., Miclea, P.T., Wenzel, J., Reisacher, E., Pfeiffer, W., and Johnson, J.A. (2010) Progress on up- and down-converted fluorescence in rare-doped fluorozirconate-based glass ceramics for high efficiency solar cells. *Proc. SPIE* (Photonics for Solar Energy Systems III, May 18, 2010) **7725**, 77250X, doi: 10.1117/12.853943.

23. Xin, F., Zhao, S., Huang, L., Deng, D., Jia, G., Wang, H., and Xu, S. (2012) Upconversion luminescence of Er^{3+}-doped glass ceramics containing β-$NaGdF_4$ nanocrystals for silicon solar cells. *Mater. Lett.*, **78**, 75–77.

24. Skrzypczak, U., Pfau, C., Bohley, C., Seifert, G., and Schweizer, S. (2013) Influence of $BaCl_2$ nanocrystal size on the optical properties of Nd^{3+} in fluorozirconate glass. *J. Phys. Chem. C*, **117**, 10630–10635.

25. Sjerps-Koomen, E.A., Alsema, E.A., and Turkenburg, W.C. (1996) A simple model for PV module reflection losses under field conditions. *Sol. Energy*, **57**, 421–432.

26. Andrews, R.W. and Pearce, J.M. (2013) The effect of spectral albedo on amorphous silicon and crystalline silicon solar photovoltaic device performance. *Sol. Energy*, **91**, 233–241.

27. Parretta, A., Sarno, A., Tortora, P., Yakubu, H., Maddalena, P., Zhao, J., and Wang, A. (1999) Angle-dependent reflectance measurements on photovoltaic materials and solar cells. *Opt. Commun.*, **172**, 139–151.

28. Parretta, A., Sarno, A., and Yakubu, H. (1999) Non-destructive optical characterization of photovoltaic modules by an integrating sphere. Part I: Mono-Si modules. *Opt. Commun.*, **161**, 297–309.

29. Maddalena, P., Parretta, A., Sarno, A., and Tortora, P. (2003) Novel techniques for the optical characterization of photovoltaic materials and devices. *Opt. Lasers Eng.*, **39**, 165–177.

30. Fontani, D., Sansoni, P., Sani, E., Coraggia, S., Jafrancesco, D., and Mercatelli, L. (2013) Solar divergence collimators for optical characterisation of solar components. *Int. J. Photoenergy*, **2013**, Article ID 610173, doi: 10.1155/2013/610173.

31. Schneider, J., Turek, M., Dyrba, M., Baumann, I., Koll, B., and Booz, T. (2014) Combined effect of light harvesting strings, anti-reflective coating, thin glass and high UV transmission

encapsulant to reduce optical losses in solar modules. *Prog. Photovolt: Res. Appl.* doi: 10.1002/pip.2470

32. Shoo, Y.S., Walsh, T.M., Lu, F., and Aberle, A.G. (2012) Method for quantifying optical parasitic absorptance loss of glass and encapsulant materials of silicon wafer based photovoltaic modules. *Sol. Energy Mater. Sol. Cells*, **102**, 153–158.

33. Fraidenraich, N. and Vilela, O.C. (2000) Exact solutions for multilayer optical structures. Application to PV modules. *Sol. Energy*, **69**, 357–362.

34. Turner, R.C. and Miles, K.E. (1957) The ultraviolet absorption spectra of the ferric ion and its first hydrolysis product in aqueous solutions. *Can. J. Chem.*, **35** (9), 1002–1009.

35. Sachot, D. and Cintora, O. (2009) Silico-sodo-calcic glass sheet. WIPO Patent Nr. 2009047463.

36. Goodyear, J.K. and Lindberg, V.L. (1980) Low absorption float glass for back surface solar reflectors. *Solar Energy Mater.*, **3** (1), 57–67.

37. Rubin, M. (1985) Optical properties of soda lime silica glasses. *Sol. Energy Mater.*, **12** (4), 275–288.

38. Fraunhofer, v.J. (1817) Versuche über die Ursachen des Anlaufens und Mattwerdens des Glases und die Mittel denselben zuvorzukommen, in *Gesammelte Schriften: Im Auftrage der mathematisch-physikalischen Classe der Königlich bayerischen Akademie der Wissenschaften* (eds E. Lommel), Verlag der K. Akademie, (1988).

39. Cathro, K.J., Constable, D.C., and Solaga, T. (1981) Durability of porous silica antireflection coatings for solar collector cover plates. *Sol. Energy*, **27** (6), 491–496.

40. Raut, H.K., Ganesh, V.A., Nair, A.S., and Ramakrishna, S. (2011) Anti-reflective coatings: a critical, in-depth review. *Energy Environ. Sci.*, **4** (10), 3779–3804.

41. Askar, K., Phillips, B.M., Fang, Y., Choi, B., Gozubenli, N., Jiang, P., and Jiang, B. (2013) Self-assembled self-cleaning broadband anti-reflection coatings. *Colloids Surf., A*, **439**, 84–100.

42. Morales, S.A. (1980) Sol-gel process for the preparation of porous coatings, using precursor solutions prepared by polymeric reactions. European Patent 1329433, Jul. 23 2003.

43. Selj, J. H., Stensrud Marstein, E., Thøgersen, A. and Foss, S. E. (2011), Porous silicon multilayer antireflection coating for solar cells; process considerations. *Phys. Status Solidi C*, **8**: 1860–1864. doi: 10.1002/pssc.201000033

44. Schmidt, H.K. (1995) in *Proceedings of the 3rd Conference of the European Society of Glass Science and Technology (ESG)* (ed H.A. Schaeffer), Verlag der Deutschen Glastechnischen Gesellschaft, Würzburg, pp. 21–32.

45. Gombert, A., Glaubitt, W., Rose, K., Dreibholz, J., Zanke, C., Bläsi, B., Heinzel, A., Horbelt, W., Sporn, D., Döll, W., Wittwer, V., and Luther, J. (1998) Glazing with very high solar transmittance. *Sol. Energy*, **62** (3), 177–188.

46. Gläser, H.J. (2008) *Vakuumbeschichtung von Architekturglas – Ein historischer Abriss Vacuum Coating of Architectural Glass – A Historic Review, Vakuum in Forschung und Praxis 20*, vol. **3**, WILEY-VCH Verlag GmbH & Co. KGaA, Weinheim.

47. Braun, S., Micard, G., and Hahn, G. (2012) Solar cell improvement by using a multi busbar design as front electrode. *Energy Procedia*, **27**, 227–233.

48. Swanson, R.M., Beckwith, S.K., Crane, R.A., Eades, W.D., Kwark, Y.H., Sinton, R.A., and Swirhun, S.E. (1984) Point-contact silicon solar cells. *IEEE Trans. Electron Devices*, **31** (5), 661–664.

49. Cousins, P.J., Smith, D.D., Luan, H.C., Manning, J., Dennis, T.D., Waldhauer, A., Wilson, K.E., Harley, G., and Mulligan, G.P. (2010) Gen III: improved performance at lower cost. 35th IEEE PVSC, Honolulu, HI, June 2010.

50. Green, M.A. and Keevers, M. (1995) Optical properties of intrinsic silicon at 300K. *Prog. Photovoltaics*, **3**, 189–192.

51. Meier, R., Kraemer, F., Wiese, S., Wolter, K., and Bagdahn, J. (2010) Reliability of copper-ribbons in photovoltaic modules under thermo-mechanical loading. 35th IEEE Photovoltaic Specialists Conference (PVSC), 2010, pp. 001283–001288, doi: 10.1109/PVSC.2010.5614220.

Index

a
air mass (AM)   329
air–silicon interface   37
alkaline/acidic chemical wet-etching   4
angularly resolved scattering (ARS)   98–99
angular restriction   186
angular-selective reflectance filters   10, 11
angular selectivity   187
angular spectrum method   98–99
anti-reflection (AR) technologies   1
– air–glass reflection loss   336
– light-trapping microstructures   338
– nano-scale technologies   336–338
– quarter-wavelength AR layer   336
anti-reflective/antireflection coating (ARC)   3–4
– EVA/Si interface   326
– nano-scale technologies   336–338
– short circuit current loss   339
atomic layer deposition (ALD)   131–132
autocorrelation length   95

b
barium borate glass
– advantages   303
– collection efficiency   305–306
– fluorescence intensity   306–307
– fluorozirconate-based glass   306–307
– optical properties   303, 305
– transmission factor   303, 305
black body emission spectrum   24
black silicon
– chemical etching   124–126
– ECE   124–126
– fabrication method   117, 119
– KOH   117, 118
– Lambertian light trapping (see light trapping)
– laser processing   122–124
– RIE   119–122
– silicon passivation (see surface passivation)
Bragg filter
– germanium solar cell   201–203
– hydrogenated amorphous silicon solar cell   197–201
Bragg stacks   11

c
Chandezon method   59, 96–97
charge carrier lifetime   42
charge transfer emission   40
chemical etching   124–126
chemical passivation   136
compound parabolic concentrator (CPC)   161–162
concentrated photovoltaic (CPV) system
– acceptance angle   154–157
– geometric concentration   154–155
– HCPV (see highly concentrating photovoltaics (HCPVs))
– low-cost optical design   153–154
– non-uniformities   158–159
– optical designs (see primary optical element (POE))
– optical efficiency   157–158
– SOG (see silicone on glass (SOG))
concentration acceptance product (CAP)   156–157
contact tabs   340–342
crystalline Si   38
crystalline silicon (c-Si)   1
– absorption depth   1, 2
– absorption spectra   2, 3
– solar cells   1, 2
$Cu(In,Ga)Se_2$ 38

d

2D gratings 8
3D photonic crystal IRLs (3D-IRL) 14–16
3D photonic crystals 9–11
dark current, solar cell 24
dielectric photonic structures 246–248, 251
differential reciprocity, Wong and Green theorem 31–33
diffractive gratings. *See* light trapping
directional selectivity. *See* light trapping
distributed Bragg reflectors (DBR) 9
donor–acceptor interface 40
doping concentration 243–251
down-conversion
 – AM1.5 standard solar spectrum 255
 – borate glasses 260
 – down-shifting 256
 – multivalent europium-doping (*see* europium-doping)
 – phonons 259
 – quantum cutting mechanism 256–257
 – rare-earth ions 257–258
 – Sm-doped borate glass (*see* samarium (Sm)-doped borate glass)
 – ZBLAN glass (*see* ZBLAN glasses)
down-shifting 256
dye-doped polymer 297
dynamic depolarization 214

e

effective acceptance angle
 – angular transmission curve 155–156
 – CAP 156–157
 – concentrator cost 155–156
 – imperfections 155
 – maximum electric power 154–155
electrochemical etching (ECE) 124–126
Electroluminescence (EL) emission 25, 36–39
electron-beam lithography 69–70, 72
emitted photon flux 23–24
emitter and collector currents, transistors 29–30
emitter wrap through (EWT) 342
energy transfer up-conversion (ETU) 231, 237
Er^{3+}-doped FZ glasses 232
 – absorption coefficient 233–234
 – absorption spectra 233–234
 – nominal compositions 232–233
 – optical density 233–234
 – up-converted fluorescence spectra 234–237
europium-doping
 – photoluminescence 274
 – trivalent europium 267
 – XANES measurement 274
 – x-ray absorption 268–270
 – XRD 270–272
exact electrodynamic theory 245
excited state absorption (ESA) 237
external quantum efficiency (EQE) 28, 105–106, 231, 237, 266, 276
 – 3D photonic crystal 194–197
 – enhancement 9, 15–17
 – front-side nanostructures 221
 – rear-side nanostructures 222
 – TCO 198–199
external radiative efficiency (ERE) 34

f

Fabry–Perot-resonator 55–56
field effect passivation 136
finite difference time-domain (FDTD) method 8, 59, 246
 – ARS function 98–99
 – black silicon interface 133–134
 – Maxwell's equation 97
 – optical simulation 99–100
 – properties 100–102
finite element method (FEM) 172–174
fluorescence concentrators
 – absorption/emission coefficient 285–286
 – barium borate glass (*see* barium borate glass)
 – bottom-mounted system 288–289
 – factors 284
 – geometrical concentration 307–308
 – IAD 310
 – LBIC 310–312
 – mono crystalline silicon solar cell 308–310
 – NanoFlukos 294
 – non-perfect filter 291–293
 – organic dyes 299
 – plasmonic nanostructures 223
 – polymer matrix material 296–298
 – principles 283–284
 – quantum dots (*see* quantum dots (QDs))
 – quantum efficiency 312
 – rugate filter 286–287
 – side-mounted system 288
 – thermodynamic efficiency 290–291
 – two- and three-dimensional photonic crystals 286–287
fluorozirconate-based glass 306–307
focused ion beam (FIB) milling 83
Förster energy-transfer processes 225
Fraunhofer diffraction 328
frequency downconverters 223

Index | 349

g

galvanostatic etching   124
germanium solar cell   203
glass ceramics. *See* ZBLAN glasses
glass-glass modules   343
gold nanoparticle   245–246

h

highly concentrating photovoltaics (HCPVs)
– concentrator design   175–179
– requirements   175–176

i

III/V based solar cells   38
inductively coupled plasma (ICP) etching   119–121
inductively coupled reactive ion etching (ICP-RIE)   5–6
inhomogeneous broadening   212, 216
interband damping   212, 215
interband transitions
– luminescence efficiencies   211
– optical properties   212
– in silver, gold and aluminum   210
interference lithography (IL)   76–77
intermediate reflective layers (IRLs)   13–16
internal quantum efficiency (IQE)   266, 329–330
ion-assisted deposition (IAD)   310–311
isolated interface   100
iterative Fourier transform algorithm (IFTA) algorithm   103, 105, 107, 109, 199

k

Kirchhoff's laws   30, 33
Köhler illumination   164
Korringa–Kohn–Rostocker (KKR) method   59

l

Lambertian scatterer   5–6, 328
laser beam induced current (LBIC)   333
laser induced periodic surface structure (LIPSS)   122–124
The Law of Entire Equilibrium   22
LED quantum efficiency   33–34, 36, 39
levelized cost of electricity (LCOE)   323
light beam induced current (LBIC)   39, 43, 310–312
light guiding efficiency (LGE)   294–295
light harvesting strings (LHSs)   333
light management, solar module
– AR treatment, PV front surface (*see* anti-reflection (AR) technologies)
– cell arrangement   324
– cell interspaces   343
– cell-to-module efficiency   323
– encapsulant materials   338–339
– experimental techniques and equipment   331–333
– front and back interface   342–343
– front glass sheets   334–336
– incidence angle variation and diffuse light   331
– metallization and contact tabs   340–342
– numerical simulations   333–334
– reflective and absorption losses   324–326
– spectral effects and solar concentration   329–331
– textured interfaces/surfaces   326–328
lightning-rod effect   212
light-scattering patterns   217
light trapping   1, 35–37
– AFM   79
– butt-coupling   55
– c-Si solar cell, textured surface of   327–328
– dielectric coatings   130–132
– electron-beam lithography   69–70, 72
– Fabry–Perot-resonator   55–56
– features   126–127
– FIB   83
– grating period   53, 55
– ICP-RIE process   128–130
– incoming radiation   56
– interference lithography   76–78
– light path   57
– limiting efficiency   132–135
– linear (1D) rear side grating   51
– micro-structured front glasses   338
– nanoimprint lithography   76–77
– optical simulation (*see* optical simulation)
– principles   49–50
– propagation constant   55
– properties   50–51
– self-organizing process   73–74
– SEM   80–81
– structures   127
– sub-micron domain   127
– substrate thickness   132–135
– textured glass surface   327–328
light-trapping
– Bragg filter (*see* Bragg filter)
– crystalline silicon cell   184
– 1D layer stack filter   192–194
– 3D photonic crystal   194–197
– directional and spectral selective absorber   191–192
– grating couplers   184

light-trapping (contd.)
- radiative efficiency limit   185–187
- radiative interaction   7, 41–50
- thin-film solar cell   184
- universal limit   188
- wafer solar cell   184

localized surface plasmons. see particle-plasmon (PP) resonances
low-iron soda-lime-silica glass   334–335
luminescence imaging   40–42
luminescent concentrators. See fluorescence concentrators

m

Maxwell's equations   96–98
metal assisted chemical etching (MACE)   125–126
metal-assisted wet-chemical etching   5
metal wrap through (MWT)   342
micromorph tandem solar cell   13–16
Mie-theory   245
multicrystalline Si solar cells   41
multifold Köhler concentrators   164–166
multi-junction cells   13
multi phonon relaxation (MPR)   240, 242–243

n

nanoimprint lithography (NIL)   76–77
nanorods (NRs)   214
nanowires   215
near-field effects   212, 217–218
near-field scanning optical microscope (NSOM)   94, 99–100, 102
normal incidence   157
normalized non-radiative lifetime   26
numerical simulations   333–334

o

Ohmic contact   23
Onsager's reciprocity   29
optical concentration   186–187
optical density   233–234
optical excitations, metal nanoparticle
- interband transitions   210–211
- PPs (see particle-plasmon (PP) resonances)

optical path enhancement   1
- energy-selective structures   13–16
- ray optical limit   4–5
- resonant structures   7–10
- scattering structures   5–7

optical simulation
- absorptance   66–67
- Auger recombination   66
- carrier conservation   66
- drift-diffusion equation   65–66
- matrix method   64
- radiative recombination: 66
- RCWA/FMM   58–61
- solar cell simulation   65
- SRH recombination   66
- surface recombination   66

opto-electronic reciprocity
- experimental verifications   37–39
- limitations   43–44
- luminescence imaging   40–42
- principle of detailed balance   21–22
- spectrally resolved luminescence analysis   39–40
- SQ limit   22–24

organic bulk heterojunction solar cells   38
ORMOCER   299

p

particle–particle coupling   215
particle-plasmon polariton   214
particle-plasmon (PP) resonances
- conduction electrons, collective oscillations of   211–212
- dielectric environment   214
- inhomogeneous broadening   212, 216
- interband damping   212, 215
- light-scattering patterns   217
- metal selection   213
- near-field effects   212, 217–218
- particle–particle coupling   215
- particle shape   214–215
- particle size   214
- radiation damping   212, 216
- scattering quantum efficiency   216–217

perfect electric conductor (PEC)   107
PERL   5
photoluminescence (PL)   256
photoluminescent (PL) emission   25, 38
photonic light trap   10
photovoltaic conversion efficiency   22
planar antireflection coating   2–4
plasma enhanced chemical vapor deposition (PECVD)   136–139
plasmon hybridization   215
plasmonic effects   8
plasmonic metal nanostructures
- advantages   226
- disadvantages   226
- fluorescence concentrators   223
- frequency downconverters   223
- front-side structures   219–221

- interband transitions (*see* interband transitions)
- intermediate reflectors, tandem cells 222–223
- near-field effects 212
- photoluminescence enhancement 224
- PPs (*see* particle-plasmon (PP) resonances)
- radiative quantum efficiency 223–224
- rear-side structures 221–222
- SPPs 209, 218–219

plasmon resonance 245–246, 251
polymethylmethacrylate (PMMA) 296, 297
polyvinyl butyral (PVB) 339
potassium hydroxide (KOH) 117, 118
primary optical element (POE)
- classical concentrators 160–161
- cost differences 166
- flat Fresnel-based architectures 166–168
- freeform SMS concentrators 163
- multifold Köhler concentrators 164–166
- nonimaging secondary optics 161–162
- spatial and spectral non-uniformity 168–169

principle of detailed balance 21–22
The Principle of Microscopic Reversibility 22
pyramidal surface structures 4–5

q

quadratic electric field 246
quantized thermal energy 259
quantum cutting mechanism 256
quantum dots (QDs)
- CdSe/ZnS-QDs types 301–303
- *ex situ* method 301
- *in situ* method 301
- properties 300
- reabsorption losses 301

quantum efficiency 12
quasi-Fermi level splitting 40, 42
quasi steady state photo conductance (QSSPC) method 140–141

r

radiation damping 212, 216
radiative quantum efficiency 223–224
randomly textured surfaces
- cell efficiency 91–92
- geometry/material properties 92
- impinging photons 92
- single-junction solar cell (*see* single-junction solar cell)

rate equation model 240–241
Rayleigh anomaly 7
ray optical limit 4–5

ray tracing approach 331
receiver antennas 223
reciprocity theorem
- derivation 24–29
- LED quantum efficiency 33–34
- light trapping 35–37
- SQ-theory 34–35
- Tellegens's network theorem 30–31
- transistors emitter and collector currents 29–30
- Wong and Green theorem 31–33
- Würfel's generalization, Kirchhoff's law 33

referential solar cell 94–96
relative emission 293–295
ribs 163
rugate filter 11–12, 192–194
rugate stacks 11

s

samarium (Sm)-doped borate glass
- barium oxide 275
- EQE 276
- fluorescence emission 276–277
- optical absorption 276–277
- quantum efficiency 278–279

scanning electron microscopy (SEM) 80–81
scanning optical microscopy (SNOM) 99–100
scattering quantum efficiency 216–217
second law of thermodynamics 22
series resistances, luminescence imaging 41, 42
Shockley–Queisser (SQ) limit 11, 22–24
Shockley–Queisser (SQ) theory 22–24, 34–35
Shockley Read Hall (SRH) recombination 66
short circuit current density 23
shunt resistances, luminescence imaging 41
silicone on glass (SOG)
- advantages 169
- disadvantages 169
- distance-temperature-map 173–174
- efficiency 173–175
- FEM 172–174
- lens temperature 170–172
- Lorentz–Lorenz equation 172

single-junction solar cell
- absorption enhancement 104–106
- angular cone 108–109
- fabrication method 109–110
- IFTA algorithm 103, 105, 109, 199
- PEC 107
- phase-only transmission function 107–108

single-junction solar cell (contd.)
- scattering cone   108–109
- short circuit current density   108
- unorthodox method   106–107
single optical surface (SILO)   164–165
solar photons   277
sol–gel glass   298
spatial distribution   98
spatial resolution   40
spectral concentration   250
spectrally resolved luminescence analysis   39–40
spines   163
stacked cells   13
sub-band-gap photons   231, 242
surface passivation
- chemical passivation   136
- disadvantages   144
- efficiency   142–144
- field effect passivation   136
- heterojunction   121, 143
- ICP-RIE processing   140–142
- intermediate structure   140
- laser treatment   142
- parameter variation   140–141
- PECVD   136–139
- porous silicon   143
- sample cleaning   139–140
- surface damage   139–140
- surface recombination   135
surface-plasmon polaritons (SPPs)   209, 218–219
surface plasmons   209–210

t

tandem solar cells   222–223
- anti-reflection effect   110–112
- current matching   110

- intermediate layer optimization   110–111
Tellegens's network theorem   30–31
thermal equilibrium   22, 188
thin-film solar cells   42, 184
- advantages   209
- disadvantages   209
- plasmonic nanostructures (see plasmonic metal nanostructures)
thinner glass (TG)   333, 339
total internal reflection   8
transparent conductive front oxide layer (TCO)
- EQE   198–199
- Lambert–Beer's law   199–200
- multi layer stack filter   197–198
- optical path length   200
- photovoltaic absorber   197

- total reflectance   198–199
- transmission and reflection   197–198
transparent conductive oxide (TCO)
- EPF   95–96
- growth condition   93
- optical properties   102–104
- sub-wavelength resolution   94
- surface profile   94–95
- wafer-based thick solar cell   93–94

u

ultra-light trapping   10–13
up-conversion (UC)
- dielectric photonic structures   246–248, 251
- doping concentration   243–244, 251
- Er^{3+}-doped ZBLAN glasses (see Er^{3+}-doped FZ glasses)
- ETU   231
- external quantum efficiency   231
- β-$NaYF_4$:20% Er^{3+} 231, 237–240
- phonon energy   242–243, 251
- plasmon resonance   245–246, 251
- rate equation model   240–241
- spectral concentration   250
- sub-band-gap photons   231, 242
- trivalent lanthanide ions   232
up-conversion quantum yield (UCQY)   231–232
- doping concentration   244
- β-$NaYF_4$:20% Er^{3+} 238–239, 240–241
- phonon energy   242–243

v

Ventana™   165

w

Wong and Green theorem   31–33

x

x-ray absorption near edge structure (XANES)   274
x-ray diffraction (XRD)   270–272

y

Yablonovitch limit   5–6, 16, 35, 189

z

ZBLAN glasses   259–260
- absorption   262–264
- Er^{3+}-doped FZ glasses (see Er^{3+}-doped FZ glasses)
- internal conversion efficiency   265
- quantum efficiency   266–267
- short circuit current   264–265
- sodium borate glass   260–261